ISNM
International Series of
Numerical Mathematics
Vol. 121

Managing Editors:
K.-H. Hoffmann, München
D. Mittelmann, Tempe

Associate Editors:
R.E. Bank, La Jolla
H. Kawarada, Chiba
R.J. LeVeque, Seattle
C. Verdi, Milano

Honorary Editor:
J. Todd, Pasadena

Stability Theory

Hurwitz Centenary Conference
Centro Stefano Franscini, Ascona, 1995

Edited by

R. Jeltsch
M. Mansour

Birkhäuser Verlag
Basel · Boston · Berlin

Editors:

Rolf Jeltsch
Seminar for Applied Mathematics
ETH Zentrum
8092 Zürich
Switzerland

Mohamed Mansour
Automatic Control Laboratory
ETH Zentrum
8092 Zürich
Switzerland

A CIP catalogue record for this book is available from the Library of Congress, Washington D.C., USA

Deutsche Bibliothek Cataloging-in-Publication Data
Stability theory / Hurwitz Centenary Conference, Centro
Stefano Franscini, Ascona, 1995. Ed. by Rolf Jeltsch ;
Mohamed Mansour. - Basel ; Boston ; Berlin : Birkhäuser, 1996
  (International series of numerical mathematics ; Vol. 121)
   ISBN 3-7643-5474-7 (Basel ...)
   ISBN 0-8176-5474-7 (Boston)
NE: Jeltsch, Rolf [Hrsg.]; Hurwitz Centenary Conference <1995,
  Ascona>; Centro Stefano Franscini; GT

This work is subject to copyright. All rights are reserved, whether the whole or part of the material is concerned, specifically the rights of translation, reprinting, re-use of illustrations, recitation, broadcasting, reproduction on microfilms or in other ways, and storage in data banks. For any kind of use the permission of the copyright owner must be obtained.

© 1996 Birkhäuser Verlag, P.O. Box 133, CH-4010 Basel, Switzerland
Printed on acid-free paper produced from chlorine-free pulp. TCF ∞
Cover design: Heinz Hiltbrunner, Basel
Printed in Germany
ISBN 3-7643-5474-7
ISBN 0-8176-5474-7

9 8 7 6 5 4 3 2 1

Contents

Preface VII

Stability theory

A. Beghi, A. Lepschy and U. Viaro
The Hurwitz Matrix and the Computation of Second-Order
Information Indices 1

B. Bernhardsson, A. Rantzer and L. Qiu
A Summary on the Real Stability Radius and Real
Perturbation Values 11

J. Garloff and B. Srinivasan
The Hadamard Factorization of Hurwitz and Schur Stable Polynomials 19

Y.V. Genin
On the Cauchy Index of a Real Rational Function and the Index Theory
of Pseudo-Lossless Rational Functions 23

H. Górecki, M. Szymkat and M. Zaczyk
A Generalization of the Orlando Formula-Symbolic
Manipulation Approach 33

L. Atanassova, D. Hinrichsen and V.L. Kharitonov
On Convex Stability Directions for Real Quasipolynomials 43

E.I. Jury
From J.J. Sylvester to Adolf Hurwitz: A Historical Review 53

F.J. Kraus, M. Mansour and M. Sebek
Hurwitz Matrix for Polynomial Matrices 67

H.C. Reddy, P.K. Rojan and G.S. Moschytz
Two-Dimensional Hurwitz Polynomials 75

E.D. Sontag and H.J. Sussmann
General Classes of Control-Lyapunov Functions 87

R. Strietzel
Towards the Stability of Fuzzy Control Systems 97

M. Vidyasagar
Discrete Optimization Using Analog Neural Networks with
Discontinuous Dynamics 107

Robust Stability

B.D.O. Anderson and S. Dasgupta
Multiplier Theory and Operator Square Roots: Application to
Robust and Time-Varying Stability 113

R. Lozano and D.A. Suáres
Adaptive Control of Non-Minimum Phase Systems Subject to
Unknown Bounded Disturbances 125

M. Mansour and B.D.O. Anderson
On the Robust Stability of Time-Varying Linear Systems 135

W. Sienel
On the Computation of Stability Profiles 151

Q.H. Wu and M. Mansour
Robust Stability of Family of Polynomials with 1-Norm-Bounded
Parameter Uncertainties 163

E. Zeheb
On the Characterization and Formation of Local Convex Directions
for Hurwitz Stability 173

Numerics

G.E. Collins
Application of Quantifier Elimination to Solotareff's
Approximation Problem 181

R. Jeltsch
Stability of Time Discretization, Hurwitz Determinants
and Order Stars 191

R. Liska and S. Steinberg
Solving Stability Problems Using Quantifier Elimination 205

M.R. Trummer
Stability of Numerical Methods for Solving Differential Equations 211

J. Sreedhar, P. Van Dooren and A.L. Tits
A Fast Algorithm to Compute the Real Structured Stability Radius 219

Some Open Problems 231

Apendix: Original Article by A. Hurwitz
A. Hurwitz
Über die Bedingungen, unter welchen eine Gleichung nur
Wurzeln mit negativen reellen Theilen besitzt 239

Preface

This book contains the historical development of the seminal paper of Adolf Hurwitz, professor in mathematics at ETH (1892–1919), and its impact on other fields. The major emphasis, however, is on modern results in stability theory and its application in the theory of control and numerics. In particular, stability of the following problems is treated: linear, nonlinear and time-dependent systems, discretizations of ordinary and partial differential equations, systems with time delay on multidimensional systems. In addition robust stability, pole placement and problems related to the stability radius are treated.

The book is an outgrowth of the international conference "Centennial Hurwitz on Stability Theory" which was held to honor Adolf Hurwitz, whose article on the location of roots of a polynomial was published one hundred years ago. The conference took place at the Centro Stefano Franscini, Monte Verità, Ascona, Switzerland, on May 21–26, 1995. This book contains a collection of the papers and open problems discussed on that occasion. Leading researchers from all over the world working on stability theory and its application were invited to present their recent results. In one paper the historic development initiated by Hurwitz's article was discussed.

The interaction between the two major groups of participants, researchers in control theory and mathematics, as well as the excellent setting of the Monte Verità strongly contributed to the success of the meeting. We thank the Centro Stefano Franscini of the Swiss Federal Institute of Technology (ETH) in Zürich and the Swiss National Science Foundation for financial support. Our thanks go also to the publisher Springer which allowed us to reproduce the original article by Adolf Hurwitz in this volume. We also thank the members of the Seminar for Applied Mathematics (ETH) for their help in the successful organization of the meeting, in particular Ms M. Krämer and M. Pfister. In addition we thank the secretaries of the Centro Stefano Franscini, Ms K. Bastianelli and F. Tewelde, for their support.

We express our deep gratitude to the staff of Birkhäuser Verlag for their excellent cooperation in producing this volume. Our biggest thanks go to our secretaries, Ms. M. Krämer and M. Pfister, as well as to our systems people, Ms. E. Copeland and Dr. P. Scherbel, who all went to great effort to produce the excellent electronic version of this book.

We hope that this volume will be helpful to engineers and mathematicians working in the area of control theory and numerics.

Rolf Jeltsch
Mohamed Mansour
Zürich, Switzerland
October 1995

The Hurwitz Matrix and the Computation of second-order Information Indices

Alessandro Beghi, Antonio Lepschy and Umberto Viaro
Department of Electronics and Informatics
University of Padova, via Gradenigo 6/A, 35131 Padova, Italy.

Abstract. An all-pole transfer function $Q(s) = 1/P(s)$, where $P(s)$ is a monic Hurwitz polynomial of degree n, is uniquely characterized by the energies (second-order information indices) of $q(t) = LT^{-1}\{Q(s)\}$ and of its first $n-1$ derivatives. These can be obtained by solving a set of linear equations whose coefficients matrix is the standard Hurwitz matrix for $P(s)$ or by using the entries of its Routh table. Any strictly proper transfer function $W(s) = N(s)/P(s)$ is characterized by n first-order information indices, e.g., Markov parameters, and by n second-order information indices, e.g., the energies of the related impulse response and its $n-1$ successive derivatives; these are simply obtainable from the energies of $q(t)$. This fact can be exploited to construct reduced-order models that retain both first- and second-order information indices of a given original system. The extension of this approach to multi-input multi-output systems described by a matrix fraction is analysed.

1. Introduction. As is known, the Hurwitz matrix as well as the Routh array were conceived for stability analysis and zeros separation problems but have found interesting applications in different fields too. In particular, they have been used to compute the integrals along the imaginary axis of the square magnitude of rational functions of a complex variable, which are related to the energies of certain signals. In this regard, we may recall Krasovskii's formula (cf. [1, p. 262]) for computing the so-called ISE (integral of the square error in the step response) which uses a matrix similar to the Hurwitz one, the procedure and related table given by Phillips [2], the method independently derived by Kac [3] and Mersman [4] which requires the computation of a number of determinants formed from the coefficients of the considered rational function, the expression suggested by Nekolný and Beneš [5], and those obtained by Effertz [6] along lines similar to those followed by Åström et al. [7] and by Åström [8]; the problem has also been considered by Lehoczky [9], by Csáki and Lehoczky [10], by Lepschy et al. [11] and by the present authors [12].

The above-mentioned techniques can usefully be applied to compute the so-called second-order information indices of a dynamic system, such as the entries of the associated impulse-response Gramian. In fact, its diagonal elements are the energies of the impulse response and of some of its derivatives, and the off-diagonal elements can be obtained by adding to such energies terms that simply depend on suitable first-order information indices, e.g., Markov parameters; for this reason the energies can be regarded as the essential second-order information.

The idea of characterizing a linear time-invariant dynamic system of order n by means of n first-order and n second-order information indices goes back to the seminal paper by Mullis and Roberts [13] in which they also suggest a procedure for constructing simplified models that retain an equal number of first- and second-order information indices of an original system. This approach has been adopted by various authors [14]÷[25] for both single-input single-output (SISO) and multi-input multi-output (MIMO) systems in discrete and continuous time.

This paper is concerned with the above topics; more precisely, it deals with the following problems: (i) the computation of the energies associated with a transfer function by using Hurwitz-type matrices or the Routh table; (ii) the use of the considered information indices for characterizing and simplifying SISO continuous-time systems; (iii) the extension to the case of MIMO systems and the comparison of the related features with those of the SISO case.

2. Computation of the energies.

Let us consider a monic Hurwitz polynomial of degree n with real coefficients

$$P(s) = s^n + a_{n-1}s^{n-1} + \ldots + a_0 \tag{1}$$

and the associated all-pole (stable) transfer function

$$Q(s) = \frac{1}{P(s)}. \tag{2}$$

The corresponding impulse response $q(t) = LT^{-1}\{1/P(s)\}$ satisfies the homogeneous differential equation

$$q^{(n)}(t) + \sum_{i=0}^{n-1} a_i q^{(i)}(t) = 0 \tag{3}$$

for the initial conditions $q^{(i)}(0_+) = 0$, $i = 0, 1, \ldots, n-2$, and $q^{(n-1)}(0_+) = 1$. Multiplying successively eqn. (3) by $(-1)^j q^{(j)}(t)$, $j = 0, 1, \ldots, n-1$, integrating from 0_+ to ∞, and denoting by e_i the energy of $q^{(i)}(t)$:

$$e_i \triangleq \int_{0_+}^{\infty} \left[q^{(i)}(t)\right]^2 dt, \qquad i = 0, 1, \ldots, n-1 \tag{4}$$

we arrive for n even at the following set of equations

$$\begin{cases} a_0 e_0 - a_2 e_1 + a_4 e_2 - \ldots - (-1)^{(n/2)-1} a_{n-2} e_{(n/2)-1} = (-1)^{(n/2)-1} e_{n/2} \\ -a_1 e_1 + a_3 e_2 - a_5 e_3 + \ldots + (-1)^{n/2} a_{n-1} e_{n/2} = 0 \\ \vdots \\ a_1 (-1)^{(n/2)-1} e_{n/2} - (-1)^{n/2} a_3 e_{(n/2)+1} + \ldots - a_{n-1} e_{n-1} = -\frac{1}{2} \end{cases} \tag{5}$$

and for n odd at a similar set. These sets are linear in the coefficients a_i and in the energies e_i. If the latter are given and coefficients a_i are unknown, these can uniquely be determined by solving the matrix equation

$$Ga = g \tag{6}$$

where G (impulse-response Gramian) is the nonsingular matrix

$$
(7) \qquad G = \begin{bmatrix} e_0 & 0 & -e_1 & \cdots & \cdots \\ 0 & e_1 & 0 & \cdots & \cdots \\ -e_1 & 0 & e_2 & \cdots & \cdots \\ \cdots & \cdots & \cdots & \cdots & \cdots \\ \cdots & \cdots & \cdots & \cdots & e_{n-1} \end{bmatrix}
$$

and

$$
(8) \quad a = \begin{bmatrix} a_0, & a_1, & \cdots & a_{n-1} \end{bmatrix}^T
$$

$$
(9) \quad g = \begin{cases} \begin{bmatrix} (-1)^{(n-2)/2} e_{n/2}, & 0, & \cdots & e_{n-1}, & \frac{1}{2} \end{bmatrix}^T & n \text{ even} \\ \begin{bmatrix} 0, & (-1)^{(n+1)/2} e_{(n+1)/2}, & 0, & \cdots & e_{n-1}, & \frac{1}{2} \end{bmatrix}^T & n \text{ odd} \end{cases}.
$$

Note that, by exploiting the structure of (7), eqn. (6) can be decoupled into two matrix equations whose unknown are the even-order coefficients a_{2i} and odd-order coefficients a_{2i+1}, respectively.

If, instead, coefficients a_i are given and energies e_i are unknown, these can uniquely be determined by solving the matrix equation

$$
(10) \qquad He = h
$$

where H is the nonsingular Hurwitz matrix associated with the stable system (2)

$$
(11) \qquad H = \begin{bmatrix} a_{n-1} & a_{n-3} & a_{n-5} & \cdots & \cdots & \cdots \\ 1 & a_{n-2} & a_{n-4} & \cdots & \cdots & \cdots \\ 0 & a_{n-1} & a_{n-3} & \cdots & \cdots & \cdots \\ 0 & 1 & a_{n-2} & \cdots & \cdots & \cdots \\ \cdots & \cdots & \cdots & \cdots & \cdots & \cdots \\ \cdots & \cdots & \cdots & \cdots & \cdots & a_0 \end{bmatrix}
$$

and

$$
(12) \qquad e = \begin{bmatrix} e_{n-1} & -e_{n-2} & e_{n-3} & -e_{n-4} & \cdots & e_0 \end{bmatrix}^T
$$

$$
(13) \qquad h = \begin{bmatrix} \frac{1}{2} & 0 & 0 & 0 & \cdots & 0 \end{bmatrix}^T.
$$

Therefore, a and e are alternative parameterizations of the all-pole transfer function (2).

Eqn. (10) can be rewritten as

$$
(14) \qquad K\tilde{e} = k
$$

where K is the matrix considered by Krasovskii for the computation of the ISE

$$
(15) \qquad K = \begin{bmatrix} a_0 & -a_2 & a_4 & \cdots & \cdots & \cdots \\ 0 & a_1 & -a_3 & \cdots & \cdots & \cdots \\ 0 & -a_0 & a_2 & \cdots & \cdots & \cdots \\ \cdots & \cdots & \cdots & \cdots & \cdots & \cdots \\ \cdots & \cdots & \cdots & \cdots & a_{n-2} & -1 \\ \cdots & \cdots & \cdots & \cdots & -a_{n-3} & a_{n-1} \end{bmatrix}
$$

and

(16) $$\tilde{e} = \begin{bmatrix} e_0 & e_1 & e_2 & \ldots & e_{n-1} \end{bmatrix}^T$$

(17) $$k = \begin{bmatrix} 0 & 0 & 0 & \ldots & \frac{1}{2} \end{bmatrix}^T.$$

The matrices in (10) are related to those in (14) via $H = TKT^{-1}$, $e = T\tilde{e}$, $h = Tk$, where

(18) $$T = \begin{bmatrix} 0 & & \ldots & 0 & 0 & 1 \\ 0 & & \ldots & 0 & -1 & 0 \\ 0 & & \ldots & 1 & 0 & 0 \\ \ldots & & \ldots & \ldots & \ldots & \ldots \\ (-1)^{n-1} & & 0 & \ldots & 0 & 0 \end{bmatrix}.$$

Relations equivalent to (10) and (14) can also be derived from the state-space realization of (2) in observability form and a Lyapunov equation. Specifically, let

(19) $$\begin{cases} \dot{x} = Ax + Bu \\ y = Cx \end{cases}$$

with

(20) $$A = \begin{bmatrix} 0 & 1 & \ldots & 0 \\ \vdots & \vdots & \vdots & \vdots \\ 0 & 0 & \ldots & 1 \\ -a_0 & -a_1 & \ldots & -a_{n-1} \end{bmatrix} \quad \begin{aligned} B^T &= \begin{bmatrix} 0 & \ldots & 0 & 1 \end{bmatrix} \\ C &= \begin{bmatrix} 1 & 0 & \ldots & 0 \end{bmatrix} \end{aligned}.$$

Now, the impulse-response Gramian (7) satisfies the Lyapunov equation

(21) $$AG + GA^T = -BB^T$$

which can be solved for $\tilde{e}$, leading to

(22) $$L\tilde{e} = -k$$

where $L = \hat{T}K$ and

(23) $$\hat{T} = \begin{bmatrix} I_{n-1} & 0_{n-1,1} \\ 0_{1,n-1} & -1 \end{bmatrix}$$

with the obvious meaning of the symbols.

As is expected, the energies can also be computed using the Routh table. To this purpose, it is convenient to normalize the coefficients of the denominator of $Q(s)$ to a_0, i.e., to refer to the polynomial

(24) $$\hat{P}(s) = 1 + \hat{a}_1 s + \hat{a}_2 s^2 + \ldots + \hat{a}_n s^n$$

where $\hat{a}_i = a_i/a_0$, $i = 0, \ldots, n$. Clearly, the energies $\hat{e}_i$ for $\hat{Q}(s) = 1/\hat{P}(s)$ are related to the energies e_i for $Q(s)$ via $\hat{e}_i = a_0^2 e_i$. By writing the Routh table for $\hat{P}(s)$ as

(25)
$$\begin{array}{ll}
r_{nn} = \hat{a}_n & r_{n,n-2} = \hat{a}_{n-2} \quad \ldots \\
r_{n-1,n-1} = \hat{a}_{n-1} & r_{n-1,n-3} = \hat{a}_{n-3} \quad \ldots \\
r_{n-2,n-2} & r_{n-2,n-4} \quad \ldots \\
r_{n-3,n-3} & r_{n-3,n-5} \quad \ldots \\
\vdots & \vdots \\
r_{11} & \\
1 &
\end{array}$$

indices $\hat{e}_i$ are obtained recursively with the formula

(26) $$\hat{e}_i = \frac{1}{r_{i+1,i+1}} \{r_{i+1,i-1}\hat{e}_{i-1} - r_{i+1,i-3}\hat{e}_{i-2} + \ldots\}$$

starting from $\hat{e}_0 = 1/2r_{11}$, which is, in general, the easiest way to compute the energies of $q(t)$ and its derivatives.

The energy $\bar{e}_0$ of the impulse response $w(t)$ of a system whose transfer function is

(27) $$W(s) = Q(s)N(s)$$

with $N(s) = b_0 + b_1 s + \ldots + b_m s^m$, $m < n$, can immediately be obtained from the energies e_i for $Q(s)$ according to

(28) $$\bar{e}_0 = \sum_{k=0}^{m} B_{2k} e_k$$

where

(29) $$B_{2k} = b_k^2 + 2\sum_{\ell=1}^{k} (-1)^\ell b_{k-\ell} b_{k+\ell}.$$

The energies $\bar{e}_i$ of the derivatives $w^{(i)}(t)$ can be evaluated in a similar way with reference to the strictly proper part of the corresponding Laplace transforms (which are not strictly proper for $i \geq n - m$).

When the numerator of $W(s)$ is not equal to 1, the system can be (completely) characterized by the first n energies $\bar{e}_i$ and n Markov parameters $m_i = w^{(i-1)}(0)$ (that are all zero except for $m_n = 1$ in the case of $Q(s)$). Specifically, from coefficients a_i and b_i we uniquely determine the information indices of the first order m_i, e.g., via the Padé procedure applied to the asymptotic series expansion of $W(s)$, and of the second order $\bar{e}_i$, e.g., via (28); conversely, from indices m_i and $\bar{e}_i$

we uniquely determine coefficients a_i by solving a set of linear equations (cf., e.g., [24]), and then coefficients b_i according to

$$b_i = \sum_{j=1}^{n-i-1} a_{i+j} m_j + m_{n-i}. \tag{30}$$

Let us finally observe that the role of first- and second-order information indices can be played by different parameters such as the energies of the transient part of the integrals of $w(t)$ and the time moments (that are related to the MacLaurin series expansion coefficients of $W(s)$) [21]; a more general characterization is provided in [16] and [25].

3. Application to model reduction. Let us consider first the problem of forming an all-pole function of order q

$$Q_q(s) = \frac{1}{P_q(s)}, \qquad q < n \tag{31}$$

in such a way that its first q energies match the first q energies for $Q(s)$. This objective can be achieved by exploiting the results of Section 2, according to either of the following procedures.

Procedure A (Hurwitz)
(i) Compute the energies e_i for $Q(s)$ by means of (10).
(ii) Form the $q \times q$ leading principal submatrix G_q from the original Gramian G and the vector g_q from the upper q entries of its $(q+1)$-th column.
(iii) Compute the vector a_q of the coefficients a_{q_i}, $i = 0, 1, \ldots, q-1$, in the monic polynomial $P_q(s)$ by solving the equation

$$G_q a_q = g_q. \tag{32}$$

(iv) The resulting $Q_q(s) = 1/P_q(s)$ matches q energies of $Q(s)$ and its Evans gain (equal to 1).

Procedure B (Routh)
(i) Form the Routh table for $P(s)$.
(ii) Using the elements of the rows of order q and $q-1$, form the polynomial

$$P_q(s) = r_{q,q} s^q + r_{q-1,q-1} s^{q-1} + r_{q,q-2} s^{q-2} + r_{q-1,q-3} s^{q-3} + \ldots. \tag{33}$$

(iii) The resulting $Q_q(s) = 1/P_q(s)$ matches q energies of $Q(s)$ and its Bode gain (equal to $1/a_0$).

Procedures similar to the preceding ones can be conceived to obtain a model of order q that matches q energies and q first-order information indices for the more general transfer function (27) [25]. In this case too, the objective can be achieved by solving linear sets of equations both for computing the considered indices from the coefficients of $W(s)$ and for determining the coefficients of the reduced model from the indices to be retained.

4. Extension to the mimo case.

This section deals with stable systems with m_i inputs and m_o outputs whose matrix fraction description (MFD) is

$$A^{-1}(s)M_n \tag{34}$$

where M_n is an $m_o \times m_i$ constant matrix and $A(s)$ is an $m_o \times m_o$ polynomial matrix given by $A(s) = Is^n + A_{n-1}s^{n-1} + \ldots + A_0$. Expression (34) can be considered as the MIMO equivalent of the all-pole transfer function (2). Its numerator is equal to the first nonzero (n-th) Markov parameter.

The inverse Laplace transform $Q(t) = \{q_{i,j}(t)\}$ of (34) solves for $t > 0$ the homogeneous matrix equation

$$\sum_{i=0}^{n-1} A_i Q^{(i)}(t) + Q^{(n)}(t) = 0 \tag{35}$$

with $Q^{(i)}(0_+) = 0$, $i = 0, 1, \ldots, n-2$, and $Q^{(n-1)}(0_+) = M_n$. Multiplying (35) on the right by $Q^{(j)^T}(t)$, $j = 0, 1, \ldots, n-1$, integrating by parts from 0_+ to ∞, and defining

$$E_{i,j} = \int_{0_+}^{\infty} Q^{(i-1)}(t) Q^{(j-1)^T}(t)\, dt = E_{j,i}^T \tag{36}$$

we get [24]

$$AE = -E_{n+1} \tag{37}$$

where

$$E = \begin{bmatrix} E_{1,1} & E_{1,2} & \ldots & E_{1,n} \\ E_{2,1} & E_{2,2} & \ldots & E_{2,n} \\ \ldots & \ldots & \ldots & \ldots \\ E_{n,1} & E_{n,2} & \ldots & E_{n,n} \end{bmatrix} = \begin{bmatrix} E_{1,1} & E_{1,2} & -E_{2,2} & \ldots \\ -E_{1,2} & E_{2,2} & E_{2,3} & \ldots \\ -E_{2,2} & -E_{2,3} & E_{3,3} & \ldots \\ \ldots & \ldots & \ldots & \ldots \end{bmatrix} \tag{38}$$

is the impulse-response Gramian (of full rank $m_0 n$) and

$$A = \begin{bmatrix} A_0 & A_1 & \ldots & A_{n-1} \end{bmatrix} \tag{39}$$
$$E_{n+1} = \begin{bmatrix} E_{1,n+1} & E_{2,n+1} & \ldots & E_{n,n+1} \end{bmatrix}. \tag{40}$$

The first $n-1$ blocks of (40) coincide, except for the sign, with blocks of E and do not depend on M_n, whereas the last block $E_{n,n+1}$ depends on M_n:

$$E_{n,n+1} = \bar{E}_{n,n+1} - \frac{1}{2} M_n M_n^T \tag{41}$$

where $\bar{E}_{n,n+1}$ does not depend on M_n. Since for $i = 1, \ldots, n$ and $j = 1, \ldots, n+1$, we have $E_{i,j} = E_{i,j}^T$ for $j - i$ even, and $E_{i,j} = -E_{i,j}^T$ for $j - i$ odd (except for

$E_{n,n+1}$), and $\bar{E}_{n,n+1} = -\bar{E}_{n,n+1}^T$, in order to form matrix (38) and block vector (40) it is enough to know M_n and determine the following $m_o^2 n$ quantities:

$$e_{i,i}^{h,h} = \sum_{\ell=1}^{m_0} \int_{0_+}^{\infty} [q_{i,\ell}^{(h-1)}(t)]^2 \, dt, \tag{42}$$

for $h = 1, 2, \ldots, n$ and $i = 1, 2, \ldots, m_0$ (corresponding to the $m_0 n$ diagonal elements of E),

$$e_{i,j}^{h,h} = \sum_{\ell=1}^{m_0} \int_{0_+}^{\infty} q_{i,\ell}^{(h-1)}(t) q_{j,\ell}^{(h-1)}(t) \, dt \tag{43}$$

for $h = 1, 2, \ldots, n$, $i = 1, 2, \ldots, m_0 - 1$ and $j = i+1, i+2, \ldots, m_0$ (corresponding to the $m_0(m_0-1)n/2$ off-diagonal entries of the diagonal blocks E_{hh} of E),

$$e_{i,j}^{h,h+1} = \sum_{\ell=1}^{m_0} \int_{0_+}^{\infty} q_{i,\ell}^{(h-1)}(t) q_{j,\ell}^{(h)}(t) \, dt \tag{44}$$

for $h = 1, 2, \ldots, n$, $i = 2, 3, \ldots, m_0$ and $j = 1, 2, \ldots, i-1$ (corresponding to the $m_0(m_0-1)n/2$ off-diagonal entries of the blocks to the right of the diagonal blocks of E and of block $\bar{E}_{n,n+1}$). These quantities have been called pseudoenergies [24] because they do not correspond to integrals of squared functions.

For example, assuming $n = 3$ and $m_o = 3$, we have

$$E = \begin{bmatrix} \begin{matrix} e_{1,1}^{1,1} & e_{1,2}^{1,1} & e_{1,3}^{1,1} \\ e_{1,2}^{1,1} & e_{2,2}^{1,1} & e_{2,3}^{1,1} \\ e_{1,3}^{1,1} & e_{2,3}^{1,1} & e_{3,3}^{1,1} \end{matrix} & \begin{matrix} 0 & -e_{2,1}^{1,2} & -e_{3,1}^{1,2} \\ e_{2,1}^{1,2} & 0 & -e_{3,2}^{1,2} \\ e_{3,1}^{1,2} & e_{3,2}^{1,2} & 0 \end{matrix} & -E_{2,2} \\ -E_{1,2} & \begin{matrix} e_{1,1}^{2,2} & e_{1,2}^{2,2} & e_{1,3}^{2,2} \\ e_{1,2}^{2,2} & e_{2,2}^{2,2} & e_{2,3}^{2,2} \\ e_{1,3}^{2,2} & e_{2,3}^{2,2} & e_{3,3}^{2,2} \end{matrix} & \begin{matrix} 0 & -e_{2,1}^{2,3} & -e_{3,1}^{2,3} \\ e_{2,1}^{2,3} & 0 & -e_{3,2}^{2,3} \\ e_{3,1}^{2,3} & e_{3,2}^{2,3} & 0 \end{matrix} \\ -E_{2,2} & -E_{2,3} & \begin{matrix} e_{1,1}^{3,3} & e_{1,2}^{3,3} & e_{1,3}^{3,3} \\ e_{1,2}^{3,3} & e_{2,2}^{3,3} & e_{2,3}^{3,3} \\ e_{1,3}^{3,3} & e_{2,3}^{3,3} & e_{3,3}^{3,3} \end{matrix} \end{bmatrix} \tag{45}$$

and

$$E_{n+1} = \begin{bmatrix} E_{2,3} & -E_{3,3} & \begin{bmatrix} 0 & -e_{2,1}^{3,4} & -e_{3,1}^{3,4} \\ e_{2,1}^{3,4} & 0 & -e_{3,2}^{3,4} \\ e_{3,1}^{3,4} & e_{3,2}^{3,4} & 0 \end{bmatrix} \end{bmatrix} + \begin{bmatrix} 0 & 0 & -\frac{1}{2} M_3 M_3^T \end{bmatrix} \tag{46}$$

By rearranging (37), it is possible to compute the pseudoenergies from the entries of matrices A_i and M_n by solving a set of linear equations which is the multivariable equivalent of eqn. (10) that uses the Hurwitz matrix; conversely, given the pseudoenergies and M_n, the MFD can be found.

This fact can be exploited to construct a reduced-order model of form (34) and McMillan degree $m_o q$, $q < n$, i.e., $\hat{A}^{-1}(s)\hat{M}_q$, in such a way that the first $m_o^2 q$ pseudoenergies of the original system are matched. Concerning $\hat{M}_q$, it is reasonable to set $\hat{M}_q = M_n$ (coincidence of the first nonzero Markov parameter). The procedure entails the following steps:
(i) Extract from E the $m_o q \times m_o q$ leading principal submatrix $\hat{E}$,
(ii) Form the block row $\hat{E}_{q+1}$ from blocks $E_{1,q+1}, \ldots, E_{q-2,q+1}$ and $E_{q-1,q+1} - \frac{1}{2}M_n M_n^T$,
(iii) Compute the reduced coefficient matrix $\hat{A}$ as

$$\hat{A} = -\hat{E}_{q+1}\hat{E}^{-1} \tag{47}$$

(iv) Form the reduced polynomial matrix $\hat{A}(s)$ from the blocks of $\hat{A}$, thus arriving at the MFD $\hat{A}^{-1}(s)M_n$.

The procedure can be generalized to the case in which the "numerator" of the MFD is not a constant. In [24] this is done by matching, besides the pseudoenergies, q (matrix) Markov parameters.

Note, finally, that the impulse response Gramian E can also be evaluated by solving the Lyapunov equation

$$F_c E + E F_c^T = -G_c G_c^T \tag{48}$$

where F_c is the $m_o n \times m_o n$ block companion matrix whose last block row is $[-A_0| - A_1| \ldots | - A_{n-1}]$ and G_c is the $m_o n \times m_i$ block column $[0 \ 0 \ldots M_n^T]^T$. Contrary to the SISO case, it is not possible to uniquely determine the parameters of the MFD from E using (48) since the resulting set of equations is underdetermined (the number of unknowns exceeds the number of equations; note that E does not contain the elements of $E_{n,n+1}$). A stable model could be determined by arbitrarily choosing $m_o(m_o - 1)/2$ entries of matrices A_i, but in this case the matching of the pseudoenergies is not ensured.

5. Conclusions. An all-pole transfer function of form (2) is characterized by the n energies (4). These can easily be computed from (10) by using the Hurwitz matrix (11) associated with the denominator of (2); other equivalent techniques have also been pointed out.

Similarly, a transfer function with zeros can be characterized by the energies of the corresponding impulse response and its $n - 1$ derivatives and by an equal number of first-order indices (Markov parameters). Again, such energies are easily evaluated, e.g., from those of the related all-pole model. These facts can be exploited to find reduced-order models that retain both first- and second-order information indices.

The results have been extended to MIMO systems by referring to MFD's. It turns out that, in this case, the approach based on the Lyapunov equation (48) is not equivalent to that based on the generalization of the Hurwitz matrix.

References

[1] A. Netushil, Ed., *Theory of Automatic Control*, Moscow:Mir Publishers, 1978.

[2] R.S. Phillips, "Rms-error criterion in servomechanism design," in *Theory of Servomechanisms*, H.M. James, N.B. Nichols, R.S. Phillips, Eds. New York: McGraw-Hill, 1947.

[3] A.M. Kac, "On the calculation of a quadratic criterion for the quality of a control system," *Prikl. Mat. Mech.*, vol. 16, pp. 362–364, 1952 (in Russian).

[4] W.A. Mersman, "Evaluation of an integral occuring in servomechanism theory," *Pacific J. Math*, vol. 3, pp. 627–632, 1953.

[5] J.Nekolný and J. Beneš, "Simultaneous control of stability and quality of adjustment - Application in statistical dynamics," in *Proc. 1960 IFAC Congress*, London: Butterworths, 1961, vol. 2, pp.734–744.

[6] F.H. Effertz, "On two coupled matrix algorithms for the evaluation of the rms error criterion of linear systems," *Proc. IEEE*, vol. 54, pp. 879–880, 1966.

[7] K.J. Åström, E.I. Jury, and R.G. Agniel, "A numerical method for the evaluation of complex integrals," *IEEE Trans. Automat. Contr.*, vol. AC-15, pp. 468–471, 1970.

[8] K.J. Åström: *Introduction to Stochastic Control Theory*, New York: Academic Press, 1970.

[9] J. Lehoczky, "The determination of simple quadratic integrals by Routh coefficients," *Periodica Polytechnica Elec. Engrg.*, vol. 10, pp. 153–166, 1966.

[10] F. Csáki and J. Lehoczky, "Comment on 'A remark on Routh's array'," *IEEE Trans. Automat. Contr.*, vol. AC-12, pp. 462–463, 1967.

[11] A. Lepschy, G.A. Mian, and U. Viaro, "Stability preservation and computational aspects of a newly proposed reduction method," *IEEE Trans. Automat. Contr.*, vol. AC-33, pp. 307–310, 1988.

[12] A. Beghi, A. Lepschy, and U. Viaro, "A property of the Routh table and its use," *IEEE Trans. Automat. Contr.*, vol. AC-39, pp. 2494–2496, 1994.

[13] C.T. Mullis and R.A. Roberts, "The use of second-order information in the approximation of discrete-time linear systems," *IEEE Trans. Acoust., Speech, Signal Processing*, vol. ASSP-24, pp. 226–238, 1976.

[14] Y. Inouye, "Approximation of multivariable linear systems with impulse response and autocorrelation sequences," *Automatica*, vol. 19, pp. 265–277, 1983.

[15] A. Youssuf, D.A. Wagie, and R.E. Skelton, "Linear system approximation via covariance equivalent realizations," *J. Math. Anal. Appl.*, vol. 106, pp. 91–115, 1985.

[16] C. de Villemagne and R.E. Skelton, "Model reduction using a projection formulation," *Int. J. Contr.*, vol. 46, pp. 2141–2169, 1987.

[17] H. Nagaoka, "Mullis-Roberts-type approximation for continuous-time linear systems," *Electronics and Communications in Japan*, part I, vol. 70, pp. 41–52, 1987.

[18] B.D.O. Anderson and R.E. Skelton, "The generation of all q-Markov covers," *IEEE Trans. Circuits Syst.*, vol. CAS-35, pp. 375–384, 1988.

[19] R.E. Skelton and B.D.O. Anderson, "Weighted q-Markov covariance equivalent realizations," *Int. J. Contr.*, vol. 49, pp. 1755–1771, 1989.

[20] P. Agathoklis and V. Sreeram, "Identification and model reduction from impulse response data," *Int. J. Systems Sci.*, vol. 21, pp. 1541–1552, 1990.

[21] V. Sreeram and P. Agathoklis, "On the computation of the Gram matrix in time domain and its applications," *IEEE Trans. Automat. Contr.*, vol. AC-38, pp. 1516–1520, 1993.

[22] V. Sreeram and P. Agathoklis, "Model reduction of linear continuous systems using weighted impulse response Gramians," *IEE Proc. D*, vol. 140, pp. 345–352, 1993.

[23] V. Sreeram and P. Agathoklis, "The discrete-time q-Markov cover models with improved low-frequency approximation," *IEEE Trans. Aut. Contr.*, vol. 39, pp. 1102–1105, 1994.

[24] W. Krajewski, A. Lepschy, and U. Viaro, "Reduction of linear continuous-time multivariable systems by matching first- and second-order information," *IEEE Trans. Automat. Contr.*, vol. AC-39, pp. 2126–2129, 1994.

[25] W. Krajewski, A. Lepschy, and U. Viaro, "Model reduction by matching Markov parameters, time moments, and impulse-response energies," *IEEE Trans. Automat. Contr.*, vol. AC-40, pp. 949–953, 1995.

A Summary on the Real Stability Radius and Real Perturbation Values

Bo Bernhardsson, Anders Rantzer, Li Qiu

1. Introduction. The stability radius problem has an interesting history in the mathematical and control theory literature. Stability radii also occur in numerical analysis in relation with the so called pseudospectra introduced by Trefethen and the analysis of stability of numerical solvers for ordinary differential equations.

There are several different definitions of so called "stability radii". The theory for the complex stability radius is equivalent to H_∞-theory and can be connected to Riccati equations. Towards the end of 1980's, attention was focused on the real stability radius. Hinrichsen, Pritchard, and associates studied various properties of the real stability radius and surveyed their results in [5]. It was also studied by people in numerical linear algebra, see [9]. Several lower bounds on the real stability radius were obtained in [12]. We will here focus on presenting the recent ideas behind the calculation of the real stability radius and the connected "real perturbation values" of a matrix.

Consider the following problem closely connected with the computation of the real stability radius: Given a complex matrix $M \in \mathbf{C}^{p \times m}$, compute the so called *"real perturbation values"* of M:

$$(1) \quad \tau_k(M) := \left[\min\{\|\Delta\| : \Delta \in \mathbf{R}^{m \times p} \text{ and } \mathrm{rank}(I_m - \Delta M) = m - k\}\right]^{-1}.$$

Note that Δ is here assumed *real*, while M is a complex matrix. The size of the matrix Δ is measured in induced 2-norm, i.e. as the largest singular value and the inverse is taken for later notational convenience. When M is real, $\tau_k(M) = \sigma_k(M)$, where $\sigma_k(M)$ denote the standard singular values of M ordered nondecreasingly.

We have recently shown the following easily computable formula for the real perturbation values:

$$(2) \quad \tau_k(M) = \inf_{\gamma \in (0,1]} \sigma_{2k}\left(\begin{bmatrix} \mathrm{Re}\, M & -\gamma \, \mathrm{Im}\, M \\ \gamma^{-1} \, \mathrm{Im}\, M & \mathrm{Re}\, M \end{bmatrix}\right).$$

This result generalizes a formula for τ_1 obtained in [10], [11] to arbitrary k. The new more general proof is quite different and sheds new light on the previous results on the real stability radius. The problem is by (2) reduced to a simple one-parametric minimization, where actually only k local minima can occur. It is also possible to give a constructive method for finding a minimizing Δ. For a discussion of numerical issues connected with the computation of (2) and the real stability radius see the contribution by van Dooren et.al. in this book.

The *(structured) real stability radius* of a matrix quadruple $(A, B, C, D) \in \mathbf{R}^{n \times n} \times \mathbf{R}^{n \times m} \times \mathbf{R}^{p \times n} \times \mathbf{R}^{p \times m}$ satisfies

$$r_R(A,B,C,D) := \inf_{\Delta \in \mathbf{R}^{m \times p}} \{\|\Delta\| : \ A + B(I - \Delta D)^{-1}\Delta C \text{ is unstable}$$
$$\text{or } \det(I - \Delta D) = 0\}.$$
$$= \left[\sup_{s \in \partial C_g} \tau_1(D + C(sI - A)^{-1}B)\right]^{-1}$$

where ∂C_g denotes the boundary of the stability domain C_g. This should be compared with the corresponding well known result for complex perturbations

$$r_C(A,B,C,D) := \inf_{\Delta \in \mathbf{C}^{m \times p}} \{\|\Delta\| : \ A + B(I - \Delta D)^{-1}\Delta C \text{ is unstable}$$
$$\text{or } \det(I - \Delta D) = 0\}.$$
$$= \left[\sup_{s \in \partial C_g} \sigma_1(D + C(sI - A)^{-1}B)\right]^{-1}$$

which connects the complex stability radius with H_∞-theory.

It is also relatively easy to show that

$$(3) \qquad \tau_k(M) = \max_{dim(S)=k} \min_{z \in S} \frac{\|\operatorname{Re}(Mz)\|}{\|\operatorname{Re}(z)\|}.$$

Compare this with the classical variational formula for the singular values

$$(4) \qquad \sigma_k(M) = \max_{dim(S)=k} \min_{z \in S} \frac{\|Mz\|}{\|z\|}.$$

The real perturbation values seem to be new interesting entities connected to a complex matrix. Many results for the singular values have counterparts for the real perturbation values. The analysis is related to other interesting areas in mathematics, such as Hermitian-symmetric inequalities, consimilarity and quaternions.

Due to page limitations it will not be possible to give all details here. An extended version of this paper will appear elsewhere.

2. Calculation of the Real Perturbation Values. Formula (2) follows from:

THEOREM 2.1. *Given a matrix $C \in \mathbf{C}^{p \times m}$ the following four conditions are equivalent, where $A = M^*M - \tau^2 I$, $B = M^T M - \tau^2 I$, $\Delta \in \mathbf{R}^{m \times p}$ and S_k is a complex matrix of rank k:*

$$(5) \qquad \exists S_k, \Delta : \quad \|\Delta\| \leq \tau^{-1} \quad \text{and} \quad (I - \Delta M)S_k = 0$$

$$(6) \qquad \exists S_k : \quad \begin{bmatrix} S_k & 0 \\ 0 & \overline{S_k} \end{bmatrix}^* \begin{bmatrix} A & \overline{B} \\ B & \overline{A} \end{bmatrix} \begin{bmatrix} S_k & 0 \\ 0 & \overline{S_k} \end{bmatrix} \geq 0$$

(7) $$\inf_{|\alpha|\leq 1} \lambda_{2k}\left(\begin{bmatrix} A & \alpha\overline{B} \\ \alpha B & \overline{A} \end{bmatrix}\right) \geq 0$$

(8) $$\inf_{\gamma\in(0,1]} \sigma_{2k}\left(\begin{bmatrix} \operatorname{Re} M & -\gamma\operatorname{Im} M \\ \gamma^{-1}\operatorname{Im} M & \operatorname{Re} M \end{bmatrix}\right) \geq \tau$$

Proof That (5) is equivalent to (6) follows from the fact that given two complex matrices U and V there exists a real contraction, i.e. a real matrix Δ with $\|\Delta\| \leq 1$, such that $\Delta U = V$ if and only if

$$\begin{bmatrix} U^* \\ U^T \end{bmatrix}\begin{bmatrix} U & \overline{U} \end{bmatrix} \geq \begin{bmatrix} V^* \\ V^T \end{bmatrix}\begin{bmatrix} V & \overline{V} \end{bmatrix}.$$

If (6) is true then it is easy to see that the same statement with B replaced with $-B$ is also true. By convexity one can then replace B by αB for any $\alpha \in [-1,1]$. Hence (6) implies (7). That (7) is equivalent to (8) can be seen as follows: Since

$$P_M(\gamma) := \begin{bmatrix} \operatorname{Re} M & -\gamma\operatorname{Im} M \\ \gamma^{-1}\operatorname{Im} M & \operatorname{Re} M \end{bmatrix} = D\begin{bmatrix} M & 0 \\ 0 & \overline{M} \end{bmatrix}D^{-1}$$

$$\text{where} \quad D = \begin{bmatrix} i\gamma I & i\gamma I \\ I & -I \end{bmatrix}$$

we can use the fact that congruence transformations do not change the sign of eigenvalues to show that

$$\lambda_{2k}(P_M(\gamma)^* P_M(\gamma) - \tau^2 I) \geq 0 \iff$$

$$\lambda_{2k}\left(\begin{bmatrix} M & 0 \\ 0 & \overline{M} \end{bmatrix}^* D^* D \begin{bmatrix} M & 0 \\ 0 & \overline{M} \end{bmatrix} - \tau^2 D^* D\right) \geq 0$$

However, because $D^* D = \begin{bmatrix} \gamma^2 + 1 & \gamma^2 - 1 \\ \gamma^2 - 1 & \gamma^2 + 1 \end{bmatrix}$ this is equivalent to

$$\lambda_{2k}\left(\begin{bmatrix} A & 0 \\ 0 & \overline{A} \end{bmatrix} + \frac{\gamma^2-1}{\gamma^2+1}\begin{bmatrix} 0 & \overline{B} \\ B & 0 \end{bmatrix}\right) \geq 0.$$

With $\alpha := (\gamma^2-1)/(\gamma^2+1)$ this proves that (7) is equivalent to (8) with the interval $(0,1]$ replaced by $(0,\infty)$. The singular values of $P_M(\gamma)$ and $P_M(\gamma^{-1})$ are however the same. This concludes the proof of the theorem except for the implication (7) $\Rightarrow$ (6). We will not prove that in full generality here. The following lemma however illustrates the main idea.

Remark The implication (7) $\Rightarrow$ (6) would be relatively easy to prove if A and B where diagonal. The existence of a simultaneous $*$- and T-block diagonalization of A and B. required in the next lemma is actually not a strong restriction. If the eigenvalues to the generalized eigenvalue problem

$$\lambda\begin{bmatrix} 0 & \overline{B} \\ B & 0 \end{bmatrix} - \begin{bmatrix} A & 0 \\ 0 & \overline{A} \end{bmatrix}.$$

are simple, S can be constructed from the corresponding eigenvectors. In the case of multiple eigenvalues the simultaneous block-diagonalization might however not be possible. One must then replace the block-diagonal matrices with matrices of a form parallel to those in the Jordan canonical form. The proof then becomes slightly more technical. For a more extensive discussion see [3]. See also the related references [1], [6], [14], [15] and [16].

LEMMA 2.2. *Let A and B a be defined as in the previous theorem. Assume that there exists a complex non-singular matrix S such that*

$$(9) \qquad S^*AS = \Lambda \quad and \quad S^T BS = I$$

where $\Lambda = diag(\Lambda_1, \ldots, \Lambda_r)$ is a block-diagonal matrix with blocks of the form $\Lambda_j = \lambda_j \in R$ or $\Lambda_j = \begin{bmatrix} 0 & \overline{\lambda_j} \\ \lambda_j & 0 \end{bmatrix}$. Then (6) and (7) are equivalent to the following condition

(*) *Let n_1, n_2 denote the number of $\lambda_j \geq 1$ and non-real λ_j respectively and let the real eigenvalues smaller than 1 be ordered so that $1 > \lambda_1 \geq \lambda_2 \geq \ldots$. Then either $n_1 + n_2 \geq k$ or else*

$$\lambda_j + \lambda_{2k-2n_1-2n_2+1-j} \geq 0, \quad j = 1, \ldots, k - n_1 - n_2.$$

Proof After a congruence transformation condition (7) means that $\begin{bmatrix} \Lambda & \alpha I \\ \alpha I & \Lambda \end{bmatrix}$ has $2k$ positive eigenvalues for all $\alpha \in [-1, 1]$. The eigenvalues can be studied block by block. Because of symmetry it is enough to study $\alpha \in [0, 1]$. The eigenvalues are given by $\lambda_j \pm \alpha$ if $\lambda_j \in R$ and $\pm(\alpha^2 + |\lambda_j|^2 \pm 2\alpha \operatorname{Re}(\lambda_j))^{1/2}$ otherwise. If the eigenvalues of the matrix above is plotted as function of α there will be $2n$ curves. The n_2 non-real blocks of Λ give rise to $4n_2$ curves, half of them above 0, half of them below 0. The real blocks will give $2n - 4n_2$ straight lines, half of them with slope 1, half with slope -1 as a function of α. The eigenvalues with $\lambda_j \geq 1$ give $2n_1$ lines above 0 for all $\alpha \in [-1, 1]$. Drawing a diagram and figuring out the condition for existence of a total of $2k$ curves above 0 for all $\alpha \in [-1, 1]$ one arrives at the somewhat involved condition (*).

From (*) it is easy to construct the nonsingular matrix S_k needed in (6) by using the blocks corresponding to the $2k$ curves above 0. In fact we construct S_k in the following way. Let S be given as above and put for each real eigenvalue ≥ 1

$$s_j = S \begin{bmatrix} 0 & \ldots & 1 & 0 & \ldots & 0 \end{bmatrix}^T, \quad j = 1, \ldots, n_1.$$

so that $s_j^* A s_j \geq 1$ and $s_j^T B s_j = 1$. For each pair of complex eigenvalues put

$$s_j = S \begin{bmatrix} 0 & \ldots & 1 & i & \ldots & 0 \end{bmatrix}^T, \quad j = n_1 + 1, \ldots, n_1 + n_2.$$

Assuming $\operatorname{Im}(\lambda_j) < 0$ this gives $s_j^* A s_j = -2\operatorname{Im}(\lambda_j) > 0$ and $s_j^T B s_j = 0$. Put for each pair with $\lambda_j + \lambda_{2k-2n_1-2n_2+1-j} \geq 0$

$$s_j = S \begin{bmatrix} 0 & \dots & 1 & i & \dots & 0 \end{bmatrix}^T, \quad j = n_1 + n_2 + 1, \dots, k.$$

This gives $s_j^* A s_j = \lambda_j + \lambda_{2k-2n_1-2n_2+1-j} \geq 0$ and $s_j^T B s_j = 0$. With $S_k = \begin{bmatrix} s_1 & \dots & s_k \end{bmatrix}^T$ we hence have

$$S_k^* A S_k = \operatorname{diag}(\Lambda_c, \Lambda_d) \quad \text{and} \quad S_k^T B S_k = \operatorname{diag}(I, 0),$$

where $\Lambda_c \geq I$ and $\Lambda_d \geq 0$. From this (6) follows. That (6) implies (7) was proved in the previous theorem.

2.1. The Courant-Fischer Type Formula.
Formula (3) follows from

LEMMA 2.3. *Let $A = M^*M - \tau^2 I$, $B = M^T M - \tau^2 I$ and assume S_k denotes a matrix of rank k. Then the following conditions are equivalent*

(6) $\exists S_k$: $\begin{bmatrix} S_k & 0 \\ 0 & \overline{S_k} \end{bmatrix}^* \begin{bmatrix} A & \overline{B} \\ B & \overline{A} \end{bmatrix} \begin{bmatrix} S_k & 0 \\ 0 & \overline{S_k} \end{bmatrix} \geq 0$

(**) $\exists S_k$: $z^* A z \geq \operatorname{Re}(z^T B z), \quad \forall z \in \operatorname{Im}(S_k)$

(3') $\exists S_k$: $\|\operatorname{Re}(Mz)\|^2 \geq \tau^2 \|\operatorname{Re}(z)\|^2, \quad \forall z \in \operatorname{Im}(S_k)$

Proof That (6) is equivalent to (**) follows from Lemma 3.1 below. Conditions (**) and (3') are equivalent since

$$\|\operatorname{Re}(MS_k w)\|^2 \geq \tau^2 \|\operatorname{Re}(S_k w)\|^2, \quad \forall w \in \mathbf{C}^k$$
$$\Leftrightarrow (MS_k w + \overline{M}\,\overline{S_k}\,\overline{w})^*(MS_k w + \overline{M}\,\overline{S_k}\,\overline{w}) \geq \tau^2(S_k w + \overline{S_k}\,\overline{w})^*(S_k w + \overline{S_k}\,\overline{w})$$
$$\Leftrightarrow w^* S_k^*(M^*M - \tau^2 I) S_k w \geq -\operatorname{Re}(w^T S_k^T (M^T M - \tau^2 I) S_k w), \quad \forall w \in \mathbf{C}^k$$
$$\Leftrightarrow w^* S_k^*(M^*M - \tau^2 I) S_k w \geq \operatorname{Re}(w^T S_k^T (M^T M - \tau^2 I) S_k w), \quad \forall w \in \mathbf{C}^k$$

2.2. Some Properties of the Real Perturbation Values.
The following properties follow directly from the definition of $\tau_k(M)$:

LEMMA 2.4.

(10) $\quad \tau_k(\alpha M) = \alpha \tau_k(M), \quad \alpha \in R$

(11) $\quad \tau_k(\overline{M}) = \tau_k(M)$

(12) $\quad \tau_k(Q_1 M Q_2) = \tau_k(M), \quad Q_i$ real orthogonal matrices

By using (2) for $\tau_1(M)$ and the fact that $\sigma_2(A) \geq \inf_{z \in S_2} \|Az\|/\|z\|$ for any subspace S_2 of dimension 2 one can also prove that

LEMMA 2.5. *$\tau_1(M)$ is continuous in M at points M where $\operatorname{Im}(M) \neq 0$.*

3. Connections to Other Parts of Mathematics.

3.1. Hermitian-Symmetric Inequalities.
Pairs (A, B) where A is Hermitian and B complex symmetric matrix occur occasionally in analysis, for instance in quadratic Hermitian-symmetric inequalities:

$$(13) \qquad z^*Az = \sum_{i,j=1}^n a_{ij}\bar{z}_i z_j \geq \left|\sum_{i,j=1}^n b_{ij} z_i z_j\right| = \left|z^T B z\right|, \quad \forall z \in \mathbf{C}^n.$$

Such inequalities are surveyed in [4] where the following theorem is proved:

LEMMA 3.1. *The following six statements are equivalent*

(i) $z^*Az \geq \left|z^T B z\right|$, $\forall z \in \mathbf{C}^n$

(ii) $x^*Ax + y^*Ay \geq 2\left|x^T By\right|$, $\forall x, y \in \mathbf{C}^n$

(iii) $x^*Ax + y^*Ay \geq 2\,\mathrm{Re}(x^T B y)$, $\forall x, y \in \mathbf{C}^n$

(iv) the $2n \times 2n$ matrix

$$\mathcal{A} = \begin{bmatrix} A & \bar{B} \\ B & \bar{A} \end{bmatrix}$$

is nonnegative definite, that is, $\zeta \mathcal{A} \zeta \geq 0$, $\forall \zeta \in \mathbf{C}^{2n}$

(v) $\zeta^* \mathcal{A} \zeta \geq 0$, for all $\zeta \in \mathbf{C}^{2n}$ of the form

$$\zeta = \begin{bmatrix} z \\ \bar{z} \end{bmatrix} \quad \text{where } z \in \mathbf{C}^n$$

(vi) $z^*Az \geq \mathrm{Re}\left(z^T B z\right)$, $\forall z \in \mathbf{C}^n$

There are several interesting instances of such inequalities, for example the Grunsky inequalities in the theory of univalent functions. Hermitian-symmetric inequalities also occur in analytic continuation, harmonic analysis and the moment problem for complex measures. Some of these applications are described below.

3.1.1. Grunsky Inequalities.
The most celebrated example of Hermitian-symmetric inequalities is probably the Grunsky inequalities in the classical theory of univalent analytic functions: If $f(z)$ is a normalized (i.e. $f(0) = 0$, $f'(0) = 1$) analytic function on the unit disc, then a necessary and sufficient condition that f be univalent, is that

$$\sum_{i,j=1}^n x_i \bar{x}_j \log \frac{1}{1 - z_i \bar{z}_j} \geq \left|\sum_{i,j=1}^n x_i x_j \log\left[\frac{z_i z_j}{f(z_i) f(z_j)} \frac{f(z_i) - f(z_j)}{z_i - z_j}\right]\right|$$

for all $x_1, \ldots, x_n \in \mathbf{C}$, all $z_1, \ldots, z_n$ in the unit disc, and all $n = 1, 2, \ldots$. Of course, the difference quotient is interpreted as $f'(z_i)$ if $z_i = z_j$.

3.1.2. The Moment Problem.
Consider a finite sequence of complex numbers a_0, a_1, ..., a_{2N} where a_0 is real and N is a positive integer. Define $a_{-n} = \bar{a}_n$, for $n = 1, 2, \ldots, 2N$. In [4] it is shown that the following conditions are equivalent:

(a) There exists an infinite sequence of complex numbers $(a_j)_{j=2N+1}^{\infty}$ such that the function $f(z) = a_0 + 2a_1 z + 2a_2 z^2 + 2a_3 z^3 + \ldots$ is analytic in the unit disc of the complex plane and satisfies
$$\operatorname{Re} f(z) \geq 0, \quad |z| \leq 1.$$

(b)
$$\sum_{i,j=0}^{N} a_{i-j} c_i \bar{c}_j \geq \left| \sum_{i,j=0}^{N} a_{i+j} c_i c_j \right|, \quad \forall c_0, c_1, \ldots, c_N \in \mathbf{C}.$$

3.1.3. Reproducing Kernel Hilbert Spaces.
Let Ω be a finite domain in the complex z-plane which is bounded by n closed analytic curves C_ν, $\nu = 1, 2, \ldots, n$. The Green function $g(z, \zeta)$ of Ω is defined by the following properties

(a) $g(z, \zeta)$ is harmonic for $\zeta \in \Omega$ fixed except for $z = \zeta$.
(b) $g(z, \zeta) + \log|z - \zeta|$ is harmonic in the neighborhood of $z = \zeta$.
(c) $g(z, \zeta) \equiv 0$ for $z \in \partial \Omega$ and $\zeta \in \Omega$.

The kernel functions are defined by
$$K(z, \bar{\zeta}) = -\frac{2}{\pi} \frac{\partial^2 g(z, \zeta)}{\partial z \partial \bar{\zeta}}, \quad L(z, \zeta) = -\frac{2}{\pi} \frac{\partial^2 g(z, \zeta)}{\partial z \partial \zeta}.$$

The following Hermitian/symmetric relations follow from the definitions:
$$\overline{K(z, \bar{\zeta})} = K(\zeta, \bar{z}), \quad L(z, \zeta) = L(\zeta, z).$$

One often also introduces the function $l(z, \zeta) = 1/\pi(z - \zeta)^2 - L(z, \zeta)$. One can show that for any Ω:
$$\sum_{i,j=1}^{n} x_i \bar{x}_j K(z_i, \bar{z}_j) \geq \left| \sum_{i,j=1}^{n} x_i x_j l(z_i, z_j) \right|, \quad \forall x_i \in \mathbf{C}, \forall z_i \in \Omega,$$

which is yet another example of a Hermitian-symmetric inequality. For a discussion of simultaneous diagonalization of K and l see [2]. See also [13] for an interesting discussion on connections to Hilbert transforms and the Fredholm integral equation.

3.2. Consimilarity.
We say that two matrices C, D are *consimilar* if there is a nonsingular P such that $\bar{P}^{-1} C P = D$. A mapping $T : V \to W$ between complex vector spaces V and W is called an *antilinear transformation* if
$$T(\alpha x + \beta y) = \bar{\alpha} T(x) + \bar{\beta} T(y), \quad \forall \alpha, \beta \in \mathbf{C}, x, y \in V.$$

Just as similar matrices are matrix representations of a linear transformation in different bases, consimilar matrices are matrix representations of an antilinear transformation in different bases. For a collection of results for consimilarity and more references see [8, Chapter 4]. Consimilarity is closely related to the simultaneous * and T-diagonalization in (9), see [8, Chapter 4.6], [7].

References

[1] R. Benedetti and P. Cragnolini. On simultaneous diagonalization of one Hermitian and one symmetric form. *Linear Algebra and its Applications*, pages 215–226, 1984.

[2] S. Bergman and M. Schiffer. Kernel functions and conformal mapping. *Compositio Math.*, 8:205–249, 1951.

[3] B. Bernhardsson. Simultaneous block-diagonalization of one Hermitian and one symmetric form. Internal Report TFRT-7520, Department of Automatic Control, Lund Institute of Technology, 1994.

[4] Carl H. Fitzgerald and Roger A. Horn. On the structure of Hermitian-symmetric inequalities. *J. London Math. Soc*, 2(15):419–430, 1977.

[5] D. Hinrichsen and A. J. Pritchard. Real and complex stability radii: a survey. In D. Hinrichsen and B. Mårtensson, editors, *Control of Uncertain Systems*. Birkhäuser, Boston, 1990.

[6] Y. P. Hong. A Hermitian canonical form for complex matrices under consimilarity. *Linear Algebra and its Applications*, 133:1–19, 1990.

[7] Y. P. Hong and R. A. Horn. A canonical form for matrices under consimilarity. *Linear Algebra and its Applications*, 102:143–168, 1988.

[8] R. A. Horn and C. A Johnson. *Matrix Analysis*. Cambridge University Press, 1985.

[9] C. Van Loan. How near is a stable matrix to an unstable one? *Contemporary Mathematics*, pages 465–478, 1985.

[10] L. Qiu, B. Bernhardsson, A. Rantzer, E. J. Davison, P. M. Young, and J. C. Doyle. On the real structured stability radius. In *Preprints IFAC 12th World Congress*, Sydney, Australia, 1993.

[11] L. Qiu, B. Bernhardsson, A. Rantzer, E. J. Davison, P. M. Young, and J. C. Doyle. A formula for computation of the real stability radius. *Automatica*, pages 879–890, June 1995.

[12] L. Qiu and E. J. Davison. The stability robustness determination of state space models with real unstructured perturbations. *Math. Control Signals Systems*, 4:247–267, 1991.

[13] M. Schiffer. Fredholm eigenvalues and Grunsky matrices. *Annales Polonici Mathmematici*, 39, 1981.

[14] F. Uhlig. Simultaneous block diagonalization of two real symmetric matrices. *Linear Algebra and its Applications*, 7:281–289, 1973.

[15] F. Uhlig. A canonical form for a pair of real symmetric matrices that generate a nonsingular pencil. *Linear Algebra and its Applications*, 14:189–209, 1976.

[16] F. Uhlig. A recurring theorem about pairs of quadratic forms and extensions: A survey. *Linear Algebra and Its Applications*, 25:219–237, 1979.

The Hadamard Factorization of Hurwitz and Schur stable Polynomials

J. Garloff
Fachhochschule Konstanz, Fachbereich Informatik
Postfach 10 05 43, D-78405 Konstanz, Germany
email: garloff@fh-konstanz.de

B. Srinivasan
EPFL Lausanne, Institut d'Automatique, DGM, CH-1015 Lausanne
email: Srinivasan@ia.epfl.ch

Abstract. We consider the Hadamard (i.e. the coefficient-wise) product of two polynomials. The set of the Hurwitz stable polynomials is closed under the Hadamard product, whereas the set of the Schur stable polynomials is not. In this note we show that each Schur stable polynomial allows a Hadamard factorization into two Schur stable polynomials, whereas there are Hurwitz stable polynomials of degree 4 which do not have a Hadamard factorization into two Hurwitz stable polynomials of degree 4.

The *Hadamard product* of two polynomials p and q with real coefficients

(1) $$p(x) = b_n x^n + b_{n-1} x^{n-1} + \ldots + b_1 x + b_0$$
(2) $$q(x) = c_n x^n + c_{n-1} x^{n-1} + \ldots + c_1 x + c_0$$

is defined to be

$$(p * q)(x) = b_n c_n x^n + b_{n-1} c_{n-1} x^{n-1} + \ldots b_1 c_1 x + b_0 c_0.$$

A polynomial p is termed *Hurwitz* (resp., *Schur*) *stable* if $p(x) \neq 0$ for $0 \leq \text{Re} x$ (resp., for $1 \leq |x|$). In a recent paper [1] it was shown that the set of the Hurwitz stable polynomials is closed under the Hadamard product, whereas the set of the Schur stable polynomials is not. Related material can be found in [2].

The purpose of this note is to show that each Schur stable polynomial f

(3) $$f(x) = a_n x^n + a_{n-1} x^{n-1} + \ldots + a_1 x + a_0$$

allows a Hadamard factorization $f = p*q$ into two Schur stable polynomials p and q, whereas a Hurwitz stable polynomial f may have no Hadamard factorization into two Hurwitz stable polynomials if $n > 3$.

Schur case: Let the polynomial f (3) be stable. Since the zeros of a polynomial vary continuously with its coefficients, the polynomial

$$p(x) = \sum_{k=0}^{n} a_k \epsilon^{-k} x^k$$

is also stable for ϵ sufficiently close to 1.

Since
$$q(x) = \sum_{k=0}^{n} \epsilon^k x^k = \frac{1 - (\epsilon x)^{n+1}}{1 - \epsilon x}$$

the polynomial q is stable if $\epsilon > 1$. Thus, the polynomials p,q provide a Hadamard factorization of f for $\epsilon > 1$ sufficiently close to 1.

Hurwitz case: For simplicity we assume that all polynomials have positive leading coefficients.

If the degree of the polynomial f in (3) is $n = 2$ stability of f is equivalent to the condition $a_0, a_1 > 0$ so that there are (infinitely many) Hadamard factorizations.

In the case $n = 3$ stability of f is equivalent to the condition $a_0, a_1, a_2 > 0$ and $a_0 a_3 < a_1 a_2$ and similarly for p and q. Therefore, we can choose as Hadamard factors polynomials p and q as

$$p(x) = \sum_{k=0}^{n} a_k^{1/\alpha} x^k \qquad q(x) = \sum_{k=0}^{n} a_k^{1/\beta} x^k,$$

where α, β are positive real numbers with $1/\alpha + 1/\beta = 1$ so that there are again (infinitely many) Hadamard factorizations of f.

We now show that there are stable polynomials f of degree 4 which do not allow a Hadamard factorization into two stable polynomials p and q of degree 4. Since $(\epsilon p) * (1/\epsilon q) = p * q$ for $0 \neq \epsilon \in R$ we can restrict the discussion to the monic case, i.e. $a_4 = b_4 = c_4 = 1$.

In the case $n = 4$ stability of f given by (3) is equivalent to the conditions $a_k > 0$, $k = 0, 1, 2, 3$, and

(4) $$a_1^2 + a_0 a_3^2 < a_1 a_2 a_3.$$

If there is a Hadamard factorization $f = p * q$ into two stable polynomials p and q given by (1) and (2) it follows that $b_k > 0$, $k = 0, 1, 2, 3$, and

(5) $$b_1^2 + b_0 b_3^2 < b_1 b_2 b_3$$

and similarly

(6) $$\left(\frac{a_1}{b_1}\right)^2 + \frac{a_0}{b_0}\left(\frac{a_3}{b_3}\right)^2 < \frac{a_1 a_2 a_3}{b_1 b_2 b_3}.$$

We shall show that it is not always possible to find positive constants b_0, b_1, b_2, b_3 such that (5) and (6) are together fulfilled.

Multiplying (6) by $b_1 b_2 b_3$, we get

(7) $$a_1^2 r + a_0 a_3^2 s < a_1 a_2 a_3,$$

where
$$r = \frac{b_2 b_3}{b_1} \quad \text{and} \quad s = \frac{b_1 b_2}{b_0 b_3}.$$

Substituting r, s into (5), we obtain $(\frac{1}{r} + \frac{1}{s})b_1 b_2 b_3 < b_1 b_2 b_3$ from which it follows that

(8) $$\frac{1}{r} + \frac{1}{s} < 1$$

and by positivity of r, s

(9) $$1 < r, s.$$

By (4) we may write

(10) $$a_1 a_2 a_3 = a_1^2 (1 + \kappa + \epsilon),$$

where $\kappa = a_0 a_3^2 / a_1^2$ and ϵ is a constant (depending on a_0, a_1, a_2, a_3).

Dividing (7) by a_1^2, we obtain from (10)

$$r + \kappa s < 1 + \kappa + \epsilon$$

hence

(11) $$\frac{1}{1 + \kappa + \epsilon - \kappa s} + \frac{1}{s} < \frac{1}{r} + \frac{1}{s}.$$

The minimum value of the left hand side of (11) as s ranges between 0 and $(1 + \kappa + \epsilon)/\kappa$ can be obtained by standard minimization, viz.

(12) $$\frac{1 + \kappa + 2\sqrt{\kappa}}{1 + \kappa + \epsilon} < \frac{1}{r} + \frac{1}{s}.$$

However, inequality (12) contradicts (8) if $\epsilon \leq 2\sqrt{\kappa}$. E.g. we may choose $a_0 = a_1 = a_3 = 1$, $a_2 = 3$, then $\kappa = 1$ and by (10) $\epsilon = 1$. Thus, we have found stable polynomials which do not allow a factorization into two stable polynomials of the same degree.

Acknowledgements. The results of this note were obtained during the conference 'Centennial Hurwitz on Stability Theory', Monte Verità, May 21–26, 1995. This conference has provided excellent conditions which we gratefully acknowledge. We also thank Nir Cohen for comments on the Schur case.

References

[1] Garloff, J. and Wagner, D.G., Hadamard products of stable polynomials are stable, Technical Report no. 9402, FH Konstanz, Fachbereich Informatik, to appear in *Journal Math. Analysis and Applications*.

[2] Garloff, J. and Wagner, D.G., Preservation of total nonnegativity under the Hadamard product and related topics, Technical Report no. 9501, FH Konstanz, Fachbereich Informatik, pp. 6–11, to appear in *Total Positivity and its Applications*, M. Gasca and C.A. Micchelli, Eds., Kluwer Acad. Publ., 1995.

On the Cauchy index of a real rational function and the index theory of pseudo-lossless rational functions

Yves V. Genin
CESAME, Université Catholique de Louvain, Bâtiment Euler,
Avenue G. Lemaître 4-6, B-1348 Louvain-La-Neuve, Belgium
e-mail: genin@auto.ucl.ac.be. *

> **Abstract.** The index theory relative to rational pseudo-lossless functions has been shown to be an interesting substitute for the Cauchy index theory and the argument principle theorem to discuss polynomial zero location problems. The reasons underlying this fact are put into light by working out the algebraic relations between these two equivalent approaches. A new simple proof of Kharitonov's theorem is proposed to further illustrate this issue.

1. Introduction. In two recent papers [1], [2], the index theory relative to rational pseudo-lossless functions has been shown to provide a particularly simple and powerful algebraic framework to approach the general problem of counting the zeros of a complex polynomial in distinguished regions of the complex plane, e.g. the unit disk or the right half-plane. The reader is especially referred to [1]–[4] for a general introduction to the question.

In particular, in [1], the author has revisited the classical problem of counting the zeros of an arbitrary complex polynomial in the right half-plane (Re $p > 0$), on the imaginary axis (Re $p = 0$) and, whence, in the left half-plane (Re $p < 0$) at the light of this index theory relative to rational pseudo-lossless functions. As a result, a generalized algorithm of the Routh-Hurwitz type has been proposed to solve the problem, which exhibits no pathological behavior (no singular case), implies no additional numerical complexity with respect to its standard version and yields, at no extra computational cost, a factorization of the given polynomial; it turns out that the resulting factors contain all the para-conjugate zeros (if any) of a fixed multiplicity and that the distribution of their zeros with respect to the imaginary axis (Re $p = 0$) can be determined from the outset. As the classical approach to the problem above [5], [6] relies on the evaluation of the Cauchy index of a real rational function which is defined from the considered polynomial in a well-defined manner, this observation suggests the existence of a close relation between the Cauchy index theory relative to real rational functions and the index theory relative to rational pseudo-lossless functions. The aim of the present short contribution is precisely to investigate this question.

In Section 2, the Cauchy index of a real rational function is first recalled to be defined on any open interval of the real axis as the difference between the number of positive and negative residues relative to its poles of odd multiplicity, that lie in

---

* This paper presents research results of the Belgian Programme on Interuniversity Poles of Attraction, initiated by the Belgian State, Prime Minister's Office for Science Technology and Culture. The scientific responsibility rests with its author.

the considered interval. The actual computation of this Cauchy index is classically made with the help of the Euclid algorithm initialized with the numerator and denominator polynomials of the given rational function [5], [6].

In Section 3, one considers the set of irreducible complex rational functions $F(p)$ satisfying the property $F(p) + \bar{F}(-\bar{p}) = 0$. Such a function $F(p)$ is termed pseudo-lossless and its index is equal, by definition, to the number of the zeros of the sum of its denominator and numerator polynomials in the right half-plane (Re $p > 0$) [1], [3]. It turns out that the value of this index can also be evaluated (although in a different manner) from a slight modification of the Euclid algorithm initialized with the denominator and the numerator polynomials of the given pseudo-lossless function. Contrarily to the Cauchy index however, this new index can be viewed as a functional defined on the space of rational pseudo-lossless functions, which is invariant for the inversion operation and which is additive under a additional mild assumption.

In Section 4, the set of real rational functions is first shown to be in one-to-one correspondence with the set of rational pseudo-lossless functions, up to an imaginary constant factor, via appropriate analytic continuation. From this property, the exact algebraic relation between the Cauchy index of a real function and the index of its associated pseudo-lossless function is then worked out. It appears that this relation is somewhat involved in the sense that it depends not only on the function degree but also on the possible existence of poles at infinity. Moreover, this relation can be viewed as a generalization of a well known property of the Cauchy index, classically derived via the argument principle theorem [5], [6].

In Section 5 and as a conclusion, it is explained why the index theory relative to rational pseudo-lossless function is an interesting substitute for the Cauchy index theory in the context of the classical polynomial zero location problem. This conclusion is illustrated by proposing a straightforward, presumably original, proof of the Kharitonov theorem relative to the stability of interval polynomials.

2. The Cauchy index of a real rational function. Let $f(x) = b(x)/a(x)$ be a rational function where $a(x)$ and $b(x)$ are coprime polynomials with real coefficients. Such a function admits a unique additive decomposition of the form

$$(1) \qquad f(x) = q(x) + f_0(x) + f_1(x)$$

where $q(x)$ is a polynomial of some degree μ and leading coefficient q_0, $f_0(x)$ and $f_1(x)$ are rational functions devoid of complex and real poles respectively, with $f_0(\infty) = f_1(\infty) = 0$.

By definition, the Cauchy index $I_\alpha^\beta f(x)$ of $f(x)$ on any fixed open interval (α, β) of the real axis is the difference between the number of jumps of $f(x)$ from $-\infty$ to $+\infty$ and its number of jumps from $+\infty$ to $-\infty$ as the argument x changes from α to β; note that one may have $\alpha = -\infty$ and/or $\beta = +\infty$ but that the extreme values of x, i.e. $x = \alpha$ and $x = \beta$, are not included in counting the number of jumps.

As the function $f_0(x)$ in (1) can be written as

$$(2) \qquad f_0(x) = \sum_{i=1}^{m} \left[\frac{h_{i,1}}{x - \omega_i} + \ldots + \frac{h_{i,n_i}}{(x - \omega_i)^{n_i}} \right],$$

where the ω_i and $h_{i,k}$ are appropriate real numbers, it appears that the Cauchy index $I_\alpha^\beta f(x)$ can be expressed by

$$(3) \qquad I_\alpha^\beta f(x) = \sum_{\substack{i=1 \\ n_i \, odd \\ \alpha < \omega_i < \beta}}^{m} \operatorname{sgn} h_{i,n_i};$$

in particular, one has

$$(4) \qquad I_{-\infty}^{+\infty} f(x) = \sum_{\substack{i=1 \\ n_i \, odd}}^{m} \operatorname{sgn} h_{i,n_i}.$$

In case $f(x)$ has no pole at infinity, i.e. $\deg a \geq \deg b$ or equivalently $\deg q = 0$, its Cauchy index is classically computed from considering a Sturm chain of polynomials $[p_i(x); i = 0, 1, \ldots, n']$ deduced from $a(x)$ and $b(x)$ with the help of the Euclid algorithm. More precisely, let us apply the polynomial recurrence relation

$$(5) \qquad p_{i-1}(x) = q_i(x)\, p_i(x) - p_{i+1}(x) \qquad \text{for } i = 1, \ldots, n'$$

initialized with $p_0(x) = a(x)$ and $p_1(x) = b(x)$ to construct the degree decreasing sequence of polynomials $p_i(x)$. As $a(x)$ and $b(x)$ are assumed to be coprime, the algorithm (5) terminates when $p_{n'}(x)$ reduces to a constant or, equivalently, when $p_{n'+1}(x) = 0$. It can then be shown that, if $V(y)$ stands for the number of sign variations in the sequence of numbers $[p_i(y); i = 0, 1, \ldots, n']$, one has the property

$$(6) \qquad I_\alpha^\beta f(x) = V(\alpha) - V(\beta).$$

In concrete situations, one is most often interested in the whole real axis ($\alpha = -\infty, \beta = +\infty$). In such cases, the Cauchy index (4) is shown from (2), (4) and (6) to take the value

$$(7) \qquad I_{-\infty}^{+\infty} f(x) = \sum_{\substack{i=1 \\ \mu_i \, odd}}^{n'} \operatorname{sgn} q_{i,0},$$

where $q_{i,0}$ is the leading coefficient of polynomial $q_i(x)$ assumed to be of degree μ_i.

It is of special interest in the context of the present contribution to consider also the situation where $f(x)$ has poles at infinity ($\deg a < \deg b$). Applying again the Euclid algorithm (5) in this case but with the initialization $p_0(x) = b(x)$, $p_1(x) = a(x)$, one obtains, in particular, the relation:

(8) $$f(x) = q_1(x) - p_2(x)/a(x)$$

so that $q(x) = q_1(x)$ in (1).

As a consequence, one has $I_{-\infty}^{+\infty} f(x) = -I_{-\infty}^{+\infty} p_2(x)/a(x)$, whence

(9) $$I_{-\infty}^{+\infty} f(x) = -\sum_{\substack{i=2 \\ \mu_i \, odd}}^{n'} \operatorname{sgn} q_{i,0},$$

which expression appears to be the appropriate substitute for (7) when $\deg a < \deg b$. Let us point out that the Cauchy indices of $f(x)$ and $1/f(x)$ assume, in general, different values.

3. The index theory of pseudo-lossless functions.

A rational function with complex coefficients $F(p) = B(p)/A(p)$, where $A(p)$ and $B(p)$ are coprime polynomials, is said to be *pseudo-lossless* if it satisfies the equality

(10) $$F(p) + \bar{F}(-\bar{p}) = 0.$$

For an appropriate normalization, $A(p)$ and $B(p)$ are thus defined as para-even or para-odd polynomials respectively, in the sense that (10) induces the relations:

(11) $$A(p) = \pm \bar{A}(-\bar{p}), \quad B(p) = \mp \bar{B}(-\bar{p}).$$

Clearly, there exists a unique decomposition of $F(p)$, similar to (1), of the form

(12) $$F(p) = Q(p) + F_0(p) + F_1(p),$$

where $Q(p)$ is a polynomial, $F_0(p)$ is a rational function with poles on the imaginary axis ($\operatorname{Re} p = 0$) and $F_1(p)$ is a rational function devoid of such poles, with $F_0(\infty) = F_1(\infty) = 0$. Note that the three elementary functions in the right hand side of (12) are themselves well-defined pseudo-lossless functions.

By definition, the index $I(F)$ of the pseudo-lossless rational function $F(p)$ counts the zeros $N(A+B)$ of the sum of the denominator and numerator polynomials of $F(p)$ in the right half-plane ($\operatorname{Re} p > 0$):

(13) $$I(F) = N(A+B).$$

To progress further, let us consider the partial fraction expansion of $F_0(p)$ whose most general form is the following (with $j = \sqrt{-1}$):

(14) $$F_0(x) = \sum_{i=1}^{m} \left[\frac{H_{i,1}}{p - j\omega_i} + \ldots + \frac{H_{i,n_i}}{(p - j\omega_i)^{n_i}} \right],$$

where the ω_i are real numbers. Let Q_0 be the leading coefficient of the polynomial $Q(p)$ assumed to be of degree μ. Note that the pseudo-losslessness constraint (10) implies the successive residues $H_{i,k}$ as well as the successive coefficients of $Q(p)$ to be alternatively real and imaginary numbers; in particular, Q_0 and H_{i,n_i} appear to be real numbers for μ odd and for n_i odd, respectively. It is then easy to prove, on the basis of the continuity property of the zeros of a polynomial as functions of its coefficients, that the the index $I(F)$ can be expressed by

$$(15) \quad 2\, I(F) = \mu + j^{\mu+1} \operatorname{sgn}_{\mu\, odd} Q_0 + \deg F_1 + \sum_{i=1}^{m} n_i + \sum_{\substack{i=1 \\ n_i\, odd}}^{m} j^{n_i+1} \operatorname{sgn} H_{i,n_i}$$

$$= \deg F + j^{\mu+1} \operatorname{sgn}_{\mu\, odd} Q_0 + \sum_{\substack{i=1 \\ n_i\, odd}}^{m} j^{n_i+1} \operatorname{sgn} H_{i,n_i},$$

where $\operatorname{sgn}_{\mu\, odd}$ stands for the usual signature function for μ odd (Q_0 real) and has, by convention, zero value otherwise (Q_0 imaginary for μ even). Moreover, it turns out that $1/F(p)$ is pseudo-lossless with $F(p)$ and that, in view of the index definition (13), one has the remarkable property

$$(16) \qquad\qquad\qquad I(F) = I(1/F).$$

The index $I(F)$ can actually be computed with the help of a slight modification of the Euclid algorithm (5), i.e.

$$(17) \qquad P_{i-1}(p) = Q_i(p)\, P_i(p) + P_{i+1}(p) \qquad \text{for } i = 1, \ldots, n',$$

with the initialization $P_0(p) = A(p)$ and $P_1(p) = B(p)$, or conversely, so as to have $\deg P_0 \geq \deg P_1$. It is noteworthy that the quotient polynomials are themselves spontaneously defined by (17) as pseudo-lossless functions and that their index values are given by

$$(18) \qquad\qquad I(Q_i) = (\mu_i + j^{\mu_i+1} \operatorname{sgn}_{\mu_i\, odd} Q_{i,0})/2$$

with μ_i and $Q_{i,0}$ the degree and the leading coefficient of $Q_i(p)$, respectively. Due to (16) and to the additivity property of the indices of rational pseudo-lossless functions devoid of common poles in $\operatorname{Re} p \geq 0$, it then appears that the index value can be evaluated from the expression

$$(19) \qquad\qquad 2I(F) = \sum_{i=1}^{n'} I(Q_i)$$

$$= \deg F + \sum_{\substack{i=1 \\ \mu_i\, odd}}^{n'} j^{\mu_i+1} \operatorname{sgn} Q_{i,0}.$$

4. From the Cauchy index to the pseudo-lossless function index and conversely.

From the preceding two sections, it is clear that the definitions of the Cauchy index of a real rational function and of the index of a pseudo-lossless rational function are intimately related. To investigate this point in detail, let us consider the invertible transformation $f(x) \rightleftharpoons F(p)$, defined by the following formulas (with x real, p complex):

$$(20) \qquad f(x) = -j\, F(jx), \quad F(p) = j\, f(p/j).$$

It is readily verified that the transformation above associates a well defined real rational function $f(x)$ to any pseudo-lossless function $F(p)$ and conversely. In other words, any pseudo-lossless function can be viewed as resulting from a real rational function by analytic continuation; conversely, any real rational function can be interpreted as the imaginary part of a rational pseudo-losslesss function whose argument is restricted to Re $p = 0$. In particular, one deduces from (1), (2), (12) and (14) the equalities:

$$(21) \qquad q(x) = -j\, Q(j\,x), \quad f_0(x) = -j\, F_0(j\,x), \quad f_1(x) = -j\, F_1(j\,x),$$

which translate in terms of the coefficients of interest into

$$(22) \qquad q_0 = j^{\mu-1}\, Q_0, \quad j^{n_i+1}\, h_{i,n_i} = H_{i,n_i}.$$

Therefore, a comparison of (4) with (15) reveals that the relation between $I_{-\infty}^{+\infty} f(x)$ and $I(F)$ can be expressed by

$$(23) \qquad I_{-\infty}^{+\infty} f(x) = 2\, I(F) - \deg F - j^{\mu+1}\, \mathrm{sgn}_{\mu\, odd}\, Q_0,$$

or, equivalently, by

$$(24) \qquad 2\, I(F) = I_{-\infty}^{+\infty} f(x) + \deg f - \mathrm{sgn}_{\mu\, odd}\, q_0.$$

Let us observe that in the particular case where $\deg A \geq \deg B$, expression (23) simplifies into $I_{-\infty}^{+\infty} f(x) = 2\, I(F) - \deg F$ or equivalently, in view of (13), into:

$$(25) \qquad I_{-\infty}^{+\infty} f(x) = 2\, N(A+B) - \deg(A+B).$$

This result can be viewed as a generalization of a well-known property, classically derived via the argument principle theorem applied to $F(p)$ in the context of the polynomial zero location problem.

The index relations (23)–(24) can alternatively be derived from the appropriate versions of the Euclid algorithm which are used for their actual computation. To see this, let us note that the polynomials $a(x)$ and $b(x)$, respectively $A(p)$ and $B(p)$, are only defined from $f(x)$, respectively $F(p)$, up to a constant factor. Let us freeze these degrees of freedom so as to have $A(p) = \bar{A}(-\bar{p})$, $B(p) = -\bar{B}(-\bar{p})$ together with

$$(26) \qquad A(p) = a(p/j), \quad B(p) = j\, b(p/j)$$

conformably with (20). In case deg $A \geq$ deg B, this induces the transformation $P_i(p) = j^i\, p_i(p/j)$ in (5), (7), whence $Q_i(p) = -j\, q_i(p/j)$ and, in particular,

(27) $$q_{i,0} = j^{\mu_i+1}\, Q_{i,0}.$$

Comparing (7) with (19) then yields relations (23)–(24) directly, since one has $Q(p) = q(p) = 0$ in the present case. The same argument carries over, mutatis mutandis, to the case deg $B >$ deg A by considering $g(x) = -p_2(x)/a(x)$ instead of $f(x)$, since one has $q(x) = q_1(x)$ by construction. Setting $G(p) = j\, g(p/j)$, one then successively derives from (9) and relations (23)–(24) applied to $G(p)$, the equalities $2\, I(G) = I_{-\infty}^{+\infty} g(x) + \deg g$ and $2\, I(F) = 2 + \mu_1 - \operatorname{sgn}_{\mu\, odd} q_{1,0}$, whence (23)–(24).

5. Conclusion.
In the context of the classical zero location problem, i.e. the problem of counting the zeros of a given polynomial $P(p)$ of degree n with complex coefficients, the rational pseudo-lossless function of interest $F(p) = B(p)/A(p)$ is defined by

(28) $$A(p) = [P(p) \pm \bar{P}(-\bar{p})]/2,\ B(p) = [P(p) \mp \bar{P}(-\bar{p})]/2.$$

As a result, one has, by definition, $I(F) = N(P)$ provided that $P(p)$ has no para-conjugate zeros (zeros on Re $p = 0$ or showing up in symmetric pairs with respect to the imaginary axis Re $p = 0$). This remarkable property together with the invertibility and additivity properties of the index of rational pseudo-lossless functions, (16) and (19) respectively, explain why the polynomial zero location problem is solved in a more direct manner via pseudo-lossless function arguments than from considering the Cauchy index of the real rational function $f(x)$ associated with $F(p)$ via (20). From this viewpoint and although the two approaches are theoretically equivalent, one can argue that the Cauchy index approach is a detour on the road to solution of the polynomial zero location problem.

It should be mentioned however that this conclusion does not hold true for the related but different problem of counting the distinct zeros of a real polynomial $p(x)$ on a *finite* interval of the real axis, whose solution is classically obtained by applying property (6) to $f(x) = p'(x)/p(x)$. It is clear that such a question cannot be discussed within the algebraic framework provided by the pseudo-lossless rational function theory.

To end up, let us illustrate the power and the conceptual simplicity of the index theory relative to pseudo-lossless functions in proposing a simple, presumably original, proof of Kharitonov's theorem [8], [9].

A polynomial $P(p)$ is said to be *Hurwitz* if it has no zero in Re $p \geq 0$; as a consequence, the rational function defined as in (28) from its para-even and para-odd parts must be a pseudo-lossless function of zero index value, in other words a lossless function in the usual sense of circuit theory [7]. The most general form of such a *real* lossless function is found from (12), (14) and (15) to have the form [1]

$$F(p) = Q_0 p + \frac{h_o}{p} + \sum_k \frac{2h_k p}{p^2 + \omega_k}, \tag{29}$$

where Q_0, h_o and the h_k must be nonnegative real numbers. Let us now consider the set of *interval* polynomials $[P(p) = \sum_i^n a_i p^i;\ a_i^- \leq a_i \leq a_i^+ \text{ for } i = 1, 2, \ldots, n]$. From the extreme values of the coefficient intervals, let us define the two even polynomials

$$P_e^+(p) = a_0 + a_2^- p^2 + a_4^+ p^4 + \ldots,\quad P_e^-(p) = a_0 + a_2^+ p^2 + a_4^- p^4 + \ldots, \tag{30}$$

and the two odd polynomials

$$P_o^+(p) = p\,[a_1^+ + a_3^- p^2 + a_5^+ p^4 + \ldots],\ P_o^-(p) = p\,[a_1^- + a_3^+ p^2 + a_5^- p^4 + \ldots]. \tag{31}$$

Note that these polynomial definitions imply that for any conformably decomposed member of the considered set, i.e. $P(p) = P_e(p) + P_o(p)$, the following inequalities are trivially satisfied:

$$\begin{array}{rcccl} P_e^-(j\omega) & \leq & P_e(j\omega) & \leq & P_e^+(j\omega) \\ P_o^-(j\omega)/j\omega & \leq & P_o(j\omega)/j\omega & \leq & P_o^+(j\omega)/j\omega \end{array} \tag{32}$$

for all ω real.

Theorem (Kharitonov)
The set of real polynomials $[P(p) = \sum_i^n a_i p^i;\ a_i^- \leq a_i \leq a_i^+ \text{ for } i = 1, 2, \ldots, n]$ *consists exclusively of Hurwitz polynomials if and only if the four corner polynomials defined by*

$$\begin{aligned} P_1(p) &= P_e^+(p) + P_o^+(p),\quad P_2(p) = P_e^-(p) + P_o^+(p), \\ P_3(p) &= P_e^+(p) + P_o^-(p),\quad P_4(p) = P_e^-(p) + P_o^-(p), \end{aligned} \tag{33}$$

are Hurwitz.

Proof. Clearly, only the *if* part of the theorem needs to be proved. Let us first observe that there is no loss of generality in assuming a_0 and all a_i^-, for $i > 0$, to be positive numbers since all the coefficients of a Hurwitz polynomial are known from the outset to have the same sign. Next, let us consider any of the rational pseudo-lossless functions $F(p) = P_e(p)/P_o^+(p)$ or $F(p) = P_e(p)/P_o^-(p)$, with $P_e(p)$ the even part of an arbitrarily selected member $P(p)$ of the polynomial set, and show that they have both zero index value. Indeed, one finds that the corresponding h_k in the partial fraction expansion (29) of $F(p)$ take the values

$$h_k = \left. \frac{P_e(j\omega_k)}{d P_o^\pm(p)/dp} \right|_{p = j\omega_k}$$

---

[1] This partial fraction expansion is known as the Foster reactance synthesis in classical network theory [7].

and are readily seen to be positive real numbers in view of (32) and of the Hurwitz property of $P_1(p)$ and $P_2(p)$; moreover, one has trivially $Q_0 > 0$ as well as $h_0 > 0$ in case $F(p)$ has a pole at infinity; as a consequence, $F(p)$ is lossless in both cases. Applying the same argument to the pseudo-lossless function $F'(p) = P_o(p)/P_e(p)$, with $P_o(p)$ the odd part of $P(p)$, and taking into account the second inequalities (32) together with the Hurwitz property of $P_3(p)$ and $P_4(p)$, one establishes similarly that $I(F') = 0$. This completes the proof since $I(F') = 0$ implies, by definition, that $P(p) = P_e(p) + P_o(p)$ is a Hurwitz polynomial.

References

[1] Y. Genin, "Euclid algorithm, orthogonal polynomials and generalized Routh-Hurwitz algorithm", to be published in *Linear Algebra Appl.*

[2] Y. Genin, "On polynomials nonnegative on the unit circle and related questions", to be published in *Linear Algebra Appl.*

[3] P. Delsarte, Y. Genin and Y. Kamp, "Pseudo-lossless functions with application to the problem of locating the zeros of a polynomial", *IEEE Trans. Circuits and Systems*, vol. CAS-32, pp. 373–381, 1985.

[4] P. Delsarte, Y. Genin and Y. Kamp, "Pseudo-Carathéodory functions and Hermitian Toeplitz matrices", *Philips J. Res.*, vol. 41, pp. 1–54, 1986.

[5] M. Marden, *Geometry of Polynomials*. Providence R.I: American Math. Soc., 1966.

[6] F.R. Gantmacher, *The Theory of Matrices*, Vol. II. New York: Chelsea, 1959.

[7] V. Belevitch, *Classical Network Theory*. San Francisco: Holden-Day, 1968.

[8] V.L. Kharitonov, "Asymptotic stability of an equilibrium position of a family of linear differential equations", *Differential Equations*, vol. 14, pp. 1483–1485, 1979.

[9] R.J. Minnichelli, J.J. Anagnost and C.A. Desoer, "An elementary proof of Kharitonov's stability theorem with extensions", *IEEE Trans. Automatic Control*, vol. AC-34, pp. 995–998, 1989.

A Generalization of the Orlando Formula — Symbolic Manipulation Approach

Henryk Górecki, Maciej Szymkat, Mieczysław Zaczyk
Institute of Automatics, St. Staszic Technical University,
al. Mickiewicza 30, 30-059 Kraków, Poland

Abstract. In this paper some new results on polynomials whose roots are linear combinations (with fixed integer coefficients) of the roots of other polynomial are presented. This research is motivated by a new method for calculation of a generalized integral criterion with the integrand function being an arbitrary polynomial of the transient error.

The well known Orlando formula establishes the useful relation between the Hurwitz determinants and the polynomial whose roots are sums of the roots of a given polynomial. However, this is only a special case of the formulae derived in this paper.

In the paper two approaches are presented. The first one uses a recursive procedure based on resultants and Hurwitz determinants, and the second exploits the theory of symmetric functions. Both methods are implemented as symbolic manipulation algorithms in Maple V. The results obtained are illustrated with an example of application.

1. Introduction. The classical Orlando formula [3] expresses the relation between roots of a polynomial and its $(n-1)$-th Hurwitz determinant being a product of sums of all different pairs of roots. In [5] the Orlando formula was generalized to the case of polynomials with complex coefficients. The result of Orlando may be directly obtained from properties of determinants of Sylvester matrix (resultants) as it is demonstrated in [2, 4].

It is well known that for two monic polynomials

$$(1) \qquad p(x) = \prod_i (x - p_i)$$

$$(2) \qquad q(x) = \prod_k (x - q_k)$$

the resultant satisfies [2]

$$(3) \qquad \det R[p,q] = \prod_{i,k}(p_i - q_k).$$

For a given polynomial

$$(4) \qquad a(s) = s^n + a_1 s^{n-1} + \ldots + a_{n-1}s + a_n = \prod_{i=1}^{n}(s - s_i)$$

let us define two 'shifted' polynomials in x with coefficients depending on s

$$(5) \qquad P(x) = a\left(\frac{s}{2} - x\right)$$

(6) $$Q(x) = a\left(\frac{s}{2} + x\right).$$

Using (3) it was proven in [2] that the resultant of $P(x)$ and $Q(x)$ satisfies

(7) $$\det R[P,Q](s) = (-1)^n \prod_{i,k} [s - (s_i + s_k)].$$

Moreover,

(8) $$\det R[P,Q](s) = (-2)^n\, a\left(\frac{s}{2}\right) R_1^2(s)$$

where $R_1(s)$ is a generalized Hurwitz matrix for $Q(x)$ defined as

(9) $$R_1(s) = \begin{bmatrix} A_1(s) & A_3(s) & A_5(s) & \cdots & 0 & 0 \\ A_0(s) & A_2(s) & A_4(s) & \cdots & 0 & 0 \\ 0 & A_1(s) & A_3(s) & \cdots & 0 & 0 \\ \vdots & \vdots & \vdots & & \vdots & \vdots \\ 0 & 0 & 0 & \cdots & A_{n-2}(s) & A_n(s) \\ 0 & 0 & 0 & \cdots & A_{n-3}(s) & A_{n-1}(s) \end{bmatrix}$$

where

(10) $$A_k(s) = \sum_{i=0}^{k} \binom{n-k+i}{n-k} a_{k-i} \left(\frac{s}{2}\right)^i.$$

Relation (8) was also pointed out in [4]. Since $A_k(0) = a_k$ we have $R_1(0)$ equal to $(n-1)$-th Hurwitz matrix $\mathcal{H}_{((n-1)\times(n-1))}$ and we may obtain the Orlando formula for monic polynomial (4)

(11) $$\det \mathcal{H}_{((n-1)\times(n-1))} = \prod_{i<k} (s_i + s_k).$$

directly from (7) and (8), [2].

Let us fix a sequence of weight coefficients $\{w_1, w_2, \ldots, w_r\}$ ($r \leq n$) and a subsequence $r_1, r_2, \ldots, r_l$ of their indices such that $w_1 = w_2 = \ldots = w_{r_1-1} < w_{r_1} = w_{r_1+1} = \ldots = w_{r_2-1} < \ldots < w_{r_l} = w_{r_l+1} = \ldots = w_r$. For the sake of notation let us define additionally $r_0 = 1$ and $r_{l+1} = r + 1$. In this paper we will derive methods to obtain polynomials

(12) $$b_{[w_1,w_2,\ldots,w_r]}(s) = \prod_{\substack{i_{r_u} < i_{r_u+1} < \ldots < i_{r_{u+1}-1} \\ u=0,1,\ldots,l}} [s - (w_1 s_{i_1} + w_2 s_{i_2} + \ldots + w_r s_{i_r})]$$

whose roots are sums of roots of $a(s)$ multiplied by fixed weight coefficients satisfying $w_1 \leq w_2 \leq \ldots \leq w_r$. Obviously $b_{[1,1]}(s) = R_1(s)$.

2. Calculation of integrals of higher powers of signals.

Integral criteria are common tool in parametric optimization of continuous and discrete linear control systems. There exist a number of closed-form representations of integrals of the square of signals. They either use formulations based on Hurwitz determinants [6] or directly use Parseval's theorem and the theory of residues of complex functions [7].

Recently, a new method for the calculation of the integral

$$(13) \qquad I_k = \int_0^\infty \varepsilon^k(t)\,dt, \quad \text{for} \quad k = 1, 2, \ldots$$

where $\varepsilon(t)$ is a stable solution of the equation

$$(14) \qquad \frac{d^n\varepsilon(t)}{dt^n} + a_1 \frac{d^{n-1}\varepsilon(t)}{dt^{n-1}} + \ldots + a_{n-1}\frac{d\varepsilon(t)}{dt} + a_n\varepsilon(t) = 0$$

has been proposed [1]. We assume that $a_1, \ldots, a_n$ are constant real coefficients and $\varepsilon^{(\mu)}(0)$ are initial values in general different from zero at least for some $\mu = 0, 1, \ldots, n-1$. The respective characteristic equation has the form

$$(15) \qquad a(s) = s^n + a_1 s^{n-1} + \ldots + a_{n-1} s + a_n = 0.$$

Assuming that the above equation has n different roots $s_i \neq s_j$ for $i \neq j$ we write the solution of equation (14) as

$$(16) \qquad \varepsilon(t) = \sum_{i=1}^n A_i e^{s_i t}$$

where the coefficients A_i may be determined using initial conditions. In order to calculate an arbitrary power of $\varepsilon(t)$ we apply the formula

$$(17) \qquad \varepsilon^k(t) = \sum \frac{k!}{w_1! w_2! \cdots w_r!} e^{(w_1 s_{i_1} + w_2 s_{i_2} + \ldots + w_r s_{i_r})t} A_{i_1}^{w_1} A_{i_2}^{w_2} \cdots A_{i_r}^{w_r}$$

where the summation is over all different r-tuples of characteristic roots and all combinations of positive integers satisfying (for various r) $w_1 + w_2 + \ldots + w_r = k$. Suppose that we have calculated the coefficients of

$$(18) \qquad \begin{aligned} B_k(s) &= s^N + B_{k,1} s^{N-1} + \ldots + B_{k,N-1} s + B_{k,N} \\ &= \prod_{\substack{w_{i_1} \leq w_{i_2} \leq \ldots \leq w_{i_r} \\ w_{i_1} + w_{i_2} + \ldots + w_{i_r} = k}} b_{[w_{i_1}, w_{i_2}, \ldots, w_{i_r}]}(s) \end{aligned}$$

where the factors in the right hand side are defined by (12). Then the integral (13) may be calculated [1] from the formula

$$(19) \qquad I_k = \frac{1}{B_{k,N}} \left[\frac{d^{N-1}[\varepsilon^k(t)]}{dt^{N-1}} + B_{k,1} \frac{d^{N-2}[\varepsilon^k(t)]}{dt^{N-2}} + \ldots + B_{k,N-1}\varepsilon^k(t) \right]_{t=0}$$

with the initial values determined recursively, with the use of the Leibniz formula

$$\frac{d^\mu \varepsilon^k(t)}{dt^\mu} = \sum_{j=0}^{\mu} \binom{\mu}{j} \frac{d^{\mu-j}\varepsilon^{k-1}(t)}{dt^{\mu-j}} \frac{d^j \varepsilon(t)}{dt^j}. \tag{20}$$

The successive values of $\frac{d^j\varepsilon(0)}{dt^j}$ are subsitituted as $\varepsilon^{(j)}(0)$ for $j = 0, 1, \ldots, n-1$ or calculated from characteristic equation (15) for $j \geq n$.

The problem we want to solve here is to express $B_{k,i}$ in (18) in terms of the original coefficients $a_1, a_2, \ldots, a_n$ avoiding in this way the explicit use of characteristic roots.

Example. Let us consider the case of

$$a(s) = s^3 + a_1 s^2 + a_2 s + a_3. \tag{21}$$

Suppose that we want to calculate

$$\begin{aligned}
\varepsilon^4(t) &= [A_1 e^{s_1 t} + A_2 e^{s_2 t} + A_3 e^{s_3 t}]^4 = \\
&\quad A_1^4 e^{4s_1 t} + A_2^4 e^{4s_2 t} + A_3^4 e^{4s_3 t} + \\
&\quad + 4A_1 A_2^3 e^{(s_1+3s_2)t} + 4A_1 A_3^3 e^{(s_1+3s_3)t} + 4A_2 A_3^3 e^{(s_2+3s_3)t} + \\
&\quad + 4A_2 A_1^3 e^{(s_2+3s_1)t} + 4A_3 A_1^3 e^{(s_3+3s_1)t} + 4A_3 A_2^3 e^{(s_3+3s_2)t} + \\
&\quad + 6A_1^2 A_2^2 e^{(2s_1+2s_2)t} + 6A_1^2 A_3^2 e^{(2s_1+2s_3)t} + 6A_2^2 A_3^2 e^{(2s_2+2s_3)t} + \\
&\quad + 12A_1 A_2 A_3^2 e^{(s_1+s_2+2s_3)t} + 12A_1 A_3 A_2^2 e^{(s_1+s_3+2s_2)t} + \\
&\quad + 12A_2 A_3 A_1^2 e^{(s_2+s_3+2s_1)t}.
\end{aligned} \tag{22}$$

We can write the characteristic polynomial corresponding to the differential equation satisfied by $\varepsilon^4(t)$ as

$$\begin{aligned}
B_4(s) &= (s-4s_1)(s-4s_2)(s-4s_3) \\
&\quad [s-(s_1+3s_2)][s-(s_1+3s_3)][s-(s_2+3s_1)] \\
&\quad [s-(s_3+3s_1)][s-(s_3+3s_2)] \\
&\quad [s-(2s_1+2s_2)][s-(2s_1+2s_3)][s-(2s_2+2s_3)] \\
&\quad [s-(s_1+s_2+2s_3)][s-(s_1+s_3+2s_2)][s-(s_2+s_3+2s_1)].
\end{aligned} \tag{23}$$

Thus, for the above example we may write

$$B_4(s) = b_{[4]} \, b_{[1,3]} \, b_{[2,2]} \, b_{[1,1,2]}. \tag{24}$$

3. Resultant based approach. We want to derive general formulae for polynomials defined by (12). In general we have:
for $r = 1$

$$b_{[w_1]}(s) = a\left(\frac{s}{w_1}\right) \tag{25}$$

for $r = 2$, $w_1 = w_2 = w$

$$b_{[w,w]}(s) = \det R_1\left(\frac{s}{w}\right) \tag{26}$$

where R_1 is the generalized Hurwitz matrix defined by (9).

For the case $r = 2$ and $w_1 \neq w_2$ we define

$$B^-_{[w]}(x) = a\left(\frac{\frac{s}{2} - x}{w}\right), \quad B^+_{[w]}(x) = a\left(\frac{\frac{s}{2} + x}{w}\right), \tag{27}$$

and

$$b_{[w_1, w_2]}(s) = \frac{\det R[B^-_{[w_1]}, B^+_{[w_2]}]}{b_{[w_1 + w_2]}} \tag{28}$$

where R is the resultant matrix for polynomials (27).

For the case $r > 2$ we define

$$B^-_{[w_1, w_2, \ldots, w_{r-1}]}(x) = b_{[w_1, w_2, \ldots, w_{r-1}]}\left(\frac{s}{2} - x\right). \tag{29}$$

Then we calculate recursively from

$$\tilde{b}_{[w_1, w_2, \ldots, w_r]}(s) = \frac{\det R[B^-_{[w_1, w_2, \ldots, w_{r-1}]}, B^+_{[w_r]}]}{\prod_{k: w_{k-1} \neq w_k} b_{[w_1, w_2, \ldots, w_k + w_r, \ldots, w_{r-1}]}(s)} \tag{30}$$

and

$$b_{[w_1, w_2, \ldots, w_r]}(s) = \frac{\tilde{b}_{[w_1, w_2, \ldots, w_r]}(s)}{\gcd\left(\tilde{b}_{[w_1, w_2, \ldots, w_r]}, \frac{d\tilde{b}_{[w_1, w_2, \ldots, w_r]}}{ds}\right)} \tag{31}$$

if $w_{r-1} = w_r$ or

$$b_{[w_1, w_2, \ldots, w_r]}(s) = \tilde{b}_{[w_1, w_2, \ldots, w_r]}(s) \tag{32}$$

otherwise.

Formulae (26), (30) and (31) have been implemented in the computer algebra system Maple V using standard algorithms for symbolic calculation of determinants, resultants and greatest common divisors [8].

4. Symmetric function approach. Again, for a given polynomial (4) we want to find the respective polynomial (12) with fixed nondecreasing sequence of weight coefficients.

The elementary symmetric function e_v is the sum of all products of v different variables s_v, such that $e_0 = 1$ and

$$e_v = \sum_{i_1 < i_2 < \ldots < i_v} s_{i_1} s_{i_2} \cdots s_{i_v}, \quad \text{for} \quad v \geq 1. \tag{33}$$

The coefficients of a monic polynomial in one variable are, disregarding the sign, the elementary symmetric polynomials of its roots (Vieta formulae).

Let us fix a sequence of weight coefficients $\{w_1, w_2, \ldots, w_r\}$ and assume the same indexing convention as when defining (12). We will use the symmetric polynomials, called power sums defined as follows

$$(34) \quad P_k = \sum_{i_{r_u} < i_{r_u+1} < \ldots < i_{r_{u+1}-1},\ u=0,1,\ldots,l} (w_1 s_{i_1} + w_2 s_{i_2} + \ldots + w_r s_{i_r})^k$$

for $k = 1, 2, \ldots, M$, where M is the degree of (12). P_k may be expressed as a linear combination of monomial symmetric functions

$$(35) \quad m_{\lambda_1,\lambda_2,\ldots,\lambda_z}(s_1, s_2, \ldots, s_n) = \sum_{i_1,i_2,\ldots,i_z} s_{i_1}^{\lambda_1} s_{i_2}^{\lambda_2} \cdots s_{i_z}^{\lambda_z}.$$

We have

$$P_k = \sum_{\lambda} \frac{\frac{n!}{(n-r)!}}{\binom{n}{z} k_1! k_2! \cdots k_q!} \frac{\binom{r}{z}}{\frac{r!}{(r-z)!}} \frac{k!}{\lambda_1! \lambda_2! \cdots \lambda_z!} \sum_{i_1,i_2,\ldots,i_z} w_{i_1}^{\lambda_1} s_{i_1}^{\lambda_1} w_{i_2}^{\lambda_2} s_{i_2}^{\lambda_2} \cdots w_{i_z}^{\lambda_z} s_{i_z}^{\lambda_z}$$

$$= \sum_{\lambda} c_\lambda m_\lambda$$

where summation is over all partitions λ, such that $\lambda_1 + \lambda_2 + \ldots + \lambda_z = k$, z - length of partition λ (assumed to be less than k), $k_1, k_2, \ldots, k_q$ - multiplicities of weights (q is the number of different weights).

It follows from the fundamental theorem on symmetric polynomials [9] that any symmetric polynomial can be expressed in terms of elementary symmetric functions. First, we represent power sums polynomials P_k (expressed in monomial functions) in the basis of power sums symmetric functions using two relation, first

$$(36) \quad h_k = \sum_{|\lambda|=k} \frac{p_{\lambda_1} p_{\lambda_2} \cdots p_{\lambda_z}}{\prod_{i \geq 1} \kappa_i! \, i^{\kappa_i}}$$

where κ_i is the multiplicity of the number i in λ, and next

$$(37) \quad h_k = \sum_{|\lambda|=k} m_\lambda, \quad |\lambda| = \lambda_1 + \lambda_2 + \ldots + \lambda_z.$$

Here h_k is the k-th complete symmetric function, which is the sum of all monomials of total degree k. The above relations enable us to express the m's in terms of the p's. Next, the conversion of power sums p into elementary symmetric functions is done using Newton's formulae

$$(38) \quad p_k = \sum_{i=1}^{k-1} (-1)^i e_i p_{k-1} + (-1)^k k e_k, \quad \text{for } k \leq r$$

or in the determinant form

(39) $$p_k = \det \begin{bmatrix} e_1 & 1 & 0 & \cdots & 0 \\ 2e_2 & e_1 & 1 & \cdots & 0 \\ \vdots & \vdots & \vdots & & \vdots \\ ke_k & e_{k-1} & e_{k-2} & \cdots & e_1 \end{bmatrix}.$$

The k-th elementary symmetric function of roots of $b(s)$ may be expressed in terms of power sums P_i using Newton's formulae again

(40) $$E_k = \frac{(-1)^k}{k}[P_k - \sum_{i=1}^{k-1}(-1)^i E_i P_{k-1}]$$

or in the determinant form

(41) $$k! E_k = \det \begin{bmatrix} P_1 & 1 & 0 & \cdots & 0 \\ P_2 & P_1 & 2 & \cdots & 0 \\ \vdots & \vdots & \vdots & & \vdots \\ P_{k-1} & P_{k-2} & \cdots & \cdots & k-1 \\ P_k & P_{k-1} & \cdots & \cdots & P_1 \end{bmatrix}.$$

In this way we can express the coefficients of the polynomial b in terms of the power sums symmetric functions. Finally, we obtain the coefficients b_i of the polynomial b in terms of coefficients a_i of the polynomial a.

The computer implementation of the above procedure uses the symmetric function library SF for Maple V package [10].

5. Application example. We will consider the transient error response of a simple third order control system with proportional feedback. Let us assume that for a given plant its Laplace transform has the following form

$$\varepsilon(s) = \varepsilon_0 \frac{s^2 + 3s + 4}{s^3 + 3s^2 + 4s + 2 + K}$$

where K is the controller gain and ε_0 is the amplitude of the step input. The corresponding initial conditions read

$$\varepsilon^{(0)}(0) = \varepsilon_0, \quad \varepsilon^{(1)}(0) = \varepsilon^{(2)}(0) = 0 .$$

The higher order derivatives at 0 may be obtained directly form the relation

$$\varepsilon^{(\mu)}(0) = -\sum_{i=1}^{i=3} a_i \varepsilon^{(\mu-i)}(0)$$

where $a_1 = 3, a_2 = 4, a_3 = 2 + K$. We want to find the formulae for

$$\int_0^\infty \varepsilon^k(t) dt, \quad \text{for} \quad k = 2, 3, 4.$$

First, we compute $\frac{d^\mu \varepsilon^k(0)}{dt^\mu}$ using (20). Next, employing one of the methods introduced in previous sections we obtain

$$\begin{aligned}
b_{[2]} &= s^3 + 2a_1 s^2 + 4a_2 s + 8a_3 \\
b_{[1,1]} &= s^3 + 2a_1 s^2 + \left(a_1^2 + a_2\right) s - a_3 + a_1 a_2 \\
b_{[3]} &= s^3 + 3a_1 s^2 + 9a_2 s + 27a_3 \\
b_{[1,2]} &= s^6 + 6a_1 s^5 + \left(6a_2 + 13a_1^2\right) s^4 + \left(12a_1^3 + 24a_1 a_2\right) s^3 + \\
&\quad \left(30a_1^2 a_2 + 9a_2^2 + 4a_1^4\right) s^2 + \left(18a_2^2 a_1 + 12a_1^3 a_2\right) s + \\
&\quad +27a_3^2 - 18a_3 a_1 a_2 + 4a_3 a_1^3 + 8a_1^2 a_2^2 + 4a_2^3 \\
b_{[1,1,1]} &= s + a_1 \\
b_{[4]} &= s^3 + 4a_1 s^2 + 16a_2 s + 64a_3 \\
b_{[1,3]} &= s^6 + 8a_1 s^5 + \left(22a_1^2 + 14a_2\right) s^4 + \left(68a_1 a_2 + 24a_1^3 + 20a_3\right) s^3 + \\
&\quad + \left(9a_1^4 + 80a_1 a_3 + 49a_2^2 + 90a_1^2 a_2\right) s^2 + \\
&\quad \left(60a_3 a_1^2 + 140a_2 a_3 + 36a_1^3 a_2 + 84a_2^2 a_1\right) s + \\
&\quad +343a_3^2 + 36a_2^3 + 36a_3 a_1^3 + 27a_1^2 a_2^2 - 42a_3 a_1 a_2 \\
b_{[2,2]} &= s^3 + 4a_1 s^2 + \left(4a_1^2 + 4a_2\right) s - 8a_3 + 8a_1 a_2 \\
b_{[1,1,2]} &= s^3 + 4a_1 s^2 + a_2 s + 5a_1^2 s + 2a_1^3 + a_3 + a_1 a_2.
\end{aligned}$$

We calculate

$$B_2(s) = b_{[2]}\, b_{[1,1]}, \quad B_3(s) = b_{[3]}\, b_{[1,2]}\, b_{[1,1,1]}, \quad B_4(s) = b_{[4]}\, b_{[1,3]}\, b_{[2,2]}\, b_{[1,1,2]}$$

and obtain from (19)

$$\begin{aligned}
I_2 &= \frac{\varepsilon_0^2}{2} \frac{5K + 58}{(K+2)(-K+10)} \\
I_3 &= \frac{\varepsilon_0^3}{3} \frac{6K^3 + 136K^2 + 2152K + 9440}{(K+2)(27K^2 + 1300)} \\
I_4 &= \frac{\varepsilon_0^4}{4} \frac{-13738K^4 - 23459K^3 + 1921008K^2 + 18705220K + 49374800}{(K+2)(-K+10)(K+68)(343K^2 + 1840K + 8500)}.
\end{aligned}$$

The above formulae are suitable, except for I_3, for parameteric optimization with respect to K. The optimal values are $K = 2.8$ for I_2 and $K = 4.854$ for I_4.

An interesting possibility arises if we admit a combined integral criterion of the form

$$I_c = I_2 - 1.5\, I_3 + 1.25\, I_4.$$

For this function we obtain the optimal gain $K = 1.826$ for $\varepsilon_0 = 1$. It is important to note that the optimal value depends on ε_0. Looking at the plots of integrand

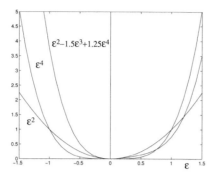

FIG. 1. *Integrands of criterion functions*

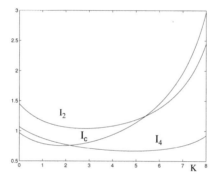

FIG. 2. *Values of criterion functions*

functions of I_2, I_4 and I_c shown in Figure 1 one clearly notices that I_c is asymmetric and introduces a larger penalty for negative values of ε.

The values of respective integrals for $K \in [0,8]$ for $\varepsilon_0 = 1$ are shown in Figure 2. The respective optimal transient responses of the error function are shown in Figure 3, the continuous line corresponds to the result of optimization using I_c criterion, the dash-dot line - I_2 and the dashed line - I_4. It is visible that optimization using I_c criterion function leads to lowest overshoots in the case under consideration.

6. Conclusion. In this paper two approaches to the computation of generalized Orlando formulae have been presented. Both methods have created sufficient base to build symbolic manipulation algorithms that have been successfully applied to find analytical solutions of certain new problems in parametric optimization of linear systems.

Acknowledgement. We would like to thank Prof. Eliahu I. Jury for valuable comments which helped to improve presentation of our results.

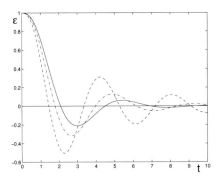

FIG. 3. *Optimal transient responses*

References

[1] H. Górecki, *A new method for calculation of the generalized integral criterion*, Bull. Pol. Acad. Sci. Tech. Sci., vol. 42, no. 4, pp. 595–603, 1994.
[2] H.Górecki, M.Szymkat, *Application of an elimination method to the study of the geometry of zeros of real polynomials*, International Journal of Control, vol. 38, no. 1, pp. 1–26, 1983.
[3] L. Orlando, *Sul problema di Hurwitz relativo alle parti reali delle radici di un'equatione algebrica*, Math. Ann., vol. 71, pp. 233–245, 1911.
[4] E.I. Jury, *From J.J. Sylvester to Adolf Hurwitz: A historical review*, these proceedings.
[5] P.C. Parks, *A new proof of Hermite's stability criterion and a generalization of Orlando's formula*, Int. J. Cont., vol. 26, no. 2, pp. 197–206, 1977.
[6] E.I. Jury, *Inners and Stability of Dynamic Systems*, Second Edition, Malabar, FL: Krieger, 1982.
[7] H. Górecki and A.Korytowski (eds.), *Advances in optimization and stability analysis of dynamical systems*, Cracow, Academy of Mining and Metallurgy Publishers, 1993.
[8] D. Redfern, *Maple Handbook*, New York, Springer Verlag, 1993.
[9] I.G. Macdonald, *Symmetric Functions and Hall Polynomials*, Oxford, Clarendon Press, 1979.
[10] J.R. Stembridge, *A Maple package for symmetric functions*, Department of Mathematics, Ann Arbor, MI, June 1989.

On convex stability directions for real quasipolynomials

L. Atanassova, D. Hinrichsen
Institut für Dynamische Systeme, Universität Bremen,
D-28334 Bremen, Germany

V.L. Kharitonov
Dept. Appl. Math. & Control Theory, St. Petersburg University,
198904 St. Petersburg, Russia

> **Abstract.** This note deals with convex directions for quasipolynomials in the real case. It gives analytic criteria ensuring Hurwitz stability of segments of real quasipolynomials of delay type. The results are derived using an approach introduced in [4].

1. Introduction. Recently, a number of publications dealing with Hurwitz stability of segments of polynomials have appeared in the literature, see e.g. [6], [1] and the references therein. In [6] Rantzer investigated under which conditions a real polynomial $q(s) \in \mathbb{R}[s]$ has the property that the implication

(1) $\qquad p, p+q$ Hurwitz stable $\quad \Rightarrow \quad [p, p+q]$ Hurwitz stable

holds true for *every* real Hurwitz polynomial p with $\deg p > \deg q$. Polynomials q with this property are called *convex directions* and have been characterized in [6]. A new interpretation of Rantzer's condition in terms of root loci was derived in [4]. The interpretation led to a new concept of *convex direction for a given Hurwitz polynomial* which was characterized by a graphical test in [4].

Some of the extreme point results for polynomials admit a natural extension to *quasipolynomials*. These are entire functions of the form

(2) $\qquad f(z) = p_0(z)e^{\tau_0 z} + p_1(z)e^{\tau_1 z} + \cdots + p_m(z)e^{\tau_m z}$

where $p_j(z)$, $j = 0, 1, \ldots, m$ are polynomials and $\tau_0 < \tau_1 < \cdots < \tau_m$ are real numbers representing "time shifts" or "delays".

Properties of quasipolynomials are discussed in [2]. Let us recall some of them. A quasipolynomial $f(z)$ with $m > 0$ has an infinite number of roots, but within any bounded region in the complex plane it has only a finite number of roots. The geometry of the roots of quasipolynomials was carefully investigated in [7]. In this note we will only deal with quasipolynomials *of delay type* (characterized by $\deg p_i < \deg p_m$, $i = 0, ..., m-1$). These quasipolynomials have only finitely many roots in each right half-plane $\{s \in \mathbb{C}; \operatorname{Re} s > \alpha\}$, $\alpha \in \mathbb{R}$. For arbitrary quasipolynomials an analogue of the Hermite-Biehler Theorem can be proved and this result is known as Pontryagin's Theorem. However, Hurwitz type stability criteria cannot be generalized to quasipolynomials. An analogue of the concept of convex directions was introduced in [5] for quasipolynomials with fixed extreme delays τ_0, τ_m.

We emphasize that convex directions for quasipolynomials are not defined relative to the whole class of quasipolynomials but to the subclass of quasipolynomials having given minimal and maximal time shifts τ_0 resp. τ_m. It was shown in [5] that this concept of convex direction can be used for testing the stability of quasipolynomial families.

In the present paper we generalize some of the convex direction results given in [4] to the case of real quasipolynomials of delay type. The main motivation for our study comes from the stability analysis of polytopes of quasipolynomials. The edge theorem [3] shows that in order to check a polytope Π of quasipolynomials for stability it is sufficient to examine the edges of Π. A stronger result can be obtained if the directions defined by the edges are convex directions satisfying the characterization given in [5]. In this case the stability of the vertices of Π ensures the stability of the whole polytope of quasipolynomials.

The paper is organized as follows. In Section 2 we formulate the convex direction problem and recall some results from the literature which we use later. Sections 3 and 4 deal with root loci of quasipolynomials. Our main result is a characterization of convex directions in terms of these root loci (Theorem 4). The characterization provides a basis for introducing the weaker concept of *convex directions for a given real Hurwitz quasipolynomial* which is briefly discussed at the end of this note.

2. Convex directions. Consider real quasipolynomials of the form

$$(3) \qquad f(z) = a_{nm} z^n e^{\tau_m z} + \sum_{j=0}^{m} \sum_{k=0}^{n-1} a_{kj} z^k e^{\tau_j z}, \quad a_{nm} > 0,$$

where the delays $\tau_0 < \tau_1 < \cdots < \tau_m$ are fixed and the coefficient vector $\mathbf{a} = [a_{00}, a_{10}, \ldots, a_{n-1\,m}, a_{nm}] \in \mathbb{R}^{1 \times [n(m+1)+1]}$ is arbitrary with $a_{nm} > 0$.

A real quasipolynomial of the form (3) is called *Hurwitz stable* or simply *stable* if all its roots have negative real parts. Let us denote the *set of all real Hurwitz stable quasipolynomials* of the form (3) (with fixed delays $\tau_0 < \tau_1 < \cdots < \tau_m$) by $\mathcal{H}$.

All the quasipolynomials under consideration in this paper are of the form (3) and therefore of delay type (as defined in [2]).

It is known that every real stable quasipolynomial $f(z)$ enjoys the following *phase increasing property* [5]

$$(4) \qquad \frac{\partial \arg(f(\imath\omega))}{\partial \omega} > \frac{\tau_0 + \tau_m}{2} + \left| \frac{\sin(2\arg(f(\imath\omega)) - (\tau_0 + \tau_m)\omega)}{2\omega} \right|$$

for all $\omega > 0$. Using this property the concept of convex direction was extended in [5] to quasipolynomials in the following way. Let $g(z)$ be a real quasipolynomial of the form

$$(5) \qquad g(z) = \sum_{j=0}^{m} \sum_{k=0}^{n-1} c_{kj} z^k e^{\tau_j z}$$

where $\mathbf{c} = [c_{00}, c_{10}, \ldots, c_{n-1\,m}] \in \mathbb{R}^{1 \times n(m+1)}$.

Definition 1 A quasipolynomial $g(z)$ of the form (5) is called a *convex direction* for the set $\mathcal{H}$ if for every stable quasipolynomial $f(z) \in \mathcal{H}$, the stability of $f(z) + g(z)$ implies the stability of the whole segment of quasipolynomials

$$[f(z), f(z) + g(z)] = \{f(z) + \mu g(z), \quad \mu \in [0,1]\}$$

The following proposition from [5] gives a characterization of the convex directions for the set $\mathcal{H}$.

PROPOSITION 2.1. *A real quasipolynomial $g(z)$ of the form (5) is a convex direction for the set $\mathcal{H}$ if and only if for all $\omega \in \{\omega > 0 \mid g(\imath\omega) \neq 0\}$ the following condition is satisfied:*

$$(6) \qquad \frac{\partial \arg(g(\imath\omega))}{\partial \omega} \leq \frac{\tau_0 + \tau_m}{2} + \left| \frac{\sin(2\arg(g(\imath\omega))) - (\tau_0 + \tau_m)\omega)}{2\omega} \right|.$$

If a quasipolynomial $g(z)$ has the property (6) then

$$[f(z), f(z) + g(z)] \subset \mathcal{H} \quad \Leftrightarrow \quad f(z), \; f(z) + g(z) \; \in \mathcal{H}$$

for *all* $f(z) \in \mathcal{H}$.

The convex directions $g(z)$ are characterized by the property that, for every Hurwitz quasipolynomial $f_0(z) \in \mathcal{H}$ the *stability set* of the pair (f_0, g)

$$(7) \qquad M(f_0, g) = \{\mu \geq 0; \; f_0(z) + \mu g(z) \in \mathcal{H}\}$$

is a real interval.

From Proposition 2.1 it follows that a quasipolynomial $g(z)$ is a convex direction if and only if $-g(z)$ is a convex direction.

For a given stable quasipolynomial $f_0(z) \in \mathcal{H}$, the condition (6) is conservative in the sense that the set (7) may be an interval for some quasipolynomial $g(z)$ although $g(z)$ does not satisfy condition (6).

3. Root loci through the origin. Consider a pair of real quasipolynomials $f_0(z)$, $g(z)$ where $f_0(z) \in \mathcal{H}$ and $g(z)$ is of the form (5).

In this section we investigate the movement of the roots of the quasipolynomial

$$(8) \qquad f_\mu(z) = f_0(z) + \mu g(z)$$

through the origin, for small variations of the parameter $\mu \geq 0$ around some parameter value $\mu_0 > 0$ such that $f_{\mu_0}(0) = 0$.

Assume that the quasipolynomial $f_\mu(z)$ has a zero root $z = 0$ of multiplicity $k \geq 1$ for some parameter value $\mu_0 > 0$. For μ varying in a small real interval $[\mu_0 - \varepsilon, \mu_0 + \varepsilon]$ we want to determine the possible changes in the number of small roots with positive real parts of $f_0(z) + \mu g(z)$. Note that $g(0) = \sum_{j=0}^{m} c_{0j} \neq 0$ since otherwise $f_0(0) = 0$, contrary to the assumption that $f_0(z) \in \mathcal{H}$. Moreover, there exists at most one value of $\mu_0 > 0$ for which $f_{\mu_0}(0) = 0$, namely $\mu_0 = -f_0(0)/g(0)$ (if $f_0(0)/g(0) < 0$).

For our analysis we need to introduce some notations and recall some results from complex analysis.

Remark 3.1 Given $k \in N$, let $u_0^k, u_1^k, \ldots, u_{k-1}^k$ denote the roots of $z^k - 1$

$$(9) \qquad u_\nu^k = \exp\{\imath \frac{2\nu\pi}{k}\}, \qquad \nu = 0, 1, \ldots, k-1$$

and $v_0^k, v_1^k, \ldots, v_{k-1}^k$ the roots of $z^k + 1$

$$(10) \qquad v_\nu^k = \exp\{\imath \frac{(2\nu+1)\pi}{k}\}, \qquad \nu = 0, 1, \ldots, k-1.$$

Then the absolute value of the difference between the number of roots of the form (9) with nonnegative real parts and the number of roots of the form (10) with nonnegative real parts does not exceed one, see [4]. □

Let $\mathbb{C}\{z, s\}$ denote the ring of convergent complex power series in two variables z, s. A function $f(z, s)$, defined by

$$f(z, s) = \sum_{i,j=0}^{\infty} A_{ij} s^i z^j = \sum_{j=0}^{\infty} A_j(s) z^j \in \mathbb{C}\{z, s\}, \quad A_j(s) = \sum_{i=0}^{\infty} A_{ij} s^i$$

is said to be of order $k \geq 1$ in z at $s = s_0$ if

$$f(z, s_0) = \sum_{j=k}^{\infty} A_j(s_0) z^j, \quad A_k(s_0) \neq 0$$

A function $f(z, s)$ is called irreducible if it cannot be represented in the form $f = gh$ where $g, h \in \mathbb{C}\{s, z\}$ are non-units in the ring $\mathbb{C}\{s, z\}$.

Let us denote by $D(\varepsilon)$ the complex disk centered at $0 \in \mathbb{C}$ with radius $\varepsilon > 0$. The following result is known, see e.g. [8].

THEOREM 3.2. Suppose that $f(z, s) \in \mathbb{C}\{z, s\}$ is irreducible and of order $k \geq 1$ in z at $s = 0$. Then there exist $\varepsilon > 0$, $\delta > 0$ such that $f(\cdot, s)$ has, for every $s \in D(\varepsilon)\backslash\{0\}$, exactly k simple zeros in the disk $D(\delta)$. Let $D = \{w \in \mathbb{C}; w^k \in D(\varepsilon)\}$. Then there exists a convergent power series $\phi(\cdot)$ on D such that
  (i) $f(\phi(w), w^k) = 0$, $w \in D$.
  (ii) For every $s = w^k \in D(\varepsilon)$, the zeros of $f(\cdot, w^k)$ in $D(\delta)$ are given by $\phi(u_0^k w), \ldots, \phi(u_{k-1}^k w)$ where the u_j^k are defined by (9).

As a consequence of the theorem the k zeros of $f(\cdot, s)$ in $D(\delta)$ can be represented by the Puiseux series for $|s| \to 0$

$$z_\nu(s) = \sum_{i=1}^{\infty} \alpha_i (u_\nu^k s^{1/k})^i, \quad s \in D(\varepsilon) \setminus \{0\}, \quad \nu = 0, \ldots, k-1$$

where $s^{1/k} = r^{1/k}e^{i\varphi/k}$ if $s = re^{i\varphi}$, $r > 0$, $0 \leq \varphi < 2\pi$. Using this result we can investigate the asymptotic behavior of "small" roots of the quasipolynomial (8) as μ tends towards μ_0. For the sake of notational simplification we make the following convention. Given $f_0(z)$, $g(z)$ of the form (3) and (5), respectively, we set $a_{ij} = 0$ and $c_{ij} = 0$ for all index pairs (i, j) for which these coefficients are not yet defined. In particular

(11) $\qquad a_{nj} = 0, \quad 0, \ldots, m-1 \quad \text{and} \quad c_{nj} = 0, \quad 0, \ldots, m.$

PROPOSITION 3.3. *Let $f_0(z)$, $g(z)$ be two real quasipolynomials of the form (3) and (5), respectively, $g(0) = \sum_{j=0}^{m} c_{0j} \neq 0$, and suppose that $f_\mu(z) = f_0(z) + \mu g(z)$ has a root of multiplicity $k \geq 1$ at $z = 0$ for $\mu = \mu_0$. Then there exist $\varepsilon > 0$, $\delta > 0$ such that, for $s \in (-\varepsilon, \varepsilon) \setminus \{0\}$, $f_{\mu_0+s}(z)$ has exactly k simple roots in the complex disk $D(\delta)$ and these roots have the following asymptotic behavior as $|s| \to 0$:*

(i) *If* $\left(\sum_{j=0}^{m} c_{0j}\right)/A_k(\mu_0) < 0$ *then*

$$z_\nu(s) = \begin{cases} s^{\frac{1}{k}} \left|\frac{\sum_{j=0}^{m} c_{0j}}{A_k(\mu_0)}\right|^{\frac{1}{k}} u_\nu^k + o\left(s^{\frac{1}{k}}\right), & \nu = 0, 1, \ldots, k-1 \quad \text{for } s > 0 \\ |s|^{\frac{1}{k}} \left|\frac{\sum_{j=0}^{m} c_{0j}}{A_k(\mu_0)}\right|^{\frac{1}{k}} v_\nu^k + o\left(|s|^{\frac{1}{k}}\right), & \nu = 0, 1, \ldots, k-1 \quad \text{for } s < 0. \end{cases}$$

(ii) *If* $\left(\sum_{j=0}^{m} c_{0j}\right)/A_k(\mu_0) > 0$ *then*

$$z_\nu(s) = \begin{cases} s^{\frac{1}{k}} \left(\frac{\sum_{j=0}^{m} c_{0j}}{A_k(\mu_0)}\right)^{\frac{1}{k}} v_\nu^k + o\left(s^{\frac{1}{k}}\right), & \nu = 0, 1, \ldots, k-1 \quad \text{for } s > 0 \\ |s|^{\frac{1}{k}} \left(\frac{\sum_{j=0}^{m} c_{0j}}{A_k(\mu_0)}\right)^{\frac{1}{k}} u_\nu^k + o\left(|s|^{\frac{1}{k}}\right), & \nu = 0, 1, \ldots, k-1 \quad \text{for } s < 0 \end{cases}$$

where

$$A_k(\mu_0) = \sum_{j=0}^{m} \left[(a_{kj} + \mu_0 c_{kj}) + (a_{k-1j} + \mu_0 c_{k-1j})\frac{\tau_j}{1!} + \cdots + (a_{0j} + \mu_0 c_{0j})\frac{\tau_j^k}{k!}\right].$$

Proof. Using (11) we have for $\mu \in [0, \infty)$:

$$\begin{aligned} f_\mu(z) &= f_0(z) + \mu g(z) \\ &= a_{nm} z^n e^{\tau_m z} + \sum_{j=0}^{m}\sum_{k=0}^{n-1}(a_{kj} + \mu c_{kj})z^k e^{\tau_j z} \\ &= a_{nm}\left[z^n + \frac{\tau_m z^{n+1}}{1!} + \frac{\tau_m^2 z^{n+2}}{2!} + \cdots\right] + \\ &\quad + \sum_{j=0}^{m}\sum_{k=0}^{n-1}(a_{kj} + \mu c_{kj})\left[z^k + \frac{\tau_j z^{k+1}}{1!} + \frac{\tau_j^2 z^{k+2}}{2!} + \cdots\right] \end{aligned}$$

$$= \sum_{j=0}^{m}\sum_{k=0}^{n}(a_{kj}+\mu c_{kj})\left[z^k + \frac{\tau_j z^{k+1}}{1!} + \frac{\tau_j^2 z^{k+2}}{2!} + \cdots\right]$$

(12)
$$= \sum_{\ell=0}^{\infty} A_\ell(\mu) z^\ell$$

where

$$A_\ell(\mu) = \sum_{j=0}^{m} \sum_{i=0}^{\min\{\ell,n\}} (a_{ij}+\mu c_{ij}) \frac{\tau_j^{\ell-i}}{(\ell-i)!}, \quad \ell \in \mathbb{N}.$$

It follows that the quasipolynomial $f_\mu(z)$ has a root of multiplicity $k \geq 1$ at $z = 0$ for $\mu = \mu_0$ if and only if

(13)
$$A_0(\mu_0) = \ldots = A_{k-1}(\mu_0) = 0, \quad A_k(\mu_0) \neq 0.$$

Suppose that (13) is satisfied. Then we obtain from (12) and (13)

(14)
$$\begin{aligned} f_{\mu_0+s}(z) &= f_{\mu_0}(z) + sg(z) \\ &= \sum_{\ell=k}^{\infty} A_\ell(\mu_0) z^\ell + s \sum_{j=0}^{m} \left[c_{0j} + (c_{1j} + c_{0j}\frac{\tau_j}{1!})z + \cdots \right. \\ &\quad + \left. (c_{k-1,j} + c_{k-2,j}\frac{\tau_j}{1!} + \cdots + c_{0j}\frac{\tau_j^{k-1}}{(k-1)!}) z^{k-1} + \cdots \right]. \end{aligned}$$

Now, if we set $f(z,s) := f_0(z) + (\mu_0 + s)g(z)$ we have

$$f(z,0) = f_0(z) + \mu_0 g(z), \qquad f(z, \mu - \mu_0) = f_0(z) + \mu g(z)$$

and therefore it follows from (14) that $f(z,s)$ is of order k in z at $s = 0$. It is easy to verify that $f(z,s)$ is irreducible in the ring $\mathbb{C}\{z,s\}$. In fact, if we assume that

(15)
$$f(z,s) = h(z,s)g(z,s), \quad h, g \in \mathbb{C}\{z,s\}$$

then $f(0,s) = h(0,s)g(0,s) = \left(\sum_{j=0}^{m} c_{0j}\right) s$ and so either $h(0,0) \neq 0$ or $g(0,0) \neq 0$, i.e. one of the factors in (15) is a unit in $\mathbb{C}\{z,s\}$.
Since $f(z,s) \in \mathbb{C}\{z,s\}$ is irreducible and of order k in z at $s = 0$ it follows from Theorem 3.2 that there exist $\varepsilon > 0$, $\delta > 0$ such that, $f(\cdot, s)$ has for every $s \in (-\varepsilon, \varepsilon) \setminus \{0\}$ exactly k roots in $D(\delta)$ and these roots can be represented by

(16)
$$z_\nu(s) = \sum_{i=1}^{\infty} \alpha_i \left(u_\nu^k s^{1/k}\right)^i, \quad s \in (-\varepsilon, \varepsilon) \setminus \{0\}, \quad \nu = 0, \ldots, k-1.$$

Substituting (16) for z in (14) and equating coefficients with the same powers of $s^{1/k}$ we obtain for α_1 the following equation

$$\left(\sum_{j=0}^{m} c_{0j}\right) + A_k(\mu_0)\alpha_1^k = 0 \tag{17}$$

where $A_k(\mu_0) \neq 0$. Therefore there exists $\nu \in \{0, 1, \ldots, k-1\}$ such that

$$\alpha_1 = \begin{cases} \left[\left(\sum_{j=0}^{m} c_{0j}\right)/A_k(\mu_0)\right]^{\frac{1}{k}} v_\nu^k, & \text{if } \left(\sum_{j=0}^{m} c_{0j}\right)/A_k(\mu_0) > 0 \\ \left|\left(\sum_{j=0}^{m} c_{0j}\right)/A_k(\mu_0)\right|^{\frac{1}{k}} u_\nu^k, & \text{if } \left(\sum_{j=0}^{m} c_{0j}\right)/A_k(\mu_0) < 0. \end{cases}$$

Substituting any of these expressions for α_1 in (16) we obtain statements (i) and (ii) of the proposition. $\square$

Let $\mathcal{N}_+(f_\mu; \delta)$ denote the number of roots of $f_\mu(z)$ in $D(\delta)$ with nonnegative real parts (counting multiplicities). As a consequence of the previous proposition and Remark 3.1 we obtain that $\mathcal{N}_+(f_\mu; \delta)$ changes at most by one as μ crosses the value μ_0. More precisely, there exist $\varepsilon > 0$, $\delta > 0$ such that

$$|\mathcal{N}_+(f_{\mu^-}; \delta) - \mathcal{N}_+(f_{\mu^+}; \delta)| \leq 1, \quad \mu^- \in (\mu_0 - \varepsilon, \mu_0), \ \mu^+ \in (\mu_0, \mu_0 + \varepsilon). \tag{18}$$

Therefore, as μ increases and some roots of $f_\mu = f(z) + \mu g(z)$ reach the closed right half-plane, not more than *one* of them may ever return to the open left half-plane through the origin $z = 0$. Note that this can only happen (at the parameter value $\mu_0 = -f(0)/g(0)$) if $f(0)/g(0) < 0$.

4. Root loci across the punctured imaginary axis. In this section we investigate how the root loci of $f_\mu(z) = f_0(z) + \mu g(z)$, $\mu \geq 0$ cross the imaginary axis if $f_0(z) \in \mathcal{H}$ and $g(z)$ is of the form (5). Let $N_+(f_\mu)$ denote the (finite) number of all unstable roots of $f_\mu(z)$ (i.e. the roots lying in $\mathbb{C}\setminus\mathbb{C}_-$) counting multiplicity. Since $f_0(z) \in \mathcal{H}$ we have $N_+(f_0) = 0$. As μ grows the roots move continuously on the complex plane and for some μ_0 the number of unstable roots $N_+(f_{\mu_0})$ may become positive. More precisely, since the quasipolynomial $f_\mu(z)$ has a countable set of roots, there exists a countable family of continuous functions $z_j(\cdot) : \mathbb{R}_+ \mapsto \mathbb{C}$, $j = 1, 2, \ldots$ such that, for every $\mu \geq 0$, $z_1(\mu), z_2(\mu), \cdots$ are the roots of $f_0(z) + \mu g(z)$. Each of the functions $z_j(\cdot)$ is analytical in μ at every value μ_0 for which $z_j(\mu_0)$ is a simple root of $f_0(z) + \mu_0 g(z)$. In this case we write $z_j'(\mu_0)$ for $(\partial z_j/\partial \mu)(\mu_0)$. We will show that, as μ increases, the roots of the quasipolynomials $f_\mu(z) = f_0(z) + \mu g(z)$ can move across the punctured imaginary axis $i\mathbb{R} \setminus \{0\}$ only from left to right. Hence it follows that $z = 0$ is the only point on the imaginary axis through which a root of $f_\mu(z)$ (at most one) can return to the open left half-plane once it has left it.

The main result in the present paper is the following characterization of convex directions for real quasipolynomials of delay type.

THEOREM 4.1. *A real quasipolynomial $g(z)$ of the form (5) is a convex direction for the set $\mathcal{H}$ if and only if it satisfies the following condition for all quasipolynomials $f_0(z) \in \mathcal{H}$:*

(CD) *If one of the roots $z_j(\mu)$ of $f_0(z) + \mu g(z)$ hits the punctured imaginary axis $\imath\mathbb{R} \setminus \{0\}$ for $\mu = \mu_0$ then $z_j(\mu_0)$ is a simple root of $f_0(z) + \mu_0 g(z)$ and $\operatorname{Re}\{z'_j(\mu_0)\} > 0$, i.e. as μ increases the roots $z_j(\mu)$ can only cross the punctured imaginary axis from left to right.*

Proof. If $g(z)$ is a convex direction for $\mathcal{H}$ then (6) is satisfied for all $w \in \{w > 0;\ g(\imath w) \neq 0\}$. In contrast every Hurwitz stable quasipolynomial $f_0(z) \in \mathcal{H}$ has the phase increasing property (4) for all $\omega > 0$. Suppose that for some $\mu_0 > 0$, $\omega_0 > 0$ the equation

$$(19) \qquad f_0(\imath\omega_0) + \mu_0 g(\imath\omega_0) = 0$$

holds. It follows from (19) and the Hurwitz stability of $f_0(z)$ that

$$(20) \qquad f_0(\imath\omega_0) \neq 0, \qquad g(\imath\omega_0) \neq 0.$$

Hence two continuous argument functions

$$\arg(f_0(\imath\omega)), \qquad \arg(g(\imath\omega))$$

can be defined in some neighborhood of ω_0 and

$$\arg(f_0(\imath\omega_0)) = \arg(g(\imath\omega_0)) + k\pi,$$

where k is an odd number. Therefore

$$\sin(2\arg(f_0(\imath\omega_0))) - (\tau_0 + \tau_m)\omega_0 = \sin(2\arg(g(\imath\omega_0))) - (\tau_0 + \tau_m)\omega_0$$

and from (4) and (6) at $\omega = \omega_0$ we obtain

$$(21) \qquad \operatorname{Re}\left[\frac{f'_0(\imath\omega_0)}{f_0(\imath\omega_0)} - \frac{g'(\imath\omega_0)}{g(\imath\omega_0)}\right] > 0.$$

Suppose by contradiction that $\imath\omega_0$ is a multiple root of $f_{\mu_0}(z)$, i.e.

$$(22) \qquad f_0(\imath\omega_0) + \mu_0 g(\imath\omega_0) = 0, \qquad f'_0(\imath\omega_0) + \mu_0 g'(\imath\omega_0) = 0$$

(20) and (22) imply

$$\frac{f'_0(\imath\omega_0)}{f_0(\imath\omega_0)} - \frac{g'(\imath\omega_0)}{g(\imath\omega_0)} = 0$$

whence a contradiction to (21). Thus $\imath\omega_0$ is a simple root of $f_{\mu_0}(z)$, and in a neighborhood of μ_0 there exists an analytic function $z_j(\mu)$ satisfying $z_j(\mu_0) = \imath\omega_0$ and $f_\mu(z_j(\mu)) = 0$. Differentiating $f_0(z_j(\mu)) + \mu g(z_j(\mu)) = 0$ with respect to μ at μ_0 we obtain

$$[f'_0(\imath\omega_0) + \mu_0 g'(\imath\omega_0)]z'_j(\mu_0) + g(\imath\omega_0) = 0$$

and dividing by $-g(\imath\omega_0) = \mu_0^{-1} f_0(\imath\omega_0)$ we get

$$\text{(23)} \quad \mu_0 \left[\frac{f_0'(\imath\omega_0)}{f_0(\imath\omega_0)} - \frac{g'(\imath\omega_0)}{g(\imath\omega_0)} \right] z_j'(\mu_0) = 1.$$

Since

$$\operatorname{Re} z > 0 \Leftrightarrow \operatorname{Re}(z^{-1}) > 0$$

for all $z \in \mathbb{C}$, $z \neq 0$ we conclude from (21) and (23) that $\operatorname{Re}\{z_j'(\mu_0)\} > 0$. Thus condition (CD) is satisfied.

Conversely, suppose that the condition (CD) is satisfied for all $f_0(z) \in \mathcal{H}$. We have to show that the stability set $M(f_0, g)$ (7) is a real interval for each $f_0(z) \in \mathcal{H}$. By continuity arguments the stability of $f_0(z)$ implies the stability of $f_\mu(z)$ for all μ in a small neighborhood of $\mu = 0$ (no root chain of the delay type quasipolynomial $f_0(z)$ converges towards the imaginary axis). Suppose that μ_0 is the smallest positive value of μ for which $f_\mu(z)$ is unstable, i.e. $f_{\mu_0}(z)$ has a root $z(\mu_0)$ on the imaginary axis. There are two possibilities, either $z(\mu_0) = \imath\omega_0$ with $\omega_0 > 0$ or $z(\mu_0) = 0$. In the first case there is the complex conjugate root $(-\imath\omega_0)$ of $f_{\mu_0}(z)$, too. By condition (CD), for μ a little bit greater than μ_0, the quasipolynomial $f_\mu(z)$ has at least two roots in the open right half-plane. Applying (18) we may conclude that at least one root must remain in the right half-plane for all $\mu > \mu_0$. Thus $M(f_0, g)$ is an interval.

Now suppose $z(\mu_0) = 0$. It follows from Proposition 3.3 that the multiplicity of the root $z(\mu_0) = 0$ cannot exceed 2 since otherwise there would exist an unstable root of $f_\mu(z)$ for $\mu < \mu_0$. By the same reason only case (i) in Proposition 3.3 can occur for $k = 1$ or $k = 2$. Hence at least one small root of $f_\mu(z)$ will enter the open left half-plane for μ a little bit greater than μ_0. But for $\mu > \mu_0$ all roots of $f_\mu(z)$ on $\imath\mathbb{R}$ are nonzero and so (CD) implies that $N_+(f_\mu) \geq 1$ for all $\mu > \mu_0$, i.e. $M(f_0, g)$ is an interval. □

We conclude this note with some comments.

Remark 4.2 Theorem 4 can be extended to the complex coefficient case. □

Remark 4.3 Following the approach in [4] we can use Theorem 4.1 as a starting point for the definition of a concept of *convex direction for a given Hurwitz quasipolynomial* f_0 of delay type. We propose the following definition:

A real quasipolynomial $g(z)$ of the form (5) is said to be a convex direction for a given Hurwitz stable quasipolynomial $f_0(z)$ of the form (3) if all the roots $z_i(\mu)$ of $f_\mu(z) = f_0(z) + \mu g(z)$ for $\mu > 0$ on the punctured imaginary axis $\imath\mathbb{R} \setminus \{0\}$ are simple and satisfy $\operatorname{Re}\{z_i'(\mu)\} > 0$.

In order to check whether $g(z)$ is a convex direction for $f_0(z)$ in the sense of this definition a similar graphical test (based on an associated Nyquist plot) can be applied as in the case of polynomials [4].

If $g(z)$ is a convex direction for $f_0(z)$ in the above sense the stability set $M(f_0, g)$ is a real interval. However, the converse does not hold. Moreover, $-g$ need not be a convex direction for $f_0(z)$ in this case. □

Remark 4.4 In this note we have fixed the delays $\tau_0 < \tau_1 < \ldots < \tau_m$ and assumed that $f_0(z)$ is of the form (3) and $g(z)$ is of the form (5). This does not mean, however, that the delays τ_j, $0 \leq j \leq m$ are actually present in $g(z)$. Some of the polynomials $\sum_{k=0}^{n-1} c_{kj}z^k$ in the expression (5) of $g(z)$ may be zero so that the term $e^{\tau_j z}$ is not present in (5). Similarly some of the polynomials $\sum_{k=0}^{n-1} a_{kj} z^k$ in (3) may be zero polynomials. Let

$$(24) \qquad g(z) = \sum_{j=0}^{m_g} \sum_{k=0}^{n-1} c_{kj} z^k e^{\sigma_j z}$$

be an arbitrary quasipolynomial with $\sigma_0 < \ldots < \sigma_{m_g}$. For any pair of real numbers $\alpha < \beta$, we denote by $\mathcal{H}(\alpha, \beta)$ the set of all Hurwitz quasipolynomials of the form (3) where $\tau_0 < \ldots < \tau_m$ are *arbitrary* delays ($m \geq 1$) with $\tau_0 = \alpha$, $\tau_m = \beta$. Suppose

$$\alpha \leq \sigma_0 \quad \text{and} \quad \sigma_{m_g} \leq \beta.$$

Applying Proposition 2.1 we conclude that $g(z)$ is a convex direction for all the quasipolynomials in $\mathcal{H}(\alpha, \beta)$ if and only if $g(z)$ satisfies condition (6). As a consequence we obtain: If $g(z)$ is of the form (24) with $\alpha \leq \sigma_0$, $\sigma_{m_g} \leq \beta$ and satisfies condition (CD) for all Hurwitz stable $f_0(z)$ of the form

$$f_0(z) = \sum_{k=0}^{n-1} a_{k0} z^k e^{\alpha z} + \sum_{k=0}^{n} a_{k1} z^k e^{\beta z}, \quad a_{n1} > 0,$$

with $a_{kj} \in \mathbb{R}$ for $k = 1, \ldots, n-1$, $j = 1, 2$ and $a_{n1} > 0$ then $g(z)$ satisfies (CD) for all $f_0(z) \in \mathcal{H}(\alpha, \beta)$. $\square$

References

[1] B.R. Barmish. *New Tools for Robustness of Linear Systems*. Macmillan Publishing Company, New York, 1994.
[2] R. Bellman and K.L. Cook. *Differential-Difference Equations*. Academic Press, New York, 1963.
[3] M. Fu, A.W. Olbrot, and M.P. Polis. Edge theorem and graphical test for robust stability of neutral time-delay system *Automatica* 17:739-742, 1991
[4] D. Hinrichsen and V.L. Kharitonov. On convex directions for stable polynomials. Technical Report 309, Inst. f. Dynamische Systeme, Universität Bremen, Bremen, 1994.
[5] V. L. Kharitonov and A.P. Zhabko. Robust Stability of Time-Delay Systems. *IEEE Transactions on Automatic Control*, AC-39:2388–2397, 1994.
[6] A. Rantzer. Stability conditions for polytopes of polynomials. *IEEE Transactions on Automatic Control*, AC-37:79–89, 1992.
[7] E. Schwengeler. *Geometrisches über die Verteilung der Nullstellen spezieller ganzer Functionen (Exponentialsummen)*. Ph.D. thesis, Zürich, 1925.
[8] H. Whitney. *Complex Analytic Varieties*. Addison-Wesley, Reading, 1972.

From J.J. Sylvester to Adolf Hurwitz: A Historical Review

Eliahu I. Jury
Research Professor Emeritus, University of Miami
Coral Gables, Florida, USA

Professor Emeritus, University of California
Berkeley, California, USA

Abstract. In this paper, a brief historical review of the works of four illustrious mathematicians in the nineteenth century – namely, Hermite (1854), Routh (1877), Lyapunov (1892), and Hurwitz (1895) – which affected research on stability in the twentieth century is mentioned.

Based on the work of Sylvester (1840), the connection between the Hurwitz criterion and resultants is established. This leads to a derivation of the generalized and specialized Orlando's formulae. The connection between Hurwitz and Routh criteria is also mentioned. Using the method of Schur complements, the equivalence between positive innerwise and positive definite stability criteria is obtained. This leads to the connection between Sylvester's resultants and Bezoutians. These formulations form the basis for obtaining all the stability criteria.

From the Hermite criterion for real polynomials, its reduced form – which in fact is the Liénard-Chipart (1918) criterion – is obtained, The connection between Hermite criterion and Lyapunov stability conditions is also indicated thus forging the link between the four stability criteria of Hermite, Routh, Lyapunov, and Hurwitz.

Finally, five important applications of the Hurwitz criterion are mentioned.

1. Introduction. In the nineteenth century, four illustrious mathematicians laid the foundation for the investigations of stability and related researches in the twentieth century.

The first of the these mathematicians was C. Hermite whose groundbreaking paper of 1854 has been translated from French into English by Parks [1] in 1977. Also, in [2], Parks had expounded another proof of the Hermite criterion. The second, E.J. Routh (1877) and his essay [3], was celebrated in various publications in 1977 [4], [5]. The third was A.M. Lyapunov whose PhD thesis of 1892 was celebrated in 1992 [6],[7], [8]. The last of the four is A. Hurwitz whose milestone paper of 1895 [9] is being celebrated this year (1995) at the Hurwitz Symposium.

In this historical review, we concentrate on the impact of Hurwitz's paper, and the many follow-up papers by several researchers.

Based on the work of Sylvester in 1840 [10], we establish the connection between the resultant determinants, or, equivalently, the inners determinants, and Hurwitz matrix minors.

In the process of this derivation, we establish the generalized as well as the well known form of Orlando's formulae [2],[11]. From the transformed Sylvester matrix, we establish its connection with the Hermite matrix [12]. From the latter, we obtain the reduced Hermite criterion [13], as well as the Liénard-Chipart criterion [14]. The connection between the Sylvester matrix and the symmetric Bezoutian matrix is also obtained. The connection between Hurwitz minors and

Routh table is then mentioned [15]. Also, the connection between Hermite criterion and Lyapunov's second method is obtained [15],[16],[17]. This forges the link between the four stability criteria of the above mentioned four illustrious mathematicians. Five important applications, among many, of the Hurwitz criterion is brought to light. Finally, some comments regarding the Liénard-Chipart criterion of 1918 and the best possible simplification of the Hurwitz criterion for stability are made. A best possible simplification for complex polynomials has already been established [18].

The composition of the paper is as follows:
2. On Orlando's formulae.
3. Connection with Hermite matrix.
4. Reduced Hermite criterion.
5. Reduced Hurwitz criterion.
6. Connection with Routh stability table.
7. Connection with Lyapunov's second method.
8. Connection between Bezoutian and Hermite matrix.
9. Applications – Five important application of Hurwitz matrix.
10. On the simplest possible property of Hurwitz criterion.
11. Conclusion.
The list of references follow.

2. On Orlando's formulae. In this section, we derive both the generalized Orlando's formula [2], as well as, the well known Orlando's formula [11]. We start with Sylvester matrix [10] which was obtained by Sylvester in 1840 in order to ascertain whether two polynomials are prime or not. The determinant of the Sylvester matrix, referred to as the *resultant*, is related to the roots of the two polynomials as indicated by Barnett [19]. This matrix is formulated for $F(s)$ and its complex conjugate $F^*(s)$ as in (3). The submatrices shown within dotted lines are referred to by Jury [20] as *inners*. If the determinants of all the inners, including the whole matrix, are positive, this matrix is said to be *positive innerwise* or *P.I.*. Also, in (3), the left triangle of zeros are also indicated by dotted lines. This fact plays an important role, as will be shown in later sections, in obtaining the half-sized Hermitian matrix as well as the Bezoutian matrix [21].

Consider the polynomial $F(s) \in \Im[s]_n$ where

$$(1) \qquad F(s) = a_0 s^n + a_1 s^{n-1} + \cdots + a_n = a_0 \prod_{i=1}^{n}(s - \lambda_i), \; a_0 \neq 0.$$

When $\operatorname{Re}\lambda_i < 0, \; \forall i = 1, \ldots, n$, $F(s)$ is said to be *stable*. Here, $\operatorname{Re}[\cdot]$ denotes the real part. Define $F^*(s)$ as

$$(2) \quad F^*(s) = (-1)^n \bar{a}_0 s^n + (-1)^{n-1} \bar{a}_1 s^{n-1} + \cdots + \bar{a}_n = (-1)^n \bar{a}_0 \prod_{j=1}^{n}(s + \bar{\lambda}_j),$$

where $\bar{[\cdot]}$ denotes complex conjugation.

2.1. Resultant. Let us denote the resultant matrix [10] of $F(s)$ and $F^*(s)$ as $R[F, F^*] \in \Im^{2n \times 2n}$ where [19]

(3)
$$R[F, F^*] = \begin{bmatrix} a_0 & a_1 & \cdots & a_n & 0 & 0 & \cdots & 0 \\ 0 & a_0 & \cdots & a_{n-1} & a_n & 0 & \cdots & 0 \\ & & \boldsymbol{R}^{(1)} & & & \boldsymbol{R}^{(2)} & \ddots & \vdots \\ & & & \ddots & & & & \vdots \\ 0 & 0 & \cdots & a_0 & a_1 & a_2 & \cdots & a_n \\ 0 & 0 & \cdots & (-1)^n \bar{a}_0 & (-1)^{n-1}\bar{a}_1 & (-1)^{n-2}\bar{a}_2 & \cdots & \bar{a}_n \\ & & \boldsymbol{R}^{(3)} & & & \boldsymbol{R}^{(4)} & & \vdots \\ 0 & (-1)^n \bar{a}_0 & \cdots & -\bar{a}_{n-1} & \bar{a}_n & 0 & \cdots & 0 \\ (-1)^n \bar{a}_0 & (-1)^{n-1}\bar{a}_1 & \cdots & \bar{a}_n & 0 & 0 & \cdots & 0 \end{bmatrix}$$

Hence [19]

(4) $$\det[R[F, F^*]] = (-1)^{n(n+1)/2}|a_0|^{2n} \prod_{i=1}^{n}\prod_{j=1}^{n}(\lambda_i + \bar{\lambda}_j),$$

where we have used the fact that $(-1)^{n(n+1)/2+n^2} = (-1)^{n(3n-1)/2} = (-1)^{n(n+1)/2}$. When $F(s)$ is real,

(5) $$\det[R[F, F^*]] = (-1)^{n(n+1)/2} 2^n a_0^{2n} \prod_{\ell=1}^{n} \lambda_\ell \prod_{1 \le i < j \le n}(\lambda_i + \lambda_j)^2.$$

The P.I. matrix $\Delta \in \Im^{2n \times 2n}$ associated with $F(s)$ is [12]

(6) $$\Delta = \begin{bmatrix} \Delta^{(1)} & \Delta^{(2)} \\ \Delta^{(3)} & \Delta^{(4)} \end{bmatrix} = R[F, F^*]M,$$

where

(7) $$M = \begin{bmatrix} I_n & 0 \\ 0 & \hat{I}_n \end{bmatrix}, \quad \hat{I}_n = \text{diag}\{(-1)^{n-1}, (-1)^{n-2}, \ldots, +1, -1, +1\}.$$

The partitioning of Δ in (6) is the same as that for R in (3). Note that, $\det[M] = (-1)^{n(n-1)/2}$ and $(-1)^{n(n+1)/2+n(n-1)/2} = (-1)^{n^2} = (-1)^n$. Hence,

(8) $$\det[\Delta] = \det[R[F, F^*]] \cdot \det[M] = (-1)^n |a_0|^{2n} \prod_{i=1}^{n}\prod_{j=1}^{n}(\lambda_i + \bar{\lambda}_j).$$

The above is a generalized version of Orlando's formula applicable to monic polynomials [2].

Note that, for the real case,

$$\det[\Delta] = (-2)^n a_0^{2n} \prod_{\ell=1}^{n} \lambda_\ell \prod_{1 \leq i < j \leq n} (\lambda_i + \lambda_j)^2. \tag{9}$$

Given matrix $[\cdot] \in \Im^{n \times n}$, its $(i \times i)$ principal minor is denoted by $[\cdot]_{(i \times i)}$. Its $(i \times i)$ inner is denoted by $[\cdot]_i \in \Im^{i \times i}$.

Now, from [12], we have, for $i = 1, 2, \ldots, n$,

$$\det[\Delta_{2i}] = 2^i a_0 \det[\mathcal{H}_{((i-1) \times (i-1))}] \cdot \det[\mathcal{H}_{(i \times i)}], \quad \mathcal{H}_{(0 \times 0)} \doteq 1, \tag{10}$$

where $a_0 > 0$ and $\mathcal{H} \in \Re^{n \times n}$ is the Hurwitz matrix given by [9]

$$\mathcal{H} = \begin{bmatrix} a_1 & a_3 & a_5 & a_7 & \cdots & 0 & 0 & 0 \\ a_0 & a_2 & a_4 & a_6 & \cdots & 0 & 0 & 0 \\ 0 & a_1 & a_3 & a_5 & \cdots & 0 & 0 & 0 \\ \vdots & \vdots & \vdots & \vdots & & \vdots & \vdots & \vdots \\ 0 & 0 & 0 & 0 & \cdots & a_{n-2} & a_n & 0 \\ 0 & 0 & 0 & 0 & \cdots & a_{n-3} & a_{n-1} & 0 \\ 0 & 0 & 0 & 0 & \cdots & a_{n-4} & a_{n-2} & a_n \end{bmatrix}. \tag{11}$$

Note that, (10) establishes a connection between Sylvester resultant determinants and Hurwitz determinants. When $i = n$, it yields

$$\det[\Delta_{2n}] = \det[\Delta] = 2^n a_0 \det[\mathcal{H}_{((n-1) \times (n-1))}] \cdot \det[\mathcal{H}_{(n \times n)}]. \tag{12}$$

We also notice from Hurwitz matrix that

$$\det[\mathcal{H}_{(n \times n)}] = a_n \det[\mathcal{H}_{(n-1) \times (n-1)}]. \tag{13}$$

Now, noting that,

$$a_n = (-1)^n a_0 \prod_{\ell=1}^{n} \lambda_\ell, \tag{14}$$

we get

$$\det[\Delta] = (-2)^n a_0^2 \prod_{\ell=1}^{n} \lambda_\ell \det[\mathcal{H}_{((n-1) \times (n-1))}]^2. \tag{15}$$

Comparing (9) and (15),

$$\det[\mathcal{H}_{((n-1) \times (n-1))}]^2 = a_0^{2(n-1)} \prod_{1 \leq i < j \leq n} (\lambda_i + \lambda_j)^2. \tag{16}$$

Hence,

$$(17) \quad \det[\mathcal{H}_{((n-1)\times(n-1))}] = \pm a_0^{n-1} \prod_{1\leq i<j\leq n} (\lambda_i + \lambda_j).$$

The factor ± 1 may be resolved from

$$(18) \quad F(s) \text{ is stable} \implies \det[\mathcal{H}_{((n-1)\times(n-1))}] > 0,$$

which gives

$$(19) \quad \det[\mathcal{H}_{((n-1)\times(n-1))}] = (-1)^{n(n-1)/2} a_0^{n-1} \prod_{1\leq i<j\leq n} (\lambda_i + \lambda_j),$$

which is Orlando's formula [11].

3. Connection with Hermite matrix. Hermite's stability criterion when applied to $F(s)$ in (1) give rise to the Hermitian matrix $H = \{h_{i,j}\} \in \Im^{n\times n}$ where [2],[12]

$$(20) \quad H = \begin{bmatrix} [a_n \bar{a}_{n-1}]_R & [a_{n-2}\bar{a}_n]_I & [a_{n-3}\bar{a}_n]_R & \cdots \\ [a_n\bar{a}_{n-2}]_I & [a_{n-1}\bar{a}_{n-2} - a_n\bar{a}_{n-3}]_R & & \cdots \\ [a_n\bar{a}_{n-3}]_R & & & \cdots \\ \vdots & \vdots & \vdots & \ddots & \vdots \\ & & & \cdots & [a_1\bar{a}_0]_R \end{bmatrix}.$$

Here, $[x]_R \doteq x + \bar{x}$ and $[x]_I \doteq x - \bar{x}$. When $F(s)$ is real, (20) becomes

$$(21) \quad H = 2 \begin{bmatrix} a_n a_{n-1} & 0 & a_n a_{n-3} & \cdots \\ 0 & a_{n-1}a_{n-2} - a_n a_{n-3} & & \cdots \\ a_n a_{n-3} & & & \cdots \\ \vdots & \vdots & \vdots & \ddots & \vdots \\ & & & \cdots & a_1 a_0 \end{bmatrix}.$$

The above matrix is symmetric in both the nonzero and zero entries. Now, form the Schur complement $[\Delta/\Delta^{(4)}]$ of Δ in (6):

$$(22) \quad [\Delta/\Delta^{(4)}] = \Delta^{(4)} - \Delta^{(3)}\Delta^{(1)^{-1}}\Delta^{(2)}.$$

The inner determinants of Δ are equal to the principal minors of $[\Delta/\Delta^{(4)}]$ (modulo a constant multiplier) [22]. Also, $[\Delta/\Delta^{(4)}]$ is related to H as follows (there is a minor error in [12] where the relationship is given is $H = S \cdot [\Delta/\Delta^{(4)}]$):

$$\hat{I}_n \cdot H \cdot \hat{I}_n = S \cdot [\Delta/\Delta^{(4)}] \cdot I_{REV}, \tag{23}$$

where

$$I_{REV} = \begin{bmatrix} 0 & 0 & 0 & \cdots & 1 \\ \vdots & & & & \vdots \\ 0 & 0 & 1 & & 0 \\ 0 & 1 & 0 & & 0 \\ 1 & 0 & 0 & \cdots & 0 \end{bmatrix} \tag{24}$$

and S is the upper triangular matrix

$$S = \begin{bmatrix} a_{n-1} & a_{n-2} & \cdots & a_0 \\ a_{n-2} & a_{n-3} & \cdots & 0 \\ \vdots & & & \\ a_0 & 0 & & 0 \end{bmatrix}. \tag{25}$$

Now,

$$\det[H] = a_0^n \det[\Delta/\Delta^{(4)}] = \det[\Delta]. \tag{26}$$

Also, the inner determinants of Δ are equal to the principal minors of H. Hence, as will be shown later,

$$F(s) \text{ is stable} \iff \Delta \in \mathfrak{I}^{2n \times 2n} \text{ is positive innerwise (P.I.)}$$
$$\iff H \in \mathfrak{I}^{n \times n} \text{ is positive definite (P.D.)}$$

In addition, for real $F(s)$, we have the following:

$$F(s) \text{ is stable} \iff \Delta \in \Re^{2n \times 2n} \text{ is positive innerwise (P.I.)}$$
$$\iff H \in \Re^{n \times n} \text{ is positive definite (P.D.)}$$
$$\iff \text{All principal minors of } \mathcal{H} \in \Re^{n \times n} \text{ are positive}$$

4. Reduced Hermite criterion. We assume all coefficients of $F(s)$ are real. Then, a permutation matrix $E \in \Re^{n \times n}$ (which also includes the factor 2 in (21)) may be found such that [23]

$$E \cdot H \cdot E^T = \begin{bmatrix} H^{(1)} & \mathbf{0} \\ \mathbf{0} & H^{(4)} \end{bmatrix}, \tag{27}$$

where $H^{(1)}$ and $H^{(4)}$ are obtained by eliminating even and odd numbered rows / columns of H the sizes of which are

$$H^{(1)} \in \begin{cases} \Re^{n/2 \times n/2} & \text{if } n \text{ is even;} \\ \Re^{(n+1)/2 \times (n+1)/2} & \text{if } n \text{ is odd, and} \end{cases} \tag{28}$$

$$H^{(4)} \in \begin{cases} \Re^{n/2 \times n/2} & \text{if } n \text{ is even;} \\ \Re^{(n-1)/2 \times (n-1)/2} & \text{if } n \text{ is odd.} \end{cases} \tag{29}$$

THEOREM 4.1. REDUCED HERMITE CRITERION [13]. $F(s)$ *is stable iff any one of the following four equivalent conditions is satisfied:*

(30) $\quad\quad\quad\quad H^{(1)} > 0,\ a_n > 0,\ a_{n-2} > 0,\ a_{n-4} > 0,\ldots$
(31) $\quad\quad\quad\quad H^{(1)} > 0,\ a_n > 0,\ a_{n-1} > 0,\ a_{n-3} > 0,\ldots$
(32) $\quad\quad\quad\quad H^{(4)} > 0,\ a_n > 0,\ a_{n-2} > 0,\ a_{n-4} > 0,\ldots$
(33) $\quad\quad\quad\quad H^{(4)} > 0,\ a_n > 0,\ a_{n-1} > 0,\ a_{n-3} > 0,\ldots$

5. Reduced Hurwitz criterion. When $F(s)$ is real, (23) reduces to

(34) $$H = S \cdot [\Delta/\Delta^{(4)}] \cdot I_{REV}.$$

Note that, (34) is in fact a symmetrizing procedure applied on the Schur complement. Using the same permutation matrix in (27), we get

(35) $$\begin{bmatrix} H^{(1)} & 0 \\ 0 & H^{(4)} \end{bmatrix} = ES \cdot [\Delta/\Delta^{(4)}] \cdot I_{REV} E^T.$$

Now, we need to find the pairs $\{S_I, \Delta_I\}$ and $\{S_{IV}, \Delta_{IV}\}$ such that the following are true:
1. The two inner matrices Δ_I and Δ_{IV} are of resultant form, and their inner determinants provide the odd or even principal minors of the Hurwitz matrix $\mathcal{H}$.
2. Both S_I and S_{IV} must be upper triangular matrices.
3. The Schur complements of Δ_I and Δ_{IV}, when symmetrized according to an appropriate version of (34), must yield the following:

$$H^{(1)} = [\,I\ \ 0\,] \cdot ES \cdot [\Delta/\Delta^{(4)}] \cdot I_{REV} E^T \cdot \begin{bmatrix} I \\ 0 \end{bmatrix}$$

(36) $\quad\quad\quad\quad \equiv S_I[\Delta_I/\Delta_I^{(4)}]I_{REV};$

$$H^{(4)} = [\,0\ \ I\,] \cdot ES \cdot [\Delta/\Delta^{(4)}] \cdot I_{REV} E^T \cdot \begin{bmatrix} 0 \\ I \end{bmatrix}$$

(37) $\quad\quad\quad\quad \equiv S_{IV}[\Delta_{IV}/\Delta_{IV}^{(4)}]I_{REV}.$

Hence, we must have

(38) $\quad\quad\quad\quad \Delta_I\ \text{P.I.}\ \iff\ H^{(1)} > 0;$
(39) $\quad\quad\quad\quad \Delta_{IV}\ \text{P.I.}\ \iff\ H^{(4)} > 0.$

THEOREM 5.1. REDUCED HURWITZ CRITERION IN INNER FORM. $F(s)$ *is stable iff any one of the following four equivalent conditions is satisfied:*

(40) $\quad\quad\quad\quad \Delta_I\ \text{P.I.},\ a_n > 0,\ a_{n-2} > 0,\ a_{n-4} > 0,\ldots$
(41) $\quad\quad\quad\quad \Delta_I\ \text{P.I.},\ a_n > 0,\ a_{n-1} > 0,\ a_{n-3} > 0,\ldots$
(42) $\quad\quad\quad\quad \Delta_{IV}\ \text{P.I.},\ a_n > 0,\ a_{n-2} > 0,\ a_{n-4} > 0,\ldots$
(43) $\quad\quad\quad\quad \Delta_{IV}\ \text{P.I.},\ a_n > 0,\ a_{n-1} > 0,\ a_{n-3} > 0,\ldots$

The above are the Liénard-Chipart conditions [14].
Historical Note. The Liénard-Chipart criterion has a long history. It was proven several times and sometimes rediscovered as indicated below:
1. Liénard in 1911 [24] obtained $3n/2$ conditions.
2. Liénard-Chipart in 1914 [14] obtained n conditions.
3. Fuller in 1957 [25] rediscovered Liénard-Chipart criterion.
4. Barnett in 1971 [23] reformulated these conditions in terms of companion matrices.
5. Anderson in 1972 [13] obtained n conditions.
6. Datta in 1978 [26] obtained $3n/2$ conditions.

Fujiwara in 1915 [27] was the first to show the connection between the Hurwitz minors and Hermite matrix.

6. Connection with routh stability table. Hurwitz, in obtaining his criterion, was not aware of the earlier work of Routh. Bompiani [15] was the first to connect both criteria. Since then, it is known as the *Routh-Hurwitz criterion*.

Notice that, from the Hurwitz matrix in (11),

$$(44) \qquad \det[\mathcal{H}_{(1\times1)}] = a_1; \quad \det[\mathcal{H}_{(2\times2)}] = a_1 a_2 - a_0 a_3; \quad \cdots$$

Then, Routh Table [3] is constructed as follows:

$$(45) \qquad \begin{array}{cccc} a_0 & a_2 & a_4 & \cdots \\ a_1 & a_3 & a_5 & \cdots \\ \dfrac{a_1 a_2 - a_0 a_3}{a_1} & \dfrac{a_1 a_4 - a_0 a_5}{a_1} & \cdots & \\ \dfrac{a_1 a_2 a_3 - a_0 a_3^2 - a_1^2 a_4 + a_0 a_1 a_5}{a_1 a_2 - a_0 a_3} & \cdots & & \\ \vdots & & & \\ a_n & & & \end{array}$$

Note that the elements of leading column are

$$(46) \qquad a_0, \quad \det[\mathcal{H}_{(1\times1)}], \quad \frac{\det[\mathcal{H}_{(2\times2)}]}{\det[\mathcal{H}_{(1\times1)}]}, \quad \cdots, \quad \frac{\det[\mathcal{H}_{(n\times n)}]}{\det[\mathcal{H}_{((n-1)\times(n-1))}]}.$$

Jeltsch in 1979 [28] obtained an optimal fraction-free Routh array thus avoiding coefficient growth.

7. Connection with Lyapunov's second method. A candidate Lyapunov function for investigation of stability of origin is the positive definite Hermitian quadratic form

$$(47) \qquad V = \bar{\mathbf{x}}^T H \mathbf{x} \quad \text{where} \quad \mathbf{x}^{(1)} = A\mathbf{x}.$$

Here, A is in its companion form. The time derivative of V is

$$(48) \qquad V^{(1)} = \bar{\mathbf{x}}^T (HA + \bar{A}^T H)\mathbf{x} = \bar{\mathbf{x}}^T \bar{\mathbf{h}} \mathbf{h}^T \mathbf{x},$$

where $\mathbf{h}^T$ is the n-th row of H itself. Then, $V^{(1)}$ is negative semidefinite and cannot be identically zero for any $\mathbf{x}$, except $\mathbf{x} = \mathbf{0}$, unless

(49) $$H\mathbf{x} = \mathbf{0} \text{ for some } \mathbf{x} \neq \mathbf{0} \Longleftrightarrow \det[H] = \det[\Delta] = 0,$$

that is, polynomial has roots on the boundary of the stability region, and thus is unstable [2],[16],[17]. This may be ascertained from the fact that $\det[\Delta] = \det[H]$ and by noting Orlando's formulae in (8) and (9).

Equivalent necessary and sufficient conditions for asymptotic stability are the following:
1. H is positive definite (for both real and complex coefficient cases): From Hermite criterion [2] and by Lyapunov's second method.
2. Δ is positive innerwise (for both real and complex coefficient cases).
3. Hurwitz determinants $\det[\mathcal{H}_{(i \times i)}]$ are all positive (for real coefficient case only): This is equivalent to the Liénard-Chipart criterion or reduced Hermite criterion.
4. First column of Routh table is of positive sign (for real coefficient case only).
Connection among the four stability criteria is hence established.

8. Connection between Bezoutian and Hermite matrix.

We have, from [22],

(50) $$I_{REV} \cdot B \cdot I_{REV} = T \cdot [R/R^{(4)}] \implies \det[B] = a_0^n \det[R/R^{(4)}],$$

where $B \in \mathfrak{I}^{n \times n}$ is the Bezoutian matrix, named after Bezout for his work on the theory of elimination in 1764 [21],

(51)
$$B = \begin{bmatrix} [a_{n-1}\bar{a}_n]_R & [a_{n-2}\bar{a}_n]_I & [a_{n-3}\bar{a}_n]_R & \cdots \\ [a_{n-2}\bar{a}_n]_I & [a_{n-3}\bar{a}_n - a_{n-2}\bar{a}_{n-1}]_R & [a_{n-4}\bar{a}_n - a_{n-3}\bar{a}_{n-1}]_I & \cdots \\ [a_{n-3}\bar{a}_n]_R & [a_{n-4}\bar{a}_n - a_{n-3}\bar{a}_{n-1}]_I & [a_{n-5}\bar{a}_n - a_{n-4}\bar{a}_{n-1} + a_{n-3}\bar{a}_{n-2}]_R & \cdots \\ \vdots & \vdots & \vdots & \ddots \end{bmatrix}$$

and $T \in \mathfrak{I}^{n \times n}$ is the lower triangular matrix

(52) $$T = \begin{bmatrix} a_0 & 0 & \cdots & 0 \\ a_1 & a_0 & \cdots & 0 \\ \vdots & \vdots & \ddots & \vdots \\ a_{n-1} & a_{n-2} & \cdots & a_0 \end{bmatrix}.$$

Now, from (6), and by taking the Schur-complement of both sides, we have

(53) $$[\Delta/\Delta^{(4)}] = [R/R^{(4)}] \cdot \hat{I}_n \implies \det[\Delta/\Delta^{(4)}] = (-1)^{n(n-1)/2} \det[R/R^{(4)}].$$

Using (50), and the fact that $S = I_{REV}T$, we get

(54)
$$\begin{aligned} B &= I_{REV}T \cdot [R/R^{(4)}] \cdot I_{REV} = S \cdot [\Delta/\Delta^{(4)}] \cdot \hat{I}_n I_{REV} \\ &\implies \det[B] = (-1)^{n(n-1)/2} a_0^n \cdot [\Delta/\Delta^{(4)}]. \end{aligned}$$

Now, comparing (23) and (50), we have

(55) $\quad \hat{I}_n H = B \check{I}_n \quad$ where $\quad \check{I}_n = \text{diag}\{(-1)^{n-1}, (-1)^{n-1}, \ldots, (-1)^{n-1}\}.$

Hence,

(56) $\quad B = \hat{I}_n \cdot H \cdot \check{I}_n \implies H = \hat{I}_n \cdot B \cdot \check{I}_n \implies \det[H] = (-1)^{n(n-1)/2} \det[B],$

where we have used the fact that $\det[\check{I}_n] = 1$. Also, from (50),

(57) $\qquad \det[B] = \det[R^{(1)}] \cdot \det[R/R^{(4)}] = \det[R].$

9. Applications.
During the century since the publication of Hurwitz's paper, Hurwitz matrix have found many uses in numerous applications. Only five important applications are mentioned now.

Consider the real polynomial

$$F(s) = \sum_{i=0}^{n} a_i s^{n-i} = a_0 \prod_{i=1}^{n} (s - \lambda_i), \ a_0 > 0.$$

1. Critical stability constraints [29]:

$$\det[\mathcal{H}_{((n-1) \times (n-1))}] \geq 0, \ a_n \geq 0.$$

These constraints are based on the Orlando's formula which is

$$\det[\mathcal{H}_{((n-1) \times (n-1))}] = \epsilon(n) a_0^{n-1} \prod_{1 \leq i < j \leq n} (\lambda_i + \lambda_j),$$

where $\epsilon(n) \doteq (-1)^{n(n-1)/2}$.

2. Abscissa of stability – Henrici's formulations [30]: If $F(s)$ is a Hurwitz polynomial and $\sigma \doteq \max_i \text{Re}[\lambda_i]$, then,

$$\sigma \leq -2^{\epsilon(n)} R^{-(n-1)(n+2)/2} \det[\mathcal{H}_{(n \times n)}];$$
$$\sigma \leq -2^{-n} R^{-2n+2} \frac{\det[\mathcal{H}_{(n \times n)}]}{\det[\mathcal{H}_{((n-2) \times (n-2))}]},$$

where $R \doteq 2 \max_i |a_i/a_0|^{1/i}$ and $\epsilon(n) \doteq (-1)^{n(n-1)/2}$.

3. Evaluation of integrals that arise in control, filtering, and communications [31]: Given the transfer function

$$G(s) = \frac{N(s)}{D(s)} \quad \text{where} \quad D(s) = a_n s^n + a_{n-1} s^{n-1} + \cdots + a_0,$$

$$\int_0^\infty g^2(t)\,dt = \frac{1}{2\pi j}\int_{-j\infty}^{j\infty} G(s)G(-s)\,ds = \frac{1}{2\pi j}\int_{-j\infty}^{j\infty}\frac{N(s)N(-s)}{D(s)D(-s)}ds$$

$$= \frac{(-1)^{n+1}}{2a_n}\frac{\det[\Delta_n^{\hat{N}}]}{\det[\Delta_n]}.$$

Here,

$$\hat{N}(s) \doteq N(s)N(-s) = b_{n-1}s^{2(n-1)} + b_{n-2}s^{2(n-2)} + \cdots + b_0,$$

and $\Delta_n^{\hat{N}}$ is formed by replacing, for n odd, the last row, and for n even, the first row, of Δ_n (written in inner form) by $[b_{n-1}, b_{n-2}, \ldots, b_0]$ [20].

4. Positive and strictly positive real functions [20]: A real rational function

$$T(s) = \frac{Q(s)}{P(s)},$$

with relatively prime polynomials, is called *positive real (p.r.)* if and only if

$$F(s) = P(s) + Q(s) \text{ is Hurwitz, and}$$
$$\text{Re}[T(j\omega)] \geq 0, \forall \omega.$$

If $\text{Re}[T(j\omega)] > 0$, $\forall \omega$, then, $T(s)$ is said to be *strictly positive real (s.p.r.)*. Applications arise in network synthesis, absolute stability, optimal control, etc.

5. Robust stability [32]: Consider the uncertain polynomial

$$F(s, q) = F_0(s) + qF_1(s),$$

where $F(s, 0) = F_0(s)$ is Hurwitz. Then, the maximum interval for robust stability is given by [32]

$$q_{max}^+ = \frac{1}{\lambda_{max}^+\left[-\mathcal{H}_{(n\times n)}^{-1}[F_0(s)]\cdot\mathcal{H}_{(n\times n)}[F_1(s)]\right]};$$

$$q_{min}^- = \frac{1}{\lambda_{min}^-\left[-\mathcal{H}_{(n\times n)}^{-1}[F_0(s)]\cdot\mathcal{H}_{(n\times n)}[F_1(s)]\right]},$$

where

$$\lambda_{max}^+ \doteq \text{maximum positive real eigenvalue;}$$
$$\lambda_{min}^- \doteq \text{minimum negative real eigenvalue.}$$

The maximum interval for robust stability is given by

$$Q_{max} = [q_{min}^-, q_{max}^+].$$

10. On the simplest possible property of Hurwitz criterion. It is not possible to reduce the *number* and *sum* of the *degrees* of the inequalities in the Liénard-Chipart criterion (which is a simpler form of Hurwitz stability criterion) if an alternate set of polynomial inequalities were used.

Though a rigorous proof is not yet available, it is believed that the above may be ascertained from [18] and [33].

11. Conclusion. In this symposium article celebrating the Hurwitz's original paper, we have attempted to connect his stability criterion with three others obtained in the nineteenth century. Namely, the Hermite, Routh, and Lyapunov criteria. Also, in this article, we obtained other important relationships motivated by Hurwitz paper, namely, Orlando's formulae and Liénard-Chipart criterion. Five important applications of Hurwitz criterion are indicated attesting to the significant and useful impact of Hurwitz paper.

A major feature in the formulation of the various stability criteria is that they are based on J.J. Sylvester's matrix. Another, or alternate, formulation is the use of the Bezoutian matrix. In this article, we connected these two formulations using the concept of inners and its extensions.

Finally, in the process of writing this paper, we were able to indicate, in a forthcoming work, that the Liénard-Chipart conditions for stability are the simplest possible.

Acknowledgement. The aid of and discussions with Dr. Kamal Premaratne in the preparation of this paper is gratefully acknowledged.

References

[1] C. Hermite, "On the number of roots of an algebraic equation between two limits," Extract of a letter from Mr. C. Hermite of Paris to Mr. Borchardt of Berlin, 1856, *J. reine angew. Math.*, vol. **52**, pp. 39–51. Translation by P.C. Parks, *Int. J. Cont.*, vol. 26, no. 2, pp. 183–196, 1977.
[2] P.C. Parks, "A new proof of Hermite's stability criterion and a generalization of Orlando's formula," *Int. J. Cont.*, vol. 26, no. 2, pp. 197–206, 1977.
[3] E.J. Routh, "A treatise on the stability of a given state of motion," Adams Prize Essay, Univ. of Cambridge, England, 1877.
[4] E.I. Jury, "Stability tests for one-, two- and multi-dimensional linear systems," Routh Centennial Lecture, *IEE Proc.*, vol. 124, no. 12, pp. 1237–1240, 1977.
[5] Routh Centenary Issue, *Int. J. Cont.*, vol. 26, no. 2, Sept. 1977.
[6] A.M. Lyapunov, "The general problem of the stability of motion," PhD Thesis, 1892. Appeared in Collected Works II, in Russian, *Kharkov. Math. Soc.*, vol. 7.
[7] A.M. Lyapunov, "The general problem of the stability of motion," 1892. Translated by A.T. Fuller, *Int. J. Cont.*, vol. 55, no. 3, pp. 531–773, Mar. 1992. Also, Guest Editorial, *ibid*, pp. 251–527.
[8] P.C. Parks, "A.M. Lyapunov's stability theory – 100 years on," *IMA J. Math. Cont. Inf.*, vol. 9, no. 4, pp. 275–303, 1992.
[9] A. Hurwitz, "Über die Bedingungen unter welchem eine Gleichung nur Wurzeln mit negativen relen Teilen besitzt," *Math. Ann.*, vol. 46, pp. 273–284, 1895.
[10] J.J. Sylvester, "A method of determining by mere inspection the derivative from two equations of an y order," *Phil. Mag.*, vol. 16, pp. 132–136, 1840. Also, Collected Mathematical Papers, *ibid*, vol. 1, pp. 54–57.
[11] L. Orlando, "Sul problema di Hurwitz relativo alle parti reali delle radici di un'equatione algebrica," *Math. Ann.*, vol. 71, pp. 233–245, 1911.
[12] E.I. Jury and M. Mansour, "A note on new inner-matrix for stability," *Proc. IEEE*, vol. 69, pp. 1579–1580, 1981.
[13] B.D.O. Anderson, "The reduced Hermite criterion and application to proof of the Liénard-Chipart criterion," *IEEE Trans. Auto. Cont.*, pp. 669–672., 1972.

[14] A. Liénard and M.H. Chipart, "Sur la signe de la partie réelle des racines d'une équation algébrique," *J. Math. Pures Appl.,* vol. 10, pp. 291–346, 1914.
[15] E. Bompiani, "Sulle conditioni sotto lequaliun-equatione a coefficienti reali ammelte solo verdici con parte reali negative," *Giornale di Mathematiche,* vol. 49, pp. 33–39, 1911.
[16] P.C. Parks, "A new proof of the Hurwitz stability criterion by the second method of Lyapunov," *Proc. Cambridge Phil. Soc.,* vol. 58, pp. 694–702, 1962.
[17] P.C. Parks, *IEEE Trans. Auto. Cont.,* vol. 11, p. 334, 1966.
[18] B.D.O. Anderson and E.I. Jury, "A simplest possible property of the generalized Hurwitz conditions," *SIAM J. Cont. Appl.,* vol. 15, no. 1, pp. 177–184, 1977.
[19] S. Barnett, *Matrices in Control Theory.* London, UK: Van Nostrand Reinhold, 1971.
[20] E.I. Jury, *Inners and Stability of Dynamic Systems.* Second Edition, Malabar, FL: Krieger, 1982.
[21] A.S. Householder, "Bezoutians, elimination and localization," *SIAM Review,* vol. 12, pp. 73–78, 1971.
[22] S. Barnett and E.I. Jury, "Inners and Schur Complement," *Lin. Alg. Appl.,* vol. 22, 57–63, 1978.
[23] S. Barnett, *Polynomials and Linear Control Systems.* New York, NY: Marcel Dekker, 1983.
[24] A. Liénard, "Conditions pour quine equation algebrique ait toutes ques racines on a partic reelle negative," *Revive Math. Speciales,* vol. 21, no. 7, pp. 153–155, Apr. 1911. Also, vol. 21, no. 8, pp. 117–180, May 1911.
[25] A.T. Fuller, "On stability criteria for linear systems and realizability criteria for RC networks," *Proc. Cambridge Phil. Soc.,* vol. 53, pp. 878–896, 1957.
[26] B.N. Datta, "An elementary proof of the stability criterion of Liénard and Chipart," *Lin. Alg. Appl.,* vol. 22, pp. 89–96, 1978.
[27] M. Fujiwara, "Über die Algebraichen Gleichungen," *Tohoku Math. J.,* vol. 8, pp. 78–85, 1915.
[28] R. Jeltsch, "An optimal fraction free Routh array," *Int. J. Cont.,* vol. 30, pp. 653–660, 1979.
[29] R.A. Frazer and W.J. Duncan, "On the criteria for the stability of small motion," *Proc. Royal Soc., Pt. A,* vol. 124, p. 642, 1929.
[30] P. Henrici, "Upper bounds for the abscissa of stability of a stable polynomial," *SIAM J. Num. Anal.,* vol. 7, pp. 538–544, 1970.
[31] W.A. Mersman, "Evaluation of an integral occuring in servomechanism theory," *Pacific J. Math.,* vol. 2, pp. 627–632, 1952.
[32] B. Barmish, *New Tools for Robustness of Linear Systems,* New York, NY: Macmillan, 1994.
[33] B.D.O. Anderson, E.I. Jury, and L.F. Chaparro, "Relations between real and complex polynomials for stability and aperiodicity conditions," *IEEE Trans. Auto. Cont.,* vol. AC-20, 244–246, 1975.

Hurwitz Matrix for Polynomial Matrices

F.J. Kraus, M. Mansour and M. Sebek
Automatic Control Laboratory, Swiss Federal Institute of Technology
ETH-Zentrum, CH-8092 Zürich, Switzerland

Abstract. If a system is given by its transfer function then the stability of the system is determined by the denominator polynomial and its corresponding Hurwitz matrix $\mathcal{H}$. Also the critical stability conditions are determined by its determinant $\det \mathcal{H}$.

The aim of this paper is to get a generalized Hurwitz matrix for polynomial matrices. In order to achieve that, we first obtain a relation between the Hurwiz matrix for a polynomial and the Lyapunov equation. Here we show how the Hurwitz matrix appears in the solution of the Lyapunov equation using the companion matrix realization and the Kronecker formulation of Lyapunov equation. Using this result we show how the generalized Hurwitz matrix for polynomial matrices can be constructed.

1. Introduction. Hurwitz gave in [1] an elegant solution of the stability problem of linear time invariant dynamic systems.

For a dynamic system given by the rational transfer function

$$(1) \qquad G(s) = \frac{b(s)}{a(s)}$$

with the denominator polynomial

$$a(s) = a_0 + a_1 s + a_2 s^2 + \cdots + a_{n-1} s^{n-1} + s^n$$

the Hurwitz matrix $\mathcal{H} \in \Re^{n \times n}$ is given by

$$(2) \qquad \mathcal{H} = \begin{bmatrix} a_{n-1} & a_{n-3} & a_{n-5} & \cdots & \cdots & 0 & 0 \\ 1 & a_{n-2} & a_{n-4} & \cdots & \cdots & 0 & 0 \\ \cdots & a_{n-1} & a_{n-3} & a_{n-5} & \cdots & 0 & 0 \\ \cdots & 1 & a_{n-2} & a_{n-4} & \cdots & 0 & 0 \\ \vdots & \vdots & & & & \vdots & \vdots \\ & & \cdots & \cdots & a_3 & a_1 & 0 \\ & & \cdots & \cdots & a_4 & a_2 & a_0 \end{bmatrix}.$$

For stability of (1) it is necessary and sufficient that all the leading principial minors of $\mathcal{H}$ are positive. Also the critical stability conditions are determined by

$$(3) \qquad \det(\mathcal{H}) \ne 0.$$

It is well known that Lyapunov theory gives an alternative method to investigate stability of (1). A realisation of (1) is given by

$$(4) \qquad \underline{\dot{x}} = A \underline{x} + \underline{b} u$$

where $a(s) = \det(sI - A)$ and the system matrix A

(5) $$A = \begin{bmatrix} -a_{n-1} & -a_{n-2} & \cdots & -a_1 & -a_0 \\ 1 & 0 & \cdots & \cdots & 0 \\ 0 & 1 & \cdots & & \vdots \\ 0 & & \ddots & & \\ 0 & \cdots & \cdots & 1 & 0 \end{bmatrix}$$

is in companion form.

The stability of (1) or (4) is given by the solution of the Lyapunov equation [4]

(6) $$A'P + PA = -Q.$$

The matrix A is stable if and only if for arbitrary positive definite $Q > 0$ there exists a positive definite solution for P in (6).

In order to obtain the critical stability conditions in this case we consider the Kronecker formulation of (6) [4], [6]:

(7) $$\underbrace{(A' \otimes I + I \otimes A')}_{\mathcal{A}} col(P) = -col(Q).$$

The eigenvalues of the operator $\mathcal{A}$ are given by

$$Eig(\mathcal{A}) = \lambda_i + \lambda_j \quad \text{for } \forall i,j$$

where λ_k are the eigenvalue of A. Therefore

(8) $$\det \mathcal{A} \neq 0$$

gives the critical stability conditions or the guardian map [5] for the stability of A. The matrix $\mathcal{A} \in \Re^{n^2 \times n^2}$ in (7) can be reduced to $\mathcal{A}_r \in \Re^{(n+1) \times n/2}$ due to symmetry of the matrix P. In the following we shall use the reduced operator $\mathcal{A}_r$ instead of $\mathcal{A}$.

We noticed that (3) and (8) are equivalent. However instead of matrix $\mathcal{H}$ of dimension n, we investigate in (8) a matrix of dimension $\nu = (n+1)n/2$ corresponding to the operator $\mathcal{A}_r$. In section 2 we shall show the relation between $\det \mathcal{H}$ and $\det \mathcal{A}_r$. In section 3 we generalize the procedure of section 2 to get a solution for polynomial matrices.

2. Relation between Hurwitz- und Lyapunov – Kronecker matrix. We consider the reduced Lyapunov-Kronecker matrix from (7). From (6) we can easily show that the first row of (5) influences only the first n columns of the matrix $\mathcal{A}_r$.

(9) $$\mathcal{A}_r = \begin{bmatrix} \mathcal{A} & \mathcal{V} \end{bmatrix}.$$

The matrix $\mathcal{V} \in \mathcal{R}^{(n+1)n/2 \times (n-1)n/2}$ results from the subdiagonal of the companion matrix and has full rank. We can find a regular transformation for (9) so that

$$\begin{bmatrix} \mathcal{N}' \\ \mathcal{W}' \end{bmatrix} \cdot \mathcal{A}_r = \begin{bmatrix} \tilde{\mathcal{H}} & 0 \\ X & I \end{bmatrix} \tag{10}$$

where $\mathcal{N}$ is the kernel of $\mathcal{V}'$ and therefore independent of a_i and $\mathcal{W}$ coincides with $\mathcal{V}$ exept for a normalization. It is clear from (10) that the regularity of $\mathcal{A}_r$ is dependent on the regularity of $\tilde{\mathcal{H}}$. To show the dependence explicitly, a possible transformation using elementary matrix operations independent of a_i is given by

$$E_r \cdot D \cdot \mathcal{A}_r \cdot E_c = \begin{bmatrix} \mathcal{H} & 0 \\ X & I \end{bmatrix} \tag{11}$$

where E_c and E_r includes the elementary row and columns operations, D a diagonal matrix with $\det(D) = 2^{-n}$ und X is a matrix which is not important for our analysis.

Therefore

$$\det(\mathcal{A}_r) = 2^n \det(\mathcal{H}) \tag{12}$$

and the elements of the matrix P are strictly proper rational functions in the parameters a_i whose denominators are the critical stability conditions. The highest element degree in these multivariable polynomials is $\frac{n}{2}$ for even polynomials and $\frac{n-1}{2}$ for odd polynomials. Similar analysis is found in [2] and [3].

3. Generalized Hurwitz Matrix for Polynomial Matrices. In this section we try to get a Hurwitz matrix for polynomial matrices following in analogous way the treatment in section 2.

Consider the polynomial matrix

$$A(s) = [a_{ij}(s)] \in \Re^{m \times m} \tag{13}$$

whose elements are polynomials in s of maximal degree n. Then we have

$$A(s) = A_0 + A_1 s + A_2 s^2 + \cdots + A_{n-1} s^{n-1} + I s^n \tag{14}$$

where $A_k \in \Re^{m \times m}$, $k = 1, 2, \ldots, n-1$ are real constant matrices.

The stability of $A(s)$ is determined by the application of Hurwitz criterion to the polynomial

$$d(s) = \det(A(s)).$$

The Hurwitz matrix thus obtained has elements which are multiaffine functions of the coefficients of the polynomials $a_{ij}(s)$. The dimension of the resulting Hurwitz matrix $\mathcal{H}$ is $\nu = n \cdot m$.

3.1. Lyapunov Stability.
To obtain an affine dependence between the stability criterion description and the coefficients of $A(s)$ we use here the state space realisation in form of the block companion matrix $A_m \in \Re^{n \cdot m \times n \cdot m}$

$$(15) \quad A_m = \begin{bmatrix} -A_{n-1} & -A_{n-2} & \cdots & -A_1 & -A_0 \\ I & 0 & \cdots & & \vdots \\ 0 & I & 0 & & \vdots \\ & & \ddots & & \\ 0 & \cdots & 0 & I & 0 \end{bmatrix}$$

The block companionmatrix A_m has only m distinguished rows. The resulting operator $\mathcal{A}_r$ using Kronecker formulation has the dimension $\mu = (n \cdot m + 1) \cdot n \cdot m/2$. For example, for $n = 4$ and $m = 2$ we get $\mathcal{H} \in \Re^{8 \cdot 8}$ and $\mathcal{A}_r \in \Re^{36 \cdot 36}$ which leads to an eigenvalue problem of considerably high dimension. Now different questions arise : Is it possible to reduce the Lyapunov-Kronecker formulation for m-companion matrices? Can we get a Hurwitz matrix for the polynomial matrix where the coefficients of the original polynomials appear explicitly in a simple structure? What are the stability characteristics which can be extracted from this generalized Hurwitz matrix?

For nonmonic (13) we can work with a generalization of the Kronecker formulation using the quardian operator

$$\mathcal{A}' \otimes \mathcal{B}' + \mathcal{B}' \otimes \mathcal{A}'$$

where

$$\mathcal{B} = \begin{bmatrix} A_n & 0 & 0 & \cdots & 0 \\ 0 & I & 0 & \cdots & \vdots \\ \vdots & & \ddots & & \vdots \\ & & & I & \\ 0 & \cdots & \cdots & 0 & I \end{bmatrix}$$

3.2. Generalized Hurwitz Matrix for m=2.
We will explain the procedure of generalization of the Hurwitz matrix for $n = 3$. But for general n the same kind of procedure is valid.

Consider the polynomial matrix

$$A(s) = \begin{bmatrix} s^3 + a_5 s^2 + a_3 s + a_1 & a_4 s^2 + a_2 s + a_0 \\ b_5 s^2 + b_3 s + b_1 & s^3 + b_4 s^2 + b_2 s + b_0 \end{bmatrix}$$

with the characteristic polynomial

$$a(s) = \det(A(s)) = s^6 + \alpha_5 s^5 + \alpha_4 s^4 + \alpha_3 s^3 + \alpha_2 s^2 + \alpha_1 s + \alpha_0$$

where

$$\begin{aligned}
\alpha_5 &= a_5 + b_4 \\
\alpha_4 &= a_3 + b_2 + a_5 b_4 - a_4 b_5 \\
\alpha_3 &= a_1 + b_0 + a_5 b_2 + a_3 b_4 - a_4 b_3 - a_2 b_5 \\
&\vdots \\
\alpha_0 &= a_1 b_0 - a_0 b_1.
\end{aligned}$$

The corresponding Hurwitz matrix $\mathcal{H} \in \Re^{n \times n}$ is

$$\mathcal{H}_6 = \begin{bmatrix} \alpha_{n-1} & \alpha_{n-3} & \alpha_{n-5} & \cdots & \cdots & 0 & 0 \\ 1 & \alpha_{n-2} & \alpha_{n-4} & \cdots & \cdots & 0 & 0 \\ \cdots & \alpha_{n-1} & \alpha_{n-3} & \alpha_{n-5} & \cdots & 0 & 0 \\ \cdots & 1 & \alpha_{n-2} & \alpha_{n-4} & \cdots & 0 & 0 \\ \vdots & \vdots & & & & \vdots & \vdots \\ & & \cdots & \cdots & \alpha_3 & \alpha_1 & 0 \\ & & \cdots & \cdots & \alpha_4 & \alpha_2 & \alpha_0 \end{bmatrix}$$

and the block companion matrix A_2 associated with $A(s)$

$$A_2 = \begin{bmatrix} -a_5 & -a_4 & -a_3 & -a_2 & -a_1 & -a_0 \\ -b_5 & -b_4 & -b_3 & -b_2 & -b_1 & -b_0 \\ 1 & 0 & & & & \\ 0 & 1 & \cdots & & & \\ & & 1 & & & \\ & & & 1 & & \end{bmatrix}.$$

Define

$$V_1 = \begin{bmatrix} 1 & a_5 & a_3 & a_1 & & & & & \\ & 1 & a_5 & a_3 & a_1 & & & & \\ & & \ddots & \ddots & \ddots & \ddots & & & \\ & & & 1 & a_5 & a_3 & a_1 & & \end{bmatrix} \in \Re^{6 \times 9}$$

$$V_2 = \begin{bmatrix} a_4 & a_2 & a_0 & & & \\ & a_4 & a_2 & a_0 & & \\ & & \ddots & \ddots & \ddots & \\ & & & a_4 & a_2 & a_0 \end{bmatrix} \in \Re^{6 \times 8}$$

$$W_1 = \begin{bmatrix} b_4 & b_0 & & & & \\ 1 & b_2 & & & & \\ & & b_4 & b_0 & & \\ & & 1 & b_2 & & \\ & & & & \ddots & \\ & & & & b_4 & b_0 \end{bmatrix} \in \Re^{9 \times 6}$$

and

$$W_2 = \begin{bmatrix} 0 & b_3 & & & & \\ 0 & b_5 & b_1 & & & \\ & & b_3 & & & \\ & & b_5 & b_1 & & \\ & & & & \ddots & \\ & & & & b_5 & b_1 \end{bmatrix} \in \Re^{8 \times 6}.$$

Then

$$\mathcal{H}_6 = \begin{bmatrix} V_1 & -V_2 \end{bmatrix} \begin{bmatrix} W_1 \\ W_2 \end{bmatrix}$$

and

(16) $$\det(\mathcal{H}_6) = \det \begin{bmatrix} I & & W_1 \\ & I & W_2 \\ V_1 & -V_2 & 0 \end{bmatrix}.$$

It is noted that the coefficients of the polynomials in $A(s)$ appear explicitly in the matrix (16) whose determinant gives the critical stability condition. Also it is obvious that the leading principal minors of the Hurwitz matrix $\mathcal{H}_6$ results also from (16). Therefore (16) gives the complete set of stability conditions. It is easy to show that for any degree n a similar construction results.

If we use the Lyapunov Kronecker formulation with elementary row operations and scaling we get

$$E \cdot D \cdot \mathcal{A}_r = \begin{bmatrix} \mathcal{W} & 0 \\ X & I \end{bmatrix}.$$

This result looks formally similar to the polynomial case. However the dimension of $\mathcal{W} \in \Re^{(4n-1) \times (4n-1)}$ is significantly higher as if we use the characteristic polynomial to compute $\mathcal{W} \in \Re^{2n \times 2n}$. We can construct $\mathcal{W}$ directly as follows. For the same example $n = 3$ define

$$K = \begin{bmatrix} a_5 & a_4 & a_3 & a_2 & a_1 & a_0 \\ b_5 & b_4 & b_3 & b_2 & b_1 & b_0 \end{bmatrix}'$$

and

$$V = \begin{bmatrix} K & I_2 & 0 \\ 0 & K & I_2 \\ 0 & 0 & K \end{bmatrix} \in \Re^{(2n-1)m \times m \cdot n}.$$

With this we obtain

$$\tilde{\mathcal{W}} = \begin{bmatrix} D_1 \begin{bmatrix} V \\ 0 \end{bmatrix}; & D_2 \begin{bmatrix} 0 \\ V \end{bmatrix} \end{bmatrix} E$$

where D_1, D_2 are periodic, diagonal weighting metrices

$$D_1 = diag[\,1,1,0,-1,-1,-1,0,1,1,1,0\,]$$

$$D_2 = diag[\,0,1,1,1,0,-1,-1,-1,0,1,1\,]$$

and the operator E cause an addition of the 2nd and 7th columns to replace column 2 and eliminate column 7. The $\tilde{\mathcal{W}}$ and $\mathcal{W}$ are identical up to elementary column operations. This gives

$$\tilde{\mathcal{W}} = \begin{bmatrix} D_1 \begin{bmatrix} a_5 & b_5 & 1 & 0 & 0 & 0 \\ a_4 & b_4 & 0 & 1 & 0 & 0 \\ a_3 & b_3 & a_5 & b_5 & 1 & 0 \\ a_2 & b_2 & a_4 & b_4 & 0 & 1 \\ a_1 & b_1 & a_3 & b_3 & a_5 & b_5 \\ a_0 & b_0 & a_2 & b_2 & a_4 & b_4 \\ 0 & 0 & a_1 & b_1 & a_3 & b_3 \\ & & a_0 & b_0 & a_2 & b_2 \\ \vdots & \vdots & \vdots & \vdots & a_1 & b_1 \\ 0 & \cdots & & \cdots & a_0 & b_0 \\ 0 & 0 & \cdots & & 0 & 0 \end{bmatrix} ; D_2 \begin{bmatrix} 0 & 0 & 0 & 0 & 0 & 0 \\ a_5 & b_5 & 1 & 0 & 0 & 0 \\ a_4 & b_4 & 0 & 1 & 0 & 0 \\ a_3 & b_3 & a_5 & b_5 & 1 & 0 \\ a_2 & b_2 & a_4 & b_4 & 0 & 1 \\ a_1 & b_1 & a_3 & b_3 & a_5 & b_5 \\ a_0 & b_0 & a_2 & b_2 & a_4 & b_4 \\ 0 & 0 & a_1 & b_1 & a_3 & b_3 \\ \vdots & & & & a_0 & b_0 & a_2 & b_2 \\ 0 & & \vdots & \vdots & a_1 & b_1 \\ 0 & 0 & & \cdots & a_0 & b_0 \end{bmatrix} \end{bmatrix} E$$

$$\tilde{\mathcal{W}} = \begin{bmatrix} a_5 & b_5 & 1 & 0 & 0 & 0 & 0 & 0 & 0 & 0 & 0 \\ a_4 & b_4 + a_5 & 0 & 1 & 0 & 0 & b_5 & 1 & 0 & 0 & 0 \\ 0 & a_4 & 0 & 0 & 0 & 0 & b_4 & 0 & 1 & 0 & 0 \\ -a_2 & -b_2 + a_3 & -a_4 & -b_4 & 0 & -1 & b_3 & a_5 & b_5 & 1 & 0 \\ -a_1 & -b_1 & -a_3 & -b_3 & -a_5 & -b_5 & 0 & 0 & 0 & 0 & 0 \\ -a_0 & -b_0 - a_1 & -a_2 & -b_2 & -a_4 & -b_4 & -b_1 & -a_3 & -b_3 & -a_5 & -b_5 \\ 0 & -a_0 & 0 & 0 & 0 & 0 & -b_0 & -a_2 & -b_2 & -a_4 & -b_4 \\ & & a_0 & b_0 & a_2 & b_2 & 0 & -a_1 & -b_1 & -a_3 & -b_3 \\ \vdots & \vdots & \vdots & \vdots & a_1 & b_1 & 0 & 0 & 0 & 0 & 0 \\ 0 & \cdots & & \cdots & a_0 & b_0 & \vdots & \vdots & \vdots & a_1 & b_1 \\ 0 & 0 & \cdots & & 0 & 0 & 0 & \cdots & & a_0 & b_0 \end{bmatrix}.$$

This $\mathcal{W}$ can be considered as the generalized Hurwitz matrix whose determinant give the critical stability condition. However, it is not yet clear how the complete set of stability conditions can be obtained. For higher values of n we have the same construction. The matrices D_1 and D_2 are thereby periodic weighting matrices. For $m > 2$ similar construction can be obtained with more complex periodicity.

4. Conclusions. It was shown that a generalized Hurwitz matrix can be obtained for polynomial matrices through construction. The determinant of this matrix gives the critical stability conditions. However how the remaining stability conditions can be retrieved from it is still an open question.

References

[1] A. Hurwitz, *Über die Bedingungen unter welchen eine Gleichung nur Wurzeln mit negativen reellen Teilen besitzt*, Math. Ann., vol. 46, pp. 273–2840, 1895.
[2] M. Mansour and B. D. O. Anderson, *On the robust stability of time-varying linear systems*, in this proceedings.
[3] A. Betser, N. Cohen and E. Zeheb, *On solving the Lyapunov and Stein Equation for a Companion Matrix*, to be published in System and control letters 1995.
[4] S. Barnett, *Polynomials and linear control systems*, Marcel Dekker Inc., 1983.
[5] L. Saydy, A. L. Tits and E. H. Abed, *Guardian Maps and the Generalized Stability of Parametrized Families of Matrices and Polynomials*, Mathematics of Control, Signals, and Systems, vol. 3, pp. 3451–371, 1990.
[6] P. Lancaster, M. Tismenetsky, *The Theory of Matrices*, Acad. Press, 1985.

Two-dimensional Hurwitz Polynomials

Hari C. Reddy*
Department of Electrical Engineering, California State University
Long Beach, California 90840, USA

P.K. Rojan
Department of Electrical Engineering, Tennessee Technological University
Cookeville, Tennessee 38505, USA

George S. Moschytz
Institute for Signal and Information Processing, Swiss Federal Institute for Technology
ETH Zentrum, CH-8092 Zurich, Switzerland

This paper is dedicated by the authors to Professor Elihau Jury who made pioneering and fundamental contributions to the theory of polynomial stability in the 20th century.

1. Introduction. The progress in two-dimensional (2-D) signal and image processing activities has stimulated active research in 2-D circuits and systems. Two-dimensional digital filters are finding applications in many areas of signal processing. Two-variable (2-V) or 2-D Hurwitz polynomial study finds application in areas such as: generation and testing of (2-V) reactance functions; bounded/positive real functions, and matrices; testing the stability of 2-D digital filters, and the generation of stable 2-D transfer functions. Stability analysis is an important aspect of the design of dynamical systems.

This analysis is often carried out by testing for the absence of zeros of the denominator polynomial of the system transfer function in some specified regions of a complex plane. One-dimensional (1-D) systems are studied through the characterization whether the denominator polynomial is Hurwitz or not. By expanding this idea we can define and study 2-D (also called bivariate, two-variable) Hurwitz polynomials. In view of the diverse needs (due to different applications) a number of 2-D Hurwitz polynomials have been defined and their test procedures developed. In this paper, a detailed presentation of various 2-D Hurwitz polynomials and their relationships to one another is given. We also study some of their relevant applications.

To highlight the relationships among the various Hurwitz polynomials, the definitions of all the Hurwitz polynomials are presented. This is done in terms of the absence of or the nature of their zerosets in specified regions such as the open or closed right half of the (S_1, S_2)-biplane. The main goal here is to make a tutorial exposition 2-D Hurwitz polynomials. The paper layout is as follows: In Section 2 some preliminaries and notations are given. Next, we present definitions

---

* H.C. Reddy is an Adjunct Professor in the Department of Electrical and Computer Engineering at the University of California, Irvine, CA 92716, USA. He is also a Guest Professor at the Institute for Signal and Information Processing, ETH, Zurich, Switzerland.

of a number of (S_1, S_2)-domain Hurwitz polynomials. The testsets for the various Hurwitz polynomials are given in Section 4. In the following section two-variable very strict Hurwitz polynomials are studied in some detail, as this is the counterpart of strict Hurwitz polynomials in the 1-D case. Some of the applications of the various Hurwitz polynomials, including the stability study of 2-D nonlinear systems, are described in Section 6.

2. Preliminaries and Notations. Infinite Distant Points: The points at infinite distances in the (S_1, S_2)-biplane play an important role in the definition of certain 2-D Hurwitz polynomials. Some of the confusion that resulted in the application of these Hurwitz polynomials can be attributed to the neglect or omission of these infinite distant points. In this paper we shall consider the extended (S_1, S_2)-biplane that includes the infinite distance points. For the sake of clarity we also explicitly indicate whether the infinite distant points are included in or excluded from the regions considered. The behavior of 2-D polynomials at infinite distant points is well described in the literature [2, 3]. Seemingly many infinite distant points in the 1-D plane such as the S_1-plane or S_2-plane may all be assumed to merge at a single point. Thus, infinity is treated as a single point. So any shift (however small it might be) from this infinite distant point leads to a finite distant point. We now deal with the notations and some passivity definitions:

Analog Biplane:

$$Re(s) = \text{Real part of the variable } s. \text{ For } i = 1, 2$$

$$S_{i+} = \{s_i | \ Re(s_i) > 0, \ |s_i| < \infty\},$$
open right half of S_i-plane

$$S_{i0} = \{s_i | \ Re(s_i) = 0, \ |s_i| \leq \infty\},$$
imaginary axis of the S_i-plane

$$S_{i\oplus} = \{s_i | \ Re(s_i) \geq 0, \ |s_i| \leq \infty\},$$
closed right half of S_i-plane

$$S^2_{+0} = \{(s_1, s_2) | \ Re(s_1) > 0, \ Re(s_2) = 0, \ |s_1| < \infty, \ |s_2| \leq \infty\},$$
open right half of S_1-plane and the imaginary axis of S_2-plane

$$S^2_{0+} = \{(s_1, s_2) | \ Re(s_1) = 0, \ Re(s_2) > 0, \ |s_1| \leq \infty, \ |s_2| < \infty\},$$
open right half of S_2-plane and the imaginary axis of S_1-plane

$$S^2_{++} = \{(s_1, s_2) | \ Re(s_1) > 0, \ Re(s_2) > 0, \ |s_1| < \infty, \ |s_2| < \infty\},$$
open right half of the (S_1, S_2)-biplane

$$S_{00}^2 = \{(s_1, s_2) |\ Re(s_1) = 0,\ Re(s_2) = 0,\ |s_1| \leq \infty,\ |s_2| \leq \infty\},$$
distinguished boundary of (S_1, S_2)-biplane

$$S_{\oplus\oplus}^2 = \{(s_1, s_2) |\ Re(s_1) \geq 0,\ Re(s_2) \geq 0,\ |s_1| \leq \infty,\ |s_2| \leq \infty\},$$
closed right half of (S_1, S_2)-biplane

Isolated and Continuum of Zeros: Some types of Hurwitz polynomials are distinguished on the basis of whether they have isolated zeros or a continuum of zeros on S_{00}^2. As a point on S_{00}^2 is characterized by $s_1 = jw_1$ and $s_2 = jw_2$ where w_1 and w_2 are real quantities, the region S_{00}^2 can be graphically represented by the (w_1, w_2)-plane. The isolated zeros on S_{00}^2 will then be points on this plane and the continuum of zeros will be represented by continuous curves.

Value of a two-variable polynomial at infinity. Because later in the paper there is a need to determine the value of two-variable polynomials at infinite distant points, the following explanation is in order here: In the one dimensional complex plane S, the infinite distant points can be represented by a single point, and the value of any function at this point is found by applying some transformation which transforms the point at infinity to some finite point s', and the value of the transformed function at s' is determined. Often $s = 1/u$ is the transformation used and infinity is mapped onto the origin. Using this transformation, the value of $B(s)$ at infinity can be defined as $B(\infty) = B_T(0)$ where $B_T(u) = B(1/u)$. In the two-dimensional biplane (S_1, S_2), consisting of two complex planes S_1 and S_2, an infinite distant point can have infinite coordinates in any one or both of these planes, and so there exist an infinite number of infinite distant points. They can be classified into three categories as follows [3].

Category 1: $s_1 = \infty$ and $s_2 =$ finite
Category 2: $s_1 =$ finite and $s_2 = \infty$
Category 3: $s_1 = \infty$ and $s_2 = \infty$

Applying the transformation method to each variable, the value of the function at each of the above points is defined as

(1)
$$B(\infty, s_2') = B_1(0, s_2')$$
where $B_1(u, s_2) = B(1/u, s_2),\ |s_2'| < \infty$

(2)
$$B(s_1', \infty) = B_2(s_1', 0)$$
where $B_2(s_1, v) = B(s_1, 1/v),\ |s_1'| < \infty$

(3)
$$B(\infty, \infty) = B_3(0, 0)$$
where $B_3(u, v) = B(1/u, 1/v)$

Nonessential Singularities of the Second Kind (NSSK). It is well known [3] that 2-V polynomials may have NSSK at some infinite distant points where the value of the polynomial is indeterminate. For the sake of notational convenience, and to indicate the type of indeterminacy involved, we write $B(s_{10}, s_{20}) = 0/0$ to say that $B(s_1, s_2)$ has a NSSK at (s_{10}, s_{20}). Of course, for a polynomial $B(s_1, s_2)$, this can occur only if s_{10}, s_{20}, or both have infinite value.

3. Various Analog Hurwitz Polynomials.

Following up the definition of one-variable Hurwitz polynomials, 2-V Hurwitz polynomials have been traditionally defined on the basis of the regions in the S_2-biplane where the 2-V polynomials have no zeros or have only some constrained types of zeros. So far, at least eight different types of Hurwitz polynomials have been defined. We next present their definitions, so as to bring out their zerofree regions and the constraints if any on their zeros in some regions. First, let us define a self-paraconjugate polynomial.

DEFINITION 3.1. **Self-Paraconjugate Polynomial.** *A 2-V polynomial $B(s_1, s_2)$ is said to be a (analog) selfparaconjugate polynomial if $B^*(s_1, s_2) = cB(s_1, s_2)$, where c is a unimodular complex constant, i.e., $|c| = 1$. When $c = +1$, $B(s_1, s_2)$ is said to be a para-even polynomial and when $c = -1$, a para-odd polynomial. $B^*(s_1, s_2)$ is defined as $B^*(s_1, s_2) = $ Paraconjugate of $B(s_1, s_2) = [B(-s_1^*, -s_2^*)]$ where s^* represents the complex conjugate of s.*

It is easily verified that the para-even and para-odd parts of a polynomial are selfparaconjugate, i.e., $B_e^*(s_1, s_2) = B_e(s_1, s_2)$ and $B_o^*(s_1, s_2) = B_o(s_1, s_2)$. In the case of real polynomials, para-even and para-odd polynomials are simply called even and odd polynomials.

Broad sense Hurwitz polynomial (BHP). BHP represents the weakest class of Hurwitz polynomials in the sense that their zerofree region in the S_2-biplane is the smallest of all the 2-V Hurwitz polynomials.

DEFINITION 3.2. *$B(s_1, s_2)$ is a BHP if $B(s_1, s_2) \neq 0$ in S_{++}^2.*

Narrow Sense Hurwitz Polynomials (NHP). A subclass of BHP is NHP introduced by Ansell [1] in the study of two-variable reactance functions (TRF)s. They may be characterized by the following equivalent definition.

DEFINITION 3.3. *$B(s_1, s_2)$ is a NHP if $B(s_1, s_2) \neq 0$ in $S_{\oplus\oplus}^2 - S_{00}^2 = S_{++}^2 + S_{+0}^2 + S_{0+}^2$, where " $-$ " is used to denote a set theoretic subtraction and "$+$" a set theoretic union.*

DEFINITION 3.4. **Scattering Hurwitz Polynomial (SHP).** *A 2-V polynomial $B(s_1, s_2)$ is a scattering Hurwitz polynomial (SHP) if*
 i) *$B(s_1, s_2)$ is a NHP and*
 ii) *$B(s_1, s_2)$ has no (onedimensional) continuum of zeros on S_{00}^2.*

DEFINITION 3.5. *A 2-V polynomial $B(s_1, s_2)$ is a scattering Hurwitz polynomial if*
 i) *$B(s_1, s_2)$ is a BHP, i.e., $B(s_1, s_2) \neq 0$ in S_{++}^2 and*
 ii) *$B(s_1, s_2)$ and $B^*(s_1, s_2)$ are relatively prime.*

Fettweis [14] identified and popularized this class and suggested the names "principal Hurwitz polynomials" and "scattering Hurwitz polynomials". Fettweis has studied the properties of SHP in depth and his pioneering efforts have established SHP as a very important class of Hurwitz polynomials in the study of stability and passivity.

Hurwitz Polynomial in the Strict Sense (HPSS). Increasing the zero free regions of a Hurwitz polynomial further by the addition of the whole finite imaginary axes of the S_1- and S_2-planes, we get the definition of a Hurwitz polynomial in the strict sense as follows.

DEFINITION 3.6. $B(s_1, s_2)$ *is a HPSS if* $B(s_1, s_2) \neq 0$ *in* $\{S^2_{\oplus\oplus} - \{\text{infinite distant points}\}\}$.

In order to get a 2-D Hurwitz polynomial that is a counter part of 1-D strict Hurwitz, we need to include the infinite distant points in the 2-D plane. (It is to be noted that in certain cases 2-D scattering Hurwitz appears to be the counter part of 1-D strict Hurwitz polynomial). Including the infinite distant points we get the really strictest Hurwitz polynomial which was named by Reddy et al. as a very strict Hurwitz polynomial in [2, 24]. Similar conclusions were also reached by Delsarte et al. [9].

Very Strict Hurwitz Polynomials (VSHP). The following equivalent definitions characterize this type of Hurwitz polynomial.

DEFINITION 3.7. $B(s_1, s_2)$ *is a VSHP if* $B(s_1, s_2) \neq 0$ *or* $0/0$ *in* $S^2_{\oplus\oplus}$.

DEFINITION 3.8. $B(s_1, s_2)$ *is a VSHP if* $1/B(s_1, s_2)$ *has no (first or second kind) singularities in* $S^2_{\oplus\oplus}$.

EXAMPLE 3.9. $B(s_1, s_2) = (5s_1 s_2 + s_1 + s_2 + 8)$ *is a VSHP*.

We shall discuss the properties of 2-D VSHP later.

DEFINITION 3.10. **Self-Paraconjugate Hurwitz Polynomial (SPHP).** $B(s_1, s_2)$ *is said to be a SPHP if*

  i) $B(s_1, s_2)$ *is a selfparaconjugate polynomial and*
  ii) $B(s_1, s_2) \neq 0$ *in* S^2_{++} .

A reactance Hurwitz polynomial (RHP) is defined as the para-even or para-odd part of a scattering Hurwitz polynomial (SHP) and an immittance Hurwitz polynomial is defined as the product of a RHP and a SHP [10].

DEFINITION 3.11. **Reactance Hurwitz Polynomial (RHP).** $B(s_1, s_2)$ *is said to be a RHP if*

  i) $B(s_1, s_2)$ *is a SPHP and*
  ii) *each zero locus of* $B(s_1, s_2)$ *on* S^2_{00} *is of multiplicity unity (i.e.* $B(s_1, s_2)$ *has no repeated factors).*

DEFINITION 3.12. **Immittance Hurwitz Polynomial (IHP).** $B(s_1, s_2)$ *is said to be an IHP if*

i) $B(s_1, s_2)$ *is a BHP and*
ii) *each continuum zero locus of* $B(s_1, s_2)$ *on* S_{00}^2 *is of multiplicity unity (i.e., $B(s_1, s_2)$ has no repeated selfparaconjugate factors.)*

We conclude this section by pointing out that the definitions of BHP, NHP, SHP, HPSS and VSHP are such that in the above sequence each Hurwitz Polynomial satisfies the conditions required for each of the preceding ones, and hence form a subset of each one of them. In other words, $\{VSHP\} \subset \{HPSS\} \subset \{SHP\} \subset \{NHP\} \subset \{BHP\}$. Also $\{IHP\} \subset \{BHP\}$ and $\{SPHP\} \subset \{BHP\}$ and $\{RHP\} = \{IHP\} \cap \{SPHP\}$.

4. Testsets for Analog Hurwitz Polynomials.

THEOREM 4.1. *A 2-D polynomial $B(s_1, s_2)$ is a BHP if and only if*

i) $B(jw_1, s_2) \neq 0$ *in S_{2+} for all real finite w_1 except possibly some isolated w_1 and*
ii) *the polynomial $B(s_1, b)$ has no zeros in S_{1+} for some $b \in S_{2+}$.*

THEOREM 4.2. *A 2-D polynomial $B(s_1, s_2)$ is a NHP if and only if*

i) *for all real finite w_1, $B(jw_1, s_2) \neq 0$ in S_{2+},*
ii) *the polynomial in s_1, $B(s_1, b) \neq 0$ in S_{1+} for some $b \in S_{2+}$, and*
iii) $B(s_1, s_2)$ *has no factor of the type $(s_2 - j\alpha)$ where α is a real constant.*

THEOREM 4.3. *A 2-D polynomial $B(s_1, s_2)$ is a SHP if and only if*

i) *for all real finite w_1, $B(jw_1, s_2)$ has no zeros in S_{2+},*
ii) *the polynomial $B(s_1, b)$ has no zeros in S_{1+} for some $b \in S_{2+}$ and*
iii) $B(s_1, s_2)$ *and* $B^*(s_1, s_2)$ *are relatively prime.*

THEOREM 4.4. *A 2-D polynomial $B(s_1, s_2)$ is a HPSS if and only if*

i) *for all real finite w_1, $B(jw_1, s_2)$ has no zeros in $\{s_2 : Re(s_2) \geq 0, |s_2| < \infty\}$ and*
ii) *the polynomial $B(s_1, b)$ has no zeros in $Re(s_1) \geq 0$, for some $b \in S_{2+}$. It is easily verified that only infinite distant points on S_{00}^2 are omitted from S_{00}^2 in testing for the zero locations of $B(s_1, s_2)$ and hence $B(s_1, s_2)$ is a HPSS.*

THEOREM 4.5. *The necessary and sufficient condition for a two-variable polynomial $B(s_1, s_2)$ to be a very strict Hurwitz polynomial are [19]:*

i) $B(s_1, s_2)$ *is a Hurwitz polynomial in the strict sense*
ii_a) $B(\infty, s_2) \neq 0/0$ *for* $Re(s_2) = 0$ *and* $|s_2| < \infty$
ii_b) $B(s_1, \infty) \neq 0/0$ *for* $Re(s_1) = 0$ *and* $|s_1| < \infty$
ii_c) $B(\infty, \infty) \neq 0/0$.

The infinite point testing method shown in equations (1) - (3) could be followed for the testing of the conditions (ii_a) to (ii_c). Let $B(s_1, s_2) = \sum_{i=0}^{M} \sum_{j=0}^{N} b_{ij} s_1^i s_2^j$, then the conditions (ii_a) to (ii_c) are equivalent to:

a) $A_M(s_2) = \sum_{j=0}^{N} b_{Mj} s_2^j \neq 0$ for $Re(s_2) = 0$
b) $B_N(s_1) = \sum_{i=0}^{M} b_{iN} s_1^i \neq 0$ for $Re(s_1) = 0$
c) $b_{MN} \neq 0$.

THEOREM 4.6.
Let $B(s_1, s_2) = \sum_{j=0}^{N} a_j(s_1) s_2^j = \sum_{i=0}^{M} b_i(s_2) s_1^i = \sum_{i=0}^{M} \sum_{j=0}^{N} c_{ij} s_1^i s_2^j$. Then $B(s_1, s_2)$ is a VSHP if and only if:

i) $B(jw_1, jw_2) \neq 0$, $-\infty < w_i < \infty$, $i = 1, 2$,
ii) $a_N(s_1) \neq 0$ in $S_{1\oplus}$ and
iii) $b_M(s_2) \neq 0$ in $S_{2\oplus}$ and
iv) $c_{MN} \neq 0$.

Testsets for SPHP, RHP and IHP can be easily formulated based on their definitions and the testset of a BHP.

5. 2-D Very Strict Hurwitz Polynomials. A brief additional discussion of Very Strict Hurwitz Polynomials (VSHP)s is given in this section. This is because this class of two-variable Hurwitz polynomials is a counter part of one-variable strict Hurwitz polynomials (at least from the domain description of the closed right half plane). We now state some of the properties of a 2-D VSH polynomial:

Let a two-variable transfer function $T(s_1, s_2)$ be expressed as

$$T(s_1, s_2) = \frac{P(s_1, s_2)}{Q(s_1, s_2)},$$

where

$$P(s_1, s_2) = \sum_i \sum_j p_{ij} s_1^i s_2^j \quad (i = 0, 1, \ldots, k;\ j = 0, 1, \ldots, l)$$

$$Q(s_1, s_2) = \sum_i \sum_j q_{ij} s_1^i s_2^j \quad (i = 0, 1, \ldots, m;\ j = 0, 1, \ldots, n) \ .$$

Applying the transformation method of (1) - (3), it can be shown that unless $m \geq k$ and $n \geq l$, polar singularities exist at a set of infinite distant points in the closed right half of the $\{S_1, S_2\}$ biplane. So let us assume that $m \geq k$ and $n \geq l$. Then the following theorem regarding the singularity in the closed right-half biplane can be stated:

THEOREM 5.1. $T(s_1, s_2)$ does not possess any singularity on the closed right half of the $\{S_1, S_2\}$ biplane defined by $S^2_{\oplus\oplus}$ if and only if $Q(s_1, s_2)$ is a VSHP.

The proof of the theorem is a straight forward one if the infinite distant points are also taken into account.

Some other useful properties of VSH polynomials are [2]:

Property 1: If $B(s_1, s_2) = B_1(s_1, s_2) \cdot B_2(s_1, s_2)$ is a VSHP if and only if $B_1(s_1, s_2)$ and $B_2(s_1, s_2)$ are VSHP's.

Property 2: If $B(s_1, s_2)$ is a VSHP then $\dfrac{\partial}{\partial s_i}[B(s_1, s_2)]$, $i = 1, 2$ are also VSHP's.
The above property is not true for other 2-D Hurwitz polynomials. Let

$$\begin{aligned}B(s_1, s_2) &= A_M(s_2)s_1^M + A_{M-1}(s_2)s_1^{M-1} + \ldots + A_1(s_2)s_1 + A_0(s_2) \\ &= C_N(s_1)s_2^N + C_{N-1}(s_1)s_2^{N-1} + \ldots + C_1(s_1)s_2 + C_0(s_1).\end{aligned}$$

Property 3: Let $B(s_1, s_2)$ be expressed as above. Then $A_i(s_2)$, $i = 0, 1, \ldots, M$ and $C_j(s_1)$, $j = 0, 1, \ldots, N$ are one-variable strict Hurwitz polynomials. This property readily follows from the partial derivative Property 2 above.

Property 4: Let $B(s_1, s_2) = \sum_i \sum_j b_{ij} s_1^i s_2^j$ be a real 2-D VSH polynomial. Then $b_{MN} b_{ij} > 0$ for all i and j ($i = 0, 1, \ldots, M$; $j = 0, 1, \ldots, N$).

THEOREM 5.2. *Let $D(s)$ be any strict Hurwitz polynomial of order n. Generate a 2-D polynomial in the following way:*

$$B(s_1, s_2) = \left[N(s_1, s_2)\right]^n * D(s) \,|s = \frac{M(s_1, s_2)}{N(s_1, s_2)}$$

where M and N are respectively the even and odd 2-D polynomials. The necessary and sufficient condition for $B(s_1, s_2)$ to be a VSHP is that $M(s_1, s_2) + N(s_1, s_2)$ be a VSHP [2].

6. Application of 2-D Hurwitz Polynomials. In this section, we would like to enumerate some properties of 2-D passive network functions with a particular reference to their Hurwitz-polynomial nature [24].

  i) Let $F(s_1, s_2) = N(s_1, s_2)/D(s_1, s_2)$ be the driving point immittance of a passive network. Then $N(s_1, s_2)$ and $D(s_1, s_2)$ are BHPs. Let the common factors of $N(s_1, s_2)$ and $D(s_1, s_2)$ be cancelled out and the resulting polynomials be called $N_1(s_1, s_2)$ and $D_1(s_1, s_2)$. Then $N(s_1, s_2)$ and $D_1(s_1, s_2)$ are immittance Hurwitz polynomials.

  ii) Let $F(s_1, s_2) = A(s_1, s_2)/B(s_1, s_2)$ be a relatively prime 2-V odd rational function. Then:
   a) $F(s_1, s_2)$ is a 2-D reactance function if $A(s_1, s_2) + B(s_1, s_2)$ is a very strict Hurwitz polynomial.
   b) $F(s_1, s_2)$ is a 2-D reactance function if and only if $A(s_1, s_2) + B(s_1, s_2)$ is a scattering Hurwitz polynomial.
   The selfparaconjugate polynomials $A(s_1, s_2)$ and $B(s_1, s_2)$ satisfy the reactance Hurwitz properties.

iii) A relatively prime 2-D odd function $F(s_1, s_2) = A(s_1, s_2)/B(s_1, s_2)$ having no second kind singularities is a reactance function if and only if $A(s_1, s_2) + B(s_1, s_2)$ is a very strict Hurwitz polynomial. Such functions are called proper reactance functions [2] and are useful as transformation functions to generate a (structurally stable) 2-D network from a stable 1-D network. This is one of the main applications of very strict Hurwitz polynomials.

iv) Let us now consider a relatively prime function $F(s_1, s_2) = N(s_1, s_2)/D(s_1, s_2)$. Then $F(s_1, s_2)$ is a TPRF only if $N(s_1, s_2) + D(s_1, s_2)$ is a SHP. Further if there are no second kind singularities for $F(s_1, s_2)$ on S_{00}^2, $N(s_1, s_2) + D(s_1, s_2)$ will be a VSH polynomial. From the above discussion we can conclude that the Hurwitz nature determines important necessary conditions (and in some cases necessary and sufficient conditions) of two-variable positive lossless functions. Now we will turn our attention to some applications concerning transfer functions.

v) Let $T(s_1, s_2) = A(s_1, s_2)/B(s_1, s_2)$ be the transfer function of a singly terminated or doubly terminated 2-V lossless (LC) network. Then $B(s_1, s_2)$ is a scattering Hurwitz polynomial. References [21, 22] provide a detailed discussion of networks with transfer functions having scattering and very strict Hurwitz polynomial denominators. It is not necessary that the denominator of all RLC 2-V network transfer functions be scattering Hurwitz. In the most general case it could be broad-sense Hurwitz.

Another interesting observation is the following. In the 1-D case the voltage transfer function cannot have a pole at the origin and infinity. Extending this to the 2-V situation we find that the 2-V voltage transfer function $T(s_1, s_2)$ cannot have first kind (polar) singularities at $s_i = 0$ or ∞ ($i = 1, 2$); however $T(s_1, s_2)$ can be 0/0 at $s_i = 0$ or ∞ ($i = 1, 2$).

vi) Let $H(s_1, s_2) = P(s_1, s_2)/Q(s_1, s_2)$ be a two-variable bounded real or lossless bounded real function. Then $Q(s_1, s_2)$ is a scattering Hurwitz polynomial (SHP). If $H(s_1, s_2)$ has no NSSK on S_{00}^2 then $Q(s_1, s_2)$ must be a VSH polynomial.

Application to 2-D Analog system stability. We shall consider the following important theorem [12].

THEOREM 6.1. *The 2-D analog transfer function $T(s_1, s_2) = A(s_1, s_2)/ B(s_1, s_2)$ is bounded input and bounded output stable only if $B(s_1, s_2)$ is a scattering Hurwitz polynomial. The sufficient condition for stability is that $B(s_1, s_2)$ be a VSH polynomial (we assume that $T(s_1, s_2)$ has no polar singularities at infinite distant points).*

The following result deals with the stability of a 2-D nonlinear system at the equilibrium point (after the linearization of 2-D nonlinear state equations):

THEOREM 6.2. *A nonlinear system exhibits asymptotic stability at the equilibrium point if the 2-D characteristic polynomial associated with the linearized model (around the equilibrium point) is a 2-D very strict Hurwitz polynomial[28].*

We conclude this section with the following unresolved problem of BIBO stability of 2-D linear continuous time systems:

Conjecture: The 2-D analog transfer function $T(s_1, s_2)$ described above is BIBO stable with no NSSK on S_{00}^2 if and only if $B(s_1, s_2)$ is a VSH polynomial. The sufficiency part of this statement is proved. The necessity has yet to be established.

7. Conclusions. The main objective for writing this paper has been to provide a comprehensive yet compact treatment of the theory of two-dimensional (analog) Hurwitz polynomials. With the help of double bilinear transformation $s_i = (1 - z_i)/(1 + z_i)$, $i = 1, 2$ the theory could easily be translated to the 2-D discrete case and thus to the stability theory and design of two-dimensional digital filters [2, 24]. As in the 1-D case the 2-D Hurwitz polynomials play a critical role in the study of 2-D circuits, systems and filters. In this paper a detailed classification and testing (theorems) of various 2-D Hurwitz polynomials is presented. Discussion of the properties of 2-D VSH polynomials is also given. The various testing procedures (algorithms) are not discussed. The test procedures could be found in the references [11, 12, 25, 26, 29]. We conclude the paper by discussing how various Hurwitz polynomials arise in passive two-variable circuit theory and 2-D analog stability.

References

[1] H. G. Ansell, "On certain two-variable generalization of circuit theory with applications to networks and transmission lines of lumped reactances," IEEE Trans.Circuit Theory, Vol. CT-11, pp.214–223, June 1964.

[2] H. C. Reddy et al., "Generation of two-dimensional digital transfer functions without nonessential singularities of the second kind," in Proc. IEEE Int. Conf. Acoust., Speech, Signal Processing, pp 13–19, April 1979. Also see P. K. Rajan et al., IEEE Trans. Acoust., Speech, Signa Processing, pp 216–223, April 1980.

[3] V. S. Valdimirov, Methods of Theory of Functions of Many Complex Variables. Cambridge, MA: M. I. T, Press, 1966, pp.36–38.

[4] M. Saito, "Synthesis of transmission line networks by multivariable techniques," in Proc. Symp. Generalized Networks, PIB, 1966, pp. 353–393.

[5] D. C. Youla, " Synthesis of networks containing lumped and distributed elements," in Proc. Symp. Generalized Networks, PIB, 1966, pp. 289–343.

[6] V. Ramachandran, "Some similarities and dissimilarities between single variable and two-variable reactance functions," IEEE Circuits and Systems Newsletter, pp. 11–14, 1976.

[7] A. Fettweis, "On the scattering matrix and the scattering transfer matrix of multidimensional lossless two-ports," Arch. El. , vol. 36, pp. 374–381, Sept. 1982.

[8] T. S. Huang, "Stability of two-dimensional recursive digital filters," IEEE Trans. Audio Electroacoust. vol. AU-20, pp. 158–163, June 1972.

[9] Ph. Delsarte, Y. Genin, and Y. Kamp, " Two-variable stability criteria, " in Proc. IEEE Int. Symp. Circuits and Syst., July 1979, pp. 495–498.

[10] A. Fettweis, "On Hurwitz Polynomials in several variables," in Proc. 1983 IEEE Int. Symp. Circuits Syst., Newport Beach, CA, 1983, pp. 382–385.

[11] N. K. Bose , Applied Multidimensional Systems Theory, New York: Van Nostrand Reinhold, 1982.

[12] E. I. Jury, "Stability of multidimensional systems and related problems," in Multidimensional Systems-Techniques and Applications, S. G. Tzafestas, (Editor) :Marcel Dekker, New York1986.

[13] A. Fettweis and S. Basu, "On discrete scattering Hurwitz polynomials," Int. J. Circuit Theory and Applications , vol. 13, Jan 1985.
[14] A. Fettweis, "Some properties of scattering Hurwitz polynomials," Arch. El. Übertragung, vol. 38, pp. 171–176, 1984.
[15] M. G. Strintzis, "Tests of stability of multidimensional filters," IEEE Trans. Circuits and Syst., vol. CAS-24, Aug 1977.
[16] D. Goodman, "Some difficulties with double bilinear transformation in 2-D filter design," Proc. IEEE, vol 66, pp. 905–914, June 1977.
[17] D. C. Youla, "The analysis and synthesis of lumped passive n-dimensional networks-Part I: Analysis," Polytechnic Inst. New York, Brooklyn, Rep. MIR-1437-84, July 1984.
[18] A. Fettweis and S. Basu, " New results on multidimensional Hurwitz polynomials," in Proc. Int. Symp. Circuits and Systems, Kyoto, Japan, June 1985, pp. 1359–1362.
[19] H. C. Reddy and P. K. Rajan, " A simpler test-set for very strict Hurwitz polynomials," Proc. IEEE, pp. 890–891, June 1986.
[20] H. C. Reddy et al., "Separability of multivariable network driving point functions," IEEE Trans. Circuits Syst., vol. CAS-29, pp. 833–840, Dec. 1982.
[21] H. C. Reddy et al., " Realization of resistively terminated two-variable lossless ladder networks," IEEE Trans. Circuits Syst., vol. CAS-29, pp. 827–832, Dec. 1982.
[22] H. C. Reddy et al., "Design of two-dimensional digital filters using analog reference filters without second kind singularities," in Proc. IEEE Int. Conf. Acoust., Speech, Signal Processing, pp. 692–695, Apr. 1981.
[23] D. Goodman, "Some stability properties of linear shift invariant digital filters," IEEE Trans. Circuits Syst., vol. 26, pp. 201–208, Apr. 1971.
[24] H. C. Reddy and P. K. Rajan, " A comprehensive study of two-variable Hurwitz polynomials," IEEE Trans. Education, vol. 32 , pp.198–209, August 1989.
[25] H. C. Reddy and P. K. Rajan, " A test procedure for the Hurwitz nature of two-dimensional analog polynomials using complex lossless function theory," Proce. of IEEE Int. symp. on circuits and systems, pp. 702–705, May 1987.
[26] P. K. Rajan and H. C. Reddy, " Hermite matrix test for very strict Hurwitz polynomials," Proc. of Midwest symp. on circuits and systems, pp.670–673, August 1986.
[27] G. A. Bliss, Algebraic functions. New York: Amer. Math. Soc., 1933.
[28] H. C. Reddy and G. M. Moschytz, "The Linearization and the Stability of Two Dimensional Analog Nonlinear Circuits", Proceeding of the 1994 IEEE-ISCAS,London, UK.
[29] H. C. Reddy, Theory of Two-Dimensional Hurwitz Polynomials, Chapter 16 in Circuits and Filters Handbook (editor W. K. Chen), CRC/ IEEE Press Book, 1995.

General Classes of Control-Lyapunov Functions

Eduardo D. Sontag*
Department of Mathematics, Rutgers University, New Brunswick, NJ 08903
e-mail: sontag@control.rutgers.edu

Héctor J. Sussmann*
Dept. of Mathematics, Rutgers University, New Brunswick, NJ 08903
e-mail: sussmann@hamilton.rutgers.edu

Abstract. The main result of this paper establishes the equivalence between null asymptotic controllability of nonlinear finite-dimensional control systems and the existence of continuous control-Lyapunov functions (CLF's) defined by means of generalized derivatives. In this manner, one obtains a complete characterization of asymptotic controllability, applying in principle to a far wider class of systems than Artstein's Theorem (which relates closed-loop feedback stabilization to the existence of *smooth* CLF's). The proof relies on viability theory and optimal control techniques.

1. Introduction. In this paper, we study systems of the general form

(1) $$\dot{x}(t) = f(x(t), u(t))$$

where the states $x(t)$ take values in a Euclidean space $\mathbb{X} = \mathbb{R}^n$, the controls $u(t)$ take values in a metric space U, and f is locally Lipschitz. A common approach for stabilization of this system to $x = 0$ relies on the use of abstract "energy" or "cost" functions that can be made to decrease in directions corresponding to possible controls. In this methodology, one starts with a "Lyapunov pair" (V, W), consisting of two positive definite functions

$$V, W : \mathbb{X} \to \mathbb{R}_{\geq 0},$$

with V continuously differentiable and proper ("radially unbounded") and W continuous, so that for each state $\xi \in \mathbb{X}$ there is some control-value $u = u_\xi$ with

(2) $$D_{f(\xi, u)} V(\xi) \leq -W(\xi).$$

We are denoting by $D_v V(\xi) = \nabla V(\xi).v$ the directional derivative of V in the direction of the vector v. This property guarantees that for each state ξ there is some control $u(\cdot)$ such that, solving the initial-value problem (1) with $x(0) = \xi$, the resulting trajectory satisfies

$$x(t) \to 0 \quad \text{as} \quad t \to +\infty \, ;$$

see for instance the textbook [11]. A function V which is part of a Lyapunov pair is generically called a *control-Lyapunov function*, henceforth abbreviated "CLF." Thus, existence of a CLF implies null asymptotic controllability.

---

* Supported in part by US Air Force Grant F49620-95-1-0101
* Supported in part by NSF Grant DMS92-02554

Besides its intrinsic theoretical interest, the CLF paradigm is extremely useful in practice, as it reduces the search for stabilizing inputs to the iterative solution of a static nonlinear programming problem: when at state ξ, find u such that Equation (2) holds. The idea underlies feedback control design (see the references in [12], and more recently the textbook [7]), the optimal control approach of Bellman, "artificial intelligence" techniques based on position evaluations in games and "critics" in learning programs, and can even be found in "neural-network" control design (see e.g. [8]).

An obvious fundamental question arises: is the existence of a continuously differentiable CLF *equivalent* to the possibility of driving every state asymptotically to zero? In other words, is this the only way, in principle, to stabilize systems?

It is the purpose of this paper to answer this question in the affirmative. But first, *the definition of* CLF *must be reformulated in a slightly weaker form*, since otherwise the answer would be negative. To see why a weakening is necessary, consider as an illustration the class of systems affine in control, that is, systems for which controls are in $\mathbb{R}^m$ and

$$f(x,u) = f_0(x) + \sum_{i=1}^{m} u_i f_i(x)$$

is affine in u. For such systems, it is well-known that the existence of a CLF in the manner defined above would imply that there is some feedback law $u = k(x)$ so that the origin is a globally asymptotically stable state for the closed-loop system $\dot{x} = f(x, k(x))$ and k is *continuous* on $\mathbb{R}^n \setminus \{0\}$. This is the content of Artstein's Theorem ([1]). More explicitely, and taking for simplicity the case $m=1$, one has the following "universal" formula for computing feedback laws (cf. [13] and also the recent textbooks [4, 9, 7] and the survey [5]): denote $a(x) := \nabla V(x).f_0(x)$ and $b(x) := \nabla V(x).f_1(x)$, for the given CLF. Then the CLF property is equivalent to:

$$b(x) \neq 0 \Rightarrow a(x) < 0$$

and the following feedback law:

$$k(x) := -\frac{a(x) + \sqrt{a(x)^2 + b(x)^2}}{b(x)}$$

(with $k(x) := 0$ when $b(x) = 0$) stabilizes the system (along closed-loop trajectories, $dV/dt = -\sqrt{a^2 + b^2} < 0$) and is smooth away from the origin. But, for most systems, even affine in control and with $m = 1$, continuous feedback may fail to exist, even for very simple controllable systems (see e.g. [11], Section 4.8, and [5]). This means that unless one weakens the definition of CLF, the converse implication "asymptotic controllability implies existence of CLF" will be false.

The main result of this paper provides such a reformulation. The critical step is to relax the differentiability assumption on V to merely *continuity*. Of course, one must then re-interpret the directional derivative appearing in Equation (2) as a *generalized* directional derivative of an appropriate sort. For this generalization, we borrow from the literature of set-valued analysis and differential inclusions, using the concept known there as "upper contingent derivative" or "contingent epiderivative." Once this generalization is allowed (and, for technical reasons, allowing derivatives in directions in the closed convex hull of the velocity set $f(\xi, U)$), the main result, Theorem 4.1, says that existence of CLF's is indeed equivalent to asymptotic controllability. This general result helps in interpreting some of the constructions for particular classes of systems which involve nondifferentiable CLF's; see for instance [10]. The proof follows easily by combining the main result in [12], which gave a necessary condition expressed in terms of Dini derivatives of trajectories, with results from [2].

REMARK 1.1. Our result shows that asymptotic controllability implies the existence of a "Lyapunov function" in the strict sense that derivatives are negative for nonzero states. In analogy with ordinary differential equations, one may ask when the existence of a "weak CLF," for which W is only required to be nonnegative, suffices for the converse. It is indeed possible to provide control theory versions of the LaSalle Invariance Principle; see [14] for details.

2. Asymptotic Controllability and CLF's.

Throughout this paper, we write $\mathbb{R}_{\geq 0} = \{r \in \mathbb{R} : r \geq 0\}$, and use $\mathcal{I}$ to denote the set of all subintervals I of $\mathbb{R}_{\geq 0}$ such that $0 \in I$; thus, $I \in \mathcal{I}$ iff either (i) $I = \mathbb{R}_{\geq 0}$, or (ii) $I = [0, a)$ for some $a > 0$, or (iii) $I = [0, a]$ for some $a \geq 0$. If μ is a map, we will use $\mathcal{D}(\mu)$ to denote the domain of μ, and $\mu | S$ to denote the restriction of μ to a subset S of $\mathcal{D}(\mu)$. For any subset S of $\mathbb{R}^n$, we use $\overline{\mathrm{co}}(S)$ to denote the closed convex hull of S.

We consider systems as in (1) and assume that a distinguished element called "0" has been chosen in the metric space U. We let U_ρ denote, for each $\rho \geq 0$, the ball $\{u \,|\, d(u, 0) \leq \rho\}$, and assume also that each set U_ρ is compact. (Typically, U is a closed subset of a Euclidean space $\mathbb{R}^m$ and 0 is the origin.) The map

$$f : \mathbb{X} \times U \to \mathbb{R}^n$$

is assumed to be locally Lipschitz with respect to (x, u) and to satisfy $f(0, 0) = 0$. (The Lipschitz property with respect to u can be weakened, but we will need to quote results from [12], where this was made as a blanket assumption.) A *control* is a bounded measurable map $u : I_u \to U$, where $I_u \in \mathcal{I}$. We use $\|u\|$ to denote the essential supremum norm of u. i.e.

$$\|u\| = \inf\{\rho \,|\, u(t) \in U_\rho \text{ for almost all } t \in I_u\}.$$

To avoid confusion with the sup norm of the controls, we will use $|\xi|$ to denote the Euclidean norm of vectors ξ in the state space $\mathbb{X}$.

We let **S** denote the class of all systems (1) that satisfy the above conditions. For a system in **S**, if $\xi \in \mathbb{X}$ and u is a control u, we let $\phi(t, \xi, u)$ denote the value at time t of the maximally defined solution $x(\cdot)$ of (1) with initial condition $x(0) = \xi$. Then $\phi(t, \xi, u)$ is defined for t in some relatively open subinterval J of I_u containing 0, and either $J = I_u$ or $\lim_{t \to \sup J} |\phi(t, \xi, u)| = +\infty$.

3. Asymptotic Controllability.

The next definition expresses the requirement that for each state ξ there should be some control driving ξ asymptotically to the origin. As for asymptotic stability of unforced systems, we require that if ξ is already close to the origin then convergence is possible without a large excursion. In addition, for technical reasons, we rule out the unnatural case in which controlling small states requires unbounded controls.

DEFINITION 3.1. The system (1) is *(null-)asymptotically controllable* (henceforth abbreviated "AC") if there exist nondecreasing functions

$$\theta, \tilde{\theta} : \mathbb{R}_{\geq 0} \to \mathbb{R}_{\geq 0}$$

such that $\lim_{r \to 0+} \tilde{\theta}(r) = 0$, with the property that, for each $\xi \in \mathbb{X}$, there exist a control $u : \mathbb{R}_{\geq 0} \to U$ and corresponding trajectory $x(\cdot) : \mathbb{R}_{\geq 0} \to \mathbb{X}$ such that $x(0) = \xi$,

$$x(t) \to 0 \text{ as } t \to +\infty,$$

$$\|u\| \leq \theta(|\xi|),$$

and

$$\sup\{|x(t)| : 0 \leq t < \infty\} \leq \tilde{\theta}(|\xi|).$$

REMARK 3.2. A routine argument involving continuity of trajectories with respect to initial states shows that the requirements of the above definition are equivalent to the following much weaker pair of conditions:
  1. For each $\xi \in \mathbb{X}$ there is a control $u : \mathbb{R}_{\geq 0} \to U$ that drives ξ asymptotically to 0 (i.e. $x(t) := \phi(t, \xi, u)$ is defined for all $t \geq 0$ and $x(t) \to 0$ as $t \to +\infty$);
  2. there exists $\rho > 0$ such that for each $\varepsilon > 0$ there is a $\delta > 0$ such that for each $\xi \in \mathbb{X}$ with $|\xi| \leq \delta$ there is a control $u : \mathbb{R}_{>0} \to U_\rho$ that drives ξ asymptotically to 0 and is such that $|\phi(t, \xi, u)| < \varepsilon$ for all $t \geq 0$.

We point out, however, that Definition 3.1, as stated, makes sense even for the more general class $\mathbf{S}^*$ of systems (1) in which f is completely arbitrary (i.e. not necessarily locally Lipschitz or even continuous), and the set of control values is state-dependent, i.e. an additional requirement $u \in \hat{U}(x)$ is imposed, where $\hat{U} : \mathbb{X} \to 2^U$ is a multifunction with values subsets of U. This includes in particular the situation when $U = \mathbb{X}$ and $f(x, u) = u$, in which case the system (1) is a *differential inclusion* $\dot{x} \in F(x)$. On the other hand, the formulation in terms of Conditions 1 and 2 above does not make sense for general systems in $\mathbf{S}^*$ (since $\phi(t, \xi, u)$ need not be well defined), and *the equivalence between the two formulations depends on the fact that each fixed control gives rise to a flow*, which is true for systems in $\mathbf{S}$ but not for systems in $\mathbf{S}^*$. Throughout the paper, systems of the form (1) are assumed to be in $\mathbf{S}$, so we will use indistinctly the two forms of the definition of AC. However in Section 8 below, and in more detail in [14], we compare systems in $\mathbf{S}$ with differential inclusions – which belong to $\mathbf{S}^*$ but not necessarily to $\mathbf{S}$ – and there one uses Definition 3.1 as stated rather than Conditions 1 and 2.

4. Directional Derivatives. We now introduce an object widely studied in Set-Valued Analysis (cf., for instance, [2], Def. 1 and Prop. 1 of Section 6.1, where it is called the "upper contingent derivative.")

DEFINITION 4.1. For a function $V : \mathbb{R}^n \to \mathbb{R} \cup \{+\infty\}$, a $\xi \in \mathbb{R}^n$ such that $F(\xi) < +\infty$, and a $v \in \mathbb{R}^n$, the *directional subderivative of V in the direction of v at ξ* is

$$D_v^- V(\xi) := \liminf_{\substack{t \to 0+ \\ w \to v}} \frac{1}{t}\left[V(\xi + tw) - V(\xi)\right].$$

(The notations $D_+ V(\xi)(v)$ and $D_\uparrow V(\xi)(v)$ are used in [2, 6] and [3] respectively, with the same meaning as our $D_v^- V(\xi)$.)

For each fixed ξ, the map $v \mapsto D_v^- V(\xi)$ is lower semicontinuous as an extended-real valued function (cf. [2], page 286); thus $\{v \mid D_v^- V(\xi) \leq \alpha\}$ is a closed set for any α. Observe that if V is Lipschitz continuous then this definition coincides with that of the classical Dini derivative, that is, $\liminf_{t \to 0+}[V(\xi + tv) - V(\xi)]/t$. However, in our results we will not assume that V is Lipschitz, so this simplification is not possible. Notice also that in the Lipschitz case $D_v^- V(\xi)$ is automatically finite, but for a general function V, even if finite-valued, it can perfectly well be the case that $D_v^- V(\xi) = +\infty$ or $D_v^- V(\xi) = -\infty$. Naturally, $D_v^- V(\xi)$ is the usual directional derivative $\nabla V(\xi).v$ if V is differentiable at ξ.

We are now ready to define what it means for a function V to be a CLF. Essentially, we want the directional derivative $D_v^- V(\xi)$ in some – ξ-dependent – control direction v to be negative for each nonzero state ξ. More precisely, we will require $D_v^- V(\xi)$ to be bounded above by a negative function of the state and, in the nonconvex case, we will allow v to belong to the convex closure of the set of control directions.

A function $V : \mathbb{X} \to \mathbb{R}_{\geq 0}$ is *positive definite* if $V(0) = 0$ and $V(\xi) > 0$ for $\xi \neq 0$, and *proper* if $V(\xi) \to \infty$ as $|\xi| \to \infty$.

DEFINITION 4.2. A *Lyapunov pair* for the system (1) is a pair (V, W) consisting of a continuous, positive definite, proper function $V : \mathbb{X} \to \mathbb{R}$ and a nonnegative continuous function $W : \mathbb{X} \to \mathbb{R}$, for which there exists a nondecreasing $\nu : \mathbb{R}_{\geq 0} \to \mathbb{R}_{\geq 0}$ with the property that for each $\xi \in \mathbb{X}$ there is a

$$v \in \overline{\mathrm{co}}(f(\xi, U_{\nu(|\xi|)}))$$

such that

(3) $$D_v^- V(\xi) \leq -W(\xi).$$

REMARK 4.3. For the special but very common case when the set of velocities $f(\xi, U_\rho)$ is convex for all ρ (for example if U is a closed convex subset of $\mathbb{R}^m$ and the system (1) is affine in the control), the condition of Definition 4.2 reduces to asking that for each $\xi \neq 0$ there be some control value $u \in U_{\nu(|\xi|)}$ such that

$$D_{f(\xi,u)}^- V(\xi) \leq -W(\xi).$$

If in addition V is differentiable at ξ, then this amounts to requiring that

$$\min_{u \in U_{\nu(|\xi|)}} [\nabla V(\xi) f(\xi, u)] \leq -W(\xi).$$

DEFINITION 4.4. A *control-Lyapunov function (*CLF*)* for the system (1) is a function $V : \mathbb{X} \to \mathbb{R}$ such that there exists a continuous positive definite $W : \mathbb{X} \to \mathbb{R}$ with the property that (V, W) is a Lyapunov pair for (1).

Our main result is as follows:

THEOREM 4.1. *A system Σ in $\mathbf{S}$ is AC if and only if it admits a* CLF.

5. A Previous Result with Relaxed Controls. We first recall the standard notion of relaxed control. If $\rho \geq 0$, a *relaxed U_ρ-valued control* is a measurable map $u : I_u \to \mathbb{P}(U_\rho)$, where $I_u \in \mathcal{I}$ and $\mathbb{P}(U_\rho)$ denotes the set of all Borel probability measures on U_ρ. An ordinary control $t \mapsto u(t)$ can be regarded as a relaxed control in the usual way, using the embedding of the space U_ρ into $\mathbb{P}(U_\rho)$ that assigns to each $w \in U_\rho$ the Dirac Delta measure at w. For $u \in \mathbb{P}(U_\rho)$, we write $f(x, u)$ for $\int_{U_\rho} f(x, w) \, du(w)$. As for ordinary controls, we also use the notation $\phi(t, \xi, u)$ for the solution of the initial value problem that obtains from initial state ξ and relaxed control u, and we denote

$$\|u\| = \inf\{\rho \mid u(t) \in \mathbb{P}(U_\rho) \text{ for almost all } t \in I_u\}.$$

The first ingredient in the proof is the following restatement of the main result in [12].

FACT 5.1. *A system Σ of the form (1) is AC if and only if there exist two continuous, positive definite functions $V, W : \mathbb{X} \to \mathbb{R}$, V proper, and a nondecreasing $\nu : \mathbb{R}_{\geq 0} \to \mathbb{R}_{\geq 0}$ so that the following property holds: for each $\xi \in \mathbb{X}$ there are a $T > 0$ and a relaxed control $\omega : [0, T) \to \mathbb{P}(U_{\nu(|\xi|)})$, so that $x(t) := \phi(t, \xi, \omega)$ is defined for all $0 \leq t < T$ and*

$$(4) \qquad V(x(t)) - V(\xi) \leq -\int_0^t W(x(\tau)) \, d\tau \quad \text{for } t \in [0, T).$$

Proof. If there are such V, W, and ν, then for each ξ we may pick a ω so that (4) holds; this implies the inequality $\liminf_{t \to 0+} t^{-1}[V(x(t)) - V(\xi)] \leq -W(\xi)$, which is the sufficient condition for AC given in [12]. Conversely, if the system is AC, then that reference shows that there exist V, W, and ν as above and such that

$$V(\xi) = \min\left\{\int_0^\infty W(\phi(\tau, \xi, \omega)) \, d\tau + \max\{\|\omega\| - k, 0\}\right\}$$

where the minimum is taken over the set of all relaxed controls

$$\omega : [0, \infty) \to \mathbb{P}(U_{\nu(|\xi|)}),$$

and k is a constant which arises from the function θ in the definition of AC. (Here we take $W(x) = N(|x|)$, where $N : \mathbb{R}_{\geq 0} \to \mathbb{R}_{\geq 0}$ from [12] is a strictly increasing, continuous function satisfying also $N(0) = 0$ and $\lim_{r \to +\infty} N(r) = +\infty$, i.e. a function of class $\mathcal{K}_\infty$. The main point of the proof was to construct an N so that the value function V is continuous and for which optimal controls exist.)

This implies property (4). Indeed, pick ξ and a minimizing ω. Let $x(\cdot) := \phi(\cdot, \xi, \omega)$ and pick any $t \geq 0$. We may consider the new initial state $x(t)$ and the control $\widetilde{\omega}$ obtained by restricting ω to the interval $[t, \infty)$. Then $V(x(t))$ is bounded above by the cost when using $\widetilde{\omega}$, that is,

$$\begin{aligned} V(x(t)) &\leq \int_t^\infty W(x(\tau)) d\tau + \max\{\|\widetilde{\omega}\| - k, 0\} \\ &\leq \int_t^\infty W(x(\tau)) d\tau + \max\{\|\omega\| - k, 0\} \\ &= V(\xi) - \int_0^t W(x(\tau)) d\tau. \end{aligned}$$

Thus property (4) holds with $T = +\infty$. $\square$

6. A Previous Result on Differential Inclusions. Next we recall some concepts from set-valued analysis. We consider set-valued maps (or "multifunctions") between two Hausdorff topological spaces X and Y. A map F from X to subsets of Y is *upper semicontinuous* (abbreviated USC) if for each open subset $V \subseteq Y$ the set $\{x \,|\, F(x) \subseteq V\}$ is open. If U is a compact topological space and $f : X \times U \to Y$ is continuous, then the set valued map $F(x) := F(x, U) = \{f(x, u), u \in U\}$ is USC (see for instance [2], Prop. 1 in Section 1.2).

We will henceforth use the abbreviations *DI* and *USCMCC* for "differential inclusion" and "upper semicontinuous multifunction with compact convex values," respectively.

Let X be a subset of $Y = \mathbb{R}^n$. A *solution* of the DI $\dot{x} \in F(x)$ is by definition a locally absolutely continuous curve $x(\cdot) : I \to X$, where I is an interval, such that $\dot{x}(t) \in F(x(t))$ for almost all $t \in I$.

The second ingredient needed to prove Theorem 4.1 is from the literature on differential inclusions and viability theory. The relevant results are as follows. (We give them in a slightly stronger form than needed, but still not in full generality: in [2], the function "W" is allowed to depend convexly on derivatives $\dot{x}(t)$, and in some implications less than continuity of V or W is required.) Theorem 1 in Section 6.3 of [2] shows that 2 implies 1 (with $T = \infty$ if X is closed and $F(X)$ is bounded), and Proposition 2 in Section 6.3 of [2] says that $1 \Rightarrow 2$. (Another good reference is [6]; see in particular Theorem 14.1 there.)

FACT 6.1. Let F be an USCMCC from X into subsets of $\mathbb{R}^n$, where X is a locally compact subset of $\mathbb{R}^n$. Assume that V and W are two continuous functions $X \to \mathbb{R}_{\geq 0}$. Let $\widetilde{V} : \mathbb{R}^n \to \mathbb{R} \cup \{+\infty\}$ be such that $\widetilde{V} \equiv V$ on X, $\widetilde{V} \equiv +\infty$ on $\mathbb{R}^n \setminus X$. Then the following properties are equivalent:

1. For each $\xi \in X$ there are a $T > 0$ and a solution of $\dot{x}(t) \in F(x(t))$ defined on $[0,T)$ with $x(0) = \xi$ which is monotone with respect to V and W, that is,

$$(5) \qquad V(x(t)) - V(x(s)) + \int_s^t W(x(\tau))d\tau \leq 0$$

for all $0 \leq s \leq t < T$.

2. For each $\xi \in X$ there is some $v \in F(\xi)$ such that $D_v^- \tilde{V}(\xi) \leq -W(\xi)$. Moreover, if X is closed and $F(X) = \bigcup_{x \in X} F(x)$ is bounded, then one can pick $T = +\infty$ in 2.

7. Proof of Theorem 4.1. Let Σ be a system of the form (1). Assume that Σ is AC. We apply Fact 5.1, and obtain V, W, and ν. Pick $\xi \in \mathbb{X}$. Let T, ω, $x(\cdot)$ be as in Fact 5.1. Then

$$x(t) - \xi = \int_0^t f(x(s), \omega(s))ds = \int_0^t f(\xi, \omega(s))ds + o(t) \in t.\overline{\mathrm{co}}(f(\xi, U_{\nu(|\xi|)})) + o(t).$$

So there is a sequence $\{t_j\}$ such that $t_j > 0$ and $t_j \to 0$, with the property that, if $v_j = t_j^{-1}(x(t_j) - \xi)$, then $v_j \to v$ for some $v \in \overline{\mathrm{co}}(f(\xi, U_{\nu(|\xi|)}))$. On the other hand, (4) implies that

$$\liminf t_j^{-1}(V(\xi + t_j v_j) - V(\xi)) \leq -W(\xi).$$

So $D_v^- V(\xi) \leq -W(\xi)$. Therefore (V,W) is a Lyapunov pair.

Conversely, assume that (V,W) is a Lyapunov pair with W continuous and positive definite, and let ν be as in the definition of Lyapunov pair. For $\xi \in \mathbb{X}$, let X_ξ be the sublevel set $\{x \mid V(x) \leq V(\xi)\}$, and write $\hat{\nu}(\xi) = \nu(r(\xi))$, where $r(\xi) = \sup\{|x| : x \in X_\xi\}$. Then let $\hat{\nu}(s) = \sup\{\tilde{\nu}(\xi) : |\xi| \leq s\}$ for $s \geq 0$. For $x \in X_\xi$, define

$$F_\xi(x) := \overline{\mathrm{co}}(f(x, U_{\hat{\nu}(|\xi|)})),$$

and let $\tilde{V}_\xi(x) = V(x)$ for $x \in X_\xi$, $\tilde{V}_\xi(x) = +\infty$ for $x \notin X_\xi$. Then it is clear that F_ξ is an USCMCC. If $x \in \mathbb{X}_\xi$, then Def. 4.2 implies that there is a $v \in \overline{\mathrm{co}}(f(x, U_{\nu(|x|)}))$ such that $D_v^- V(x) \leq -W(x)$. Since $|x| \leq r(\xi)$, we have $\nu(|x|) \leq \tilde{\nu}(\xi) \leq \hat{\nu}(|\xi|)$. So v belongs to $F_\xi(x)$. If $v_j \to v$, $t_j > 0$, $t_j \to 0$, and

$$\frac{1}{t_j}(V(x + t_j v_j) - V(x)) \to w \leq -W(x),$$

then $V(x + t_j v_j)$ must be finite for all large j. Therefore $V(x + t_j v_j) = \tilde{V}_\xi(x + t_j v_j)$ for large j. So $D_v^- \tilde{V}_\xi(x) \leq -W(x)$. This shows that Condition 2 of Fact 6.1 holds with $X = X_\xi$, $F = F_\xi$, and $V_\xi = V|X$ in the role of V. Fact 6.1 – together with standard measurable selection theorems – then implies that there is a control

$\omega : [0,+\infty) \to \mathbb{P}(U_{\hat\nu(\xi)})$ such that Equation (4) holds with $T = +\infty$, $x(t) = \phi(t,x,\omega)$. Since this is true for every ξ, we see that the condition of Fact 5.1 holds (with $\hat\nu$ in the role of ν), so Σ is AC.

REMARK 7.1. The proof actually shows that in the AC case one has trajectories, corresponding to relaxed controls, which are monotone with respect to V and W, and are defined on the entire $[0,+\infty)$. (Observe that the cost function used in [12] is not additive, because of the term "$\max\{\|\omega\|-k,0\}$", so the dynamic programming principle does not apply, and hence we cannot conclude that *optimal* trajectories are monotone. If desired, this situation could be remedied by redefining the optimal control problem as follows: drop the term $\max\{\|\omega\|-k,0\}$ but instead add a state-dependent control constraint forcing $u(t)$ to be bounded by $\theta(x(t))$.)

8. Comparison with differential inclusions.

The purpose of this Section is basically to remark that the necessary and sufficient condition presented here is truly a result about control systems as opposed to about abstract differential inclusions. Notice first that, as explained in Remark 3.2, the systems corresponding to DI's are in $\mathbf{S}^*$, so the concept of asymptotic controllability given by Def. 3.1 makes sense for them. Moreover, there is an obvious definition of CLF in this case as well. It is easy to see that it is still true that the existence of a CLF implies AC (indeed, the proof of the "if" part of Theorem 4.1 applies in this case as well). We now show that the converse implication can fail, that is, we provide an example of an AC system for which there is no CLF. (It is proved in [6] that an AC DI arising from an USCMCC always has a *lower semicontinuous* "CLF." Our definition requires the CLF to be continuous.)

We let $f : \mathbb{R}^2 \to \mathbb{R}^2$ be given by $f(x,y) = (-y,x)$. Let $S = \mathbb{R}_{\geq 0} \times \{0\}$. Define an USCMCC F on $\mathbb{R}^2$ by letting

$$F(x,y) = \begin{cases} \{f(x,y)\} & \text{if } (x,y) \notin S \\ \text{co}(\{f(x,y),(-1,0)\}) & \text{if } (x,y) \in S \,. \end{cases}$$

Then for every $p \in \mathbb{R}^2$ we can construct a trajectory

$$\gamma_p : [0,T_p] \to \mathbb{R}^2$$

of the DI $\dot\xi \in F(\xi)$ such that $\gamma(0) = p$, $\gamma(T_p) = 0$, and $t \mapsto |\gamma_p(t)|$ is nonincreasing. So our DI is AC. However, *there is no continuous function $V : \mathbb{R}^2 \to \mathbb{R}$ such that*

$$\inf_{v \in F(\xi)} D_v^- V(\xi) < 0 \text{ for all } \xi \neq 0 \,.$$

Indeed, let V be such a function. Then $D_{f(x,y)}^- V(x,y) < 0$ if $(x,y) \notin S$. If $r > 0$, then Fact 6.1 – with $W \equiv 0$ – easily implies that the function

$$[0,2\pi] \ni t \mapsto h_r(t) = V(r\cos t, r\sin t)$$

is nonincreasing on $(0,2\pi)$. Since V is continuous, and $h_r(0) = h_r(2\pi)$, we conclude that h_r is constant. So V is in fact a radial function, i.e. $V(\xi) = \hat V(|\xi|)$ for some continuous $\widehat{V} : \mathbb{R}_{\geq 0} \to \mathbb{R}$. Given $r > 0$, let $\xi = (0, r)$, so that $f(\xi) = (-r, 0)$, and find $w_n \to (-r, 0)$, $h_n \to 0+$, such that

$$V(\xi + h_n w_n) - V(\xi) \leq -ch_n$$

for some $c > 0$. Let $r_n = |\xi + h_n w_n|$. Then $|r_n - r| = o(h_n)$ as $n \to \infty$. Pick any $L > 0$, and define $k_n = \frac{h_n}{L}$. Write $r_n = r + k_n s_n$. Then $s_n \to 0$, $k_n \to 0+$, and

$$\widehat{V}(r + k_n s_n) - \widehat{V}(r) \leq -cLk_n .$$

Therefore $D_0^- \widehat{V}(r) \leq -cL$. So $D_0^- \widehat{V}(r) = -\infty$. Since this is true for all $r > 0$, Fact 6.1 – with $W \equiv -1$ – yields the existence, for each r, of an $a > 0$ and a solution $\rho : [0, a] \to \mathbb{R}$ of $\dot{\rho} = 0$, such that $\rho(0) = r$ and $\widehat{V}(\rho(a)) < \widehat{V}(r)$. Since $\rho(a) = r$, we have reached a contradiction.

References

[1] Artstein, Z., "Stabilization with relaxed controls," *Nonlinear Analysis, Theory, Methods & Applications* **7**(1983): 1163-1173.

[2] Aubin, J.-P., and A. Cellina, *Differential Inclusions: Set-Valued Maps and Viability Theory*, Springer-Verlag, Berlin, 1984.

[3] Aubin, J.-P., *Viability Theory*, Birkhäuser, Boston, 1991.

[4] Bacciotti, A., *Local Stabilizability of Nonlinear Control Systems*, World Scientific, London, 1991.

[5] Coron, J.M., L. Praly, and A. Teel, "Feedback stabilization of nonlinear systems: sufficient conditions and Lyapunov and input-output techniques," in *Trends in Control: A European Perspective* (A. Isidori, Ed.), Springer, London, 1995 (pp. 293-348).

[6] Deimling, K., *Multivalued Differential Equations*, de Gruyter, Berlin, 1992.

[7] Krstic, M., I. Kanellakopoulos, and P. Kokotovic, *Nonlinear and adaptive control design*, John Wiley & Sons, New York, 1995.

[8] Long, Y., and M. M. Bayoumi, "Feedback stabilization: Control Lyapunov functions modeled by neural networks," in *Proc. IEEE Conf. Decision and Control*, San Antonio, Dec. 1993, IEEE Publications, 1993, pp. 2812–2814.

[9] Isidori, A., *Nonlinear Control Systems: An Introduction*, Springer-Verlag, Berlin, third ed., 1995.

[10] Lafferriere, G. A., "Discontinuous stabilizing feedback using partially defined Lyapunov functions," in *Proc. IEEE Conf. Decision and Control*, Lake Buena Vista, Dec. 1994, IEEE Publications, 1994, pp. 3487–3491.

[11] Sontag, E.D., *Mathematical Control Theory: Deterministic Finite Dimensional Systems*, Springer, New York, 1990.

[12] Sontag, E.D., "A Lyapunov-like characterization of asymptotic controllability," *SIAM J. Control & Opt.* **21**(1983): 462-471. (See also "A characterization of asymptotic controllability," in *Dynamical Systems II* (A. Bednarek and L. Cesari, eds.), Academic Press, NY, 1982, pp. 645-648.)

[13] Sontag, E.D., "A 'universal' construction of Artstein's theorem on nonlinear stabilization," *Systems and Control Letters*, **13**(1989): 117-123.

[14] Sontag, E.D., and H.J. Sussmann, "Non-smooth control-Lyapunov functions," *Proc. IEEE Conf. Decision and Control, New Orleans, Dec. 1995*, IEEE Publications, 1995.

Towards the stability of fuzzy control systems

Roland Strietzel
Technische Universität Dresden, Fakultät Elektrotechnik
LS für Regelungs- und Steuerungstheorie, Mommsenstr. 13
D-01062 Dresden, Germany

Abstract. There are some different methods to evaluate the stability behaviour of fuzzy control systems depending on the availability of a mathematical process model. Two possibilities are discussed, stability by linearization and stability supervision by an additional fuzzy unit. The stability supervision is very adequate to the application of fuzzy control because a mathematical process model seems to be not necessary. In many cases of stability proof, relations to linear systems can be exploited.

1. Introduction. Many practical examples show, that fuzzy control enlarges the possibilities to solve control problems. The advantages of fuzzy control especially consist in the combination of analogue and digital systems, the design of control systems by means of linguistic models without mathematical models, the possibility of robust control by using the situation oriented calculation of control variables and the design of nonlinear controls.

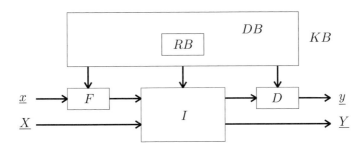

Fig. 1 Fuzzy unit, F fuzzification, I inference, D defuzzification, KB knowledge base, DB data base, RB rule base, $\underline{x}$, $\underline{X}$ input vectors, $\underline{y}$, $\underline{Y}$, output vectors.

The main part of fuzzy controller is the fuzzy unit of Fig. 1 containing modules of fuzzification, inference and defuzzification and the knowledge base formed by a data base and a rule base. The properties of this fuzzy unit are the following:
- fuzzy unit is a single- or multi-variable static nonlinear unique transfer element (excluding dead time of computation, sampling and holding),
- it can also process linguistic variables (e.g. medical diagnoses),
- it has a high realizable variety and flexibility of transfer behaviour,
- it can be completed by dynamic linear components at input and output.

The variety and flexibility of the transfer behaviour results from a high number of input variables of different types (binary, multivalent, continuous), different membership functions for fuzzification and defuzzification (database), the high number of possible rule bases and different logic operations, inference and defuzzification methods used in fuzzy units. In order to get smooth non-linear or sometimes linear behaviour, the logic operations of algebraic sum and algebraic product, the sumprod inference and the center-of-gravity defuzzification by means of singletons according to Table 1 should be applied.

Logical operations	Inference method	Defuzzification
Calculation of membership functions: **And** $C = A \cap B$ $\mu_C(x) = \mu_A(x)\,\mu_B(x)$, $x \in X$ (algebraic product) **Or** $C = A \cup B$ $\mu_C(x) = \mu_A(x) +$ $+ \mu_B(x) - \mu_A(x)\,\mu_B(x)$ (algebraic sum)	Truth value of the i-th premise v_{pi} and the i-th conclusion v_{ci}: $v_{ci} = v_{pi}$, Modification of rule output fuzzy sets (singletons): $\mu'_i(y) = v_{ci}\mu_i(y)$ Aggregation of modified fuzzy sets: $\mu_Y(y) = \sum_{i=1}^{m} v_{ci}\mu_i(y)$	Center-of-gravity method output: $y = \dfrac{\Sigma y_k \mu_k}{\Sigma \mu_k}$, $\Sigma \mu_k = 1$

Table 1: Fuzzy unit operator system

This methodology allows
- the determination of nonlinear behaviour only by fuzzification and rule base,
- the realization of smooth nonlinear functions,
- the implementation of linear input-output relations in special cases,
- the differentiability and the linearization near the operating point,
- the application of the back propagation principle in neuro-fuzzy nets.

Fig. 2 demonstrates the specification of a (linear) PI controller with normalized input and output variables and the transmission behaviour $\Delta y = (e + \Delta e)/2$ as an example of applying this operator system. e is the control error, y the controller output and Δ the backward difference.

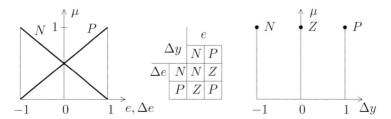

Fig. 2 Specification of a linear PI controller by membership functions of fuzzification, rule base and rule output singletons for defuzzification

Fuzzy units according to Fig. 1 can be used in all hierarchical levels of a control system, in all places of the level where a nonlinear processing of signals is necessary (control, multiplexing, smooth switching, adaptation, supervision, ...).

2. Life cycle and stability. According to its modules the fuzzy unit of Fig. 1 allows the implementation of non-linear transfer elements by a nonlinear fuzzification, a special form of rule base and by a suitable placement of the rule output singletons for defuzzification.

The aim of the design procedure, the beginning part of the life cycle, is to form a fuzzy unit of Fig. 1 solving the planned control problem. In general the following tasks must be treated:
- formulation of the control objective and the general system structure (including stability requirements),
- study of the process to be controlled (evaluation of dangerousness of instability),
- investigation of possible control structures and determination of variables (stability criterion),
- analysis of the process and forming a mathematical or a linguistic model (stability criterion, stabilizing methods),
- decision over the structure of fuzzy controller (stability criterion, stability region),
- determination of the rule base, the membership functions,
- parameterization (stability reserve),
- simulation, testing (stability reserve),
- technical implementation,
- start-up, tuning (stability check),
- operation, long time optimization (including stability behaviour).

Practically all these steps influence the stability behaviour of the system. But the character of stability investigations essentially depends on the type of process model (mathematical or linguistic model).

3. Stability investigation of fuzzy control systems with mathematical process model.

In the case of availability of a mathematical process model, the stability check of a fuzzy control system can use the methods for nonlinear systems. Following methods are applied for stability investigations of fuzzy systems [2], [3], [4], [5]:
- Harmonic balance, method of describing function
- Popov criterion of absolute stability
- Hyperstability method
- Method of Lyapunov functions.

The designed fuzzy unit can be tested by simulation and improved by using parameter optimization with process model.

Supposing the general structure of Fig. 1, it is possible to extend the listed methods for more than one input and output variable.

Using a single-input-single-output fuzzy unit the method of harmonic balance can be applied to investigate the stability behaviour of a PI control according to Fig. 3. The plant is modelled by a Hammerstein approach with the invertible static nonlinear function $y_s = \phi(y)$, $y_s \in [y_{sl}, y_{su}]$, $y \in [y_l, y_u]$, and the linear transfer function $G(s)$. The controller consists of a fuzzy unit and a linear input part (k_p, k_I) with integrator I.

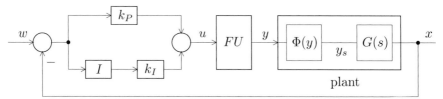

Fig. 3 Control loop with PI module, fuzzy unit FU and plant with Hammerstein model

The fuzzy unit compensates the non-linear part of the plant model by using a singleton placement method before defuzzification in order to get an approximately linear control system. Fig. 4 gives the specification of the fuzzy unit for $m = 5$ linguistic values $(-2, -1, 0, 1, 2)$ of input and output. Sum-prod inference according to Table 1 is applied.

The parameters of the output singletons (y_i, μ_i), $i = 1, \ldots, m$ for defuzzification are calculated in the following way:

$$\Delta y_s = \frac{(y_{su} - y_{sl})}{(m-1)}, \tag{1}$$

$$y_i = \phi^{-1}[y_{sl} + (i-1)\Delta y_s], \; i = 1, \ldots, m, \; y_{sl} = \phi(y_l), \tag{2}$$

$$y'_i = \phi^{-1}\left[y_{sl} + \left(i - \frac{1}{2}\right)\Delta y_s\right], \; i = 1, \ldots, m-1, \tag{3}$$

(4) $$\mu_{i+1} = \frac{\mu_i(y'_i - y_i)}{(y_{i+1} - y'_i)}, \ \mu_1 = 1, \ i = 1, \ldots, m-1.$$

$2m - 1$ measurement points of the nonlinear function $y_s = \phi(y)$ are necessary for compensation.

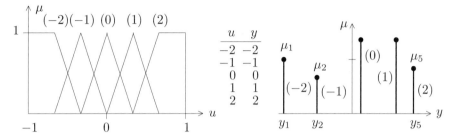

Fig. 4 Specification of the fuzzy unit: fuzzification, rule base, distribution of output singletons

The describing function for evaluating the stability behaviour results from the nonlinear functions of the fuzzy unit and the nonlinear part of the plant. It depends on the current operating point influenced by the reference value and disturbances. The PI unit is adapted to the linear dynamic part of the plant, to obtain the desired control performance. The resulting linear part of the control loop has the frequency response $G^*(jw)$. Both form a resulting locus diagram, sketched in Fig. 5. If there is no point of intersection between the loci of the frequency response and the negative inverse of the describing function the system is asymptotically stable.

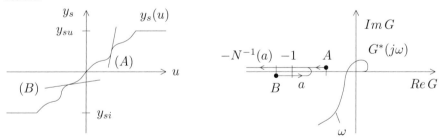

Fig. 5 Characteristic of the resultant nonlinear part and locus diagram

The negative inverse of the describing function $N(a)$ has two extreme start points A and B for the amplitude $a = 0$ corresponding to the maximal amplification and the minimal amplification in $[y_{sl}, y_{su}]$ respectively.

This example shows an interesting design method, which suggests to compensate the static nonlinearity of the plant by means of single input single output fuzzy units. The compensation is possible, if the inverse of $\phi(y)$ can be calculated.

The necessary pairs (y, y_s) can be also obtained by measurement. A higher number m of linguistic values brings a closer compensation of the plant nonlinearity. Amplitude and phase margin are calculable.

The quality of compensation is represented by the distance of the extreme points of the describing function for minimal excitation $(a > 0)$. In the case of ideal compensation the locus of describing function shrinks to the Nyquist point -1, if saturation is not considered.

In a similar way a plant with Wiener model can be treated to get an approximately linear control loop. Another possibility to design a control system with fuzzy controllers by given plant model is the parameter optimization of a sufficiently general fuzzy controller.

4. Stability investigation of fuzzy control systems with a linguistic process model.
The main application field of fuzzy control is based on linguistic models. Linguistic models are more or less complete verbal expressions in form of production rules describing the necessary manipulations for controlling plants.

The control algorithm principally depends on the rule base built from the linguistic model, the membership functions of the fuzzification and the parameterization of the fuzzy unit and the additional blocks to realize dynamic transfer functions.

To guarantee the stability of the system the following methods can be used:
- tuning of the controller at starting-up,
- adaptation of the fuzzy controller,
- passive stability supervision,
- active stability supervision.

The methods can also be used in combination.

Because a mathematical model of the process is not available, the manipulations to observe stability or to realize a stable system must be done with the real plant.

4.1. Tuning of fuzzy controllers. Tuning is a parameter optimization at the process [7]. Therefore a method is necessary, which consumes little time. The direct method of Ziegler and Nichols is not usable because of the dangerousness of increasing oscillations in nonlinear systems. A useful alternative is a parameter optimization with the real plant. The closed loop control system is tested by means of the response of sufficiently small reference input steps at different operating points. The linearization near the operating point is used. The controller parameters can be adjusted in order to optimize the step responses. In a so-called direct method the controller parameter also can be calculated from the course of the step response. This is possible in an easy way, if the step response has no overshoots and by using linear methods.

To guarantee the stability under different working conditions the step response check has to be realized at a sufficient number of different operating points. Fig. 6 shows an example of tuning fuzzy controllers for a bicycle [6].

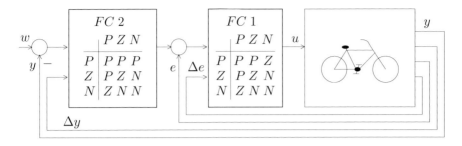

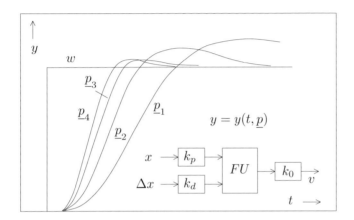

Fig. 6 Simulation of tuning the cascaded controller for a bicycle model, y deviation of the gravity center from the straight-line track, w reference step to a parallel track, $\underline{p}$ parameter vector of scaling factors of the fuzzy controller (e.g. k_p, k_d, k_0), u steering angle, e tilt angle.

4.2. Stability supervision by fuzzy unit. Fuzzy logic allows the implementation of stability criteria [1]. A possible way is shown by Fig. 7. The stability supervision unit uses the Lyapunov criterion.

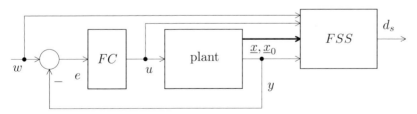

Fig. 7 Control system with fuzzy controller FC and fuzzy stability supervisor FSS, w reference variable, e control difference, u control variable, $\underline{x}$ state vector, $\underline{x}_0$ state vector at the actual operating point, d_s degree of stability.

The fuzzy stability supervisor performs the following tasks
- calculation of a Lyapunov function of the complete or incomplete state vector,
- calculation of the tendency of Lyapunov function,
- application of production rules for programming the fuzzy unit to calculate a measure of stability.

The production rules define the stability behaviour. A Lyapunov function of the form $V = (\underline{x}-\underline{x}_0)^T P(\underline{x}-\underline{x}_0)$ is used. If the Lyapunov function $V(\underline{x}-\underline{x}_0)$ is limited at a prescribed value depending on the motion of the reference variable or it is decreasing, then the system is stable ($d_s = 1$). Instability exists ($d_s = 0$), if the Lyapunov function is increasing at constant reference signal or if it is exceeding the limitation. In this case a excessively high distortion which gives high values of the Lyapunov function is equivalent with an instability. The calculated value $d_s \in [0,1]$ serves for alarming, shutting down the system or correcting the controller parameter.

4.3. Appended stabilizer. By means of a linguistic model a fuzzy controller is designed. This controller realizes the desired control action but does not meet the stability requirements.

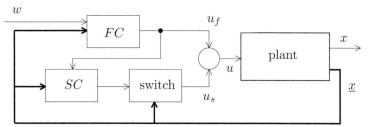

Fig. 8 Plant with fuzzy controller and stabilizing controller smoothly switched in dependence on state $\underline{x}$. FC fuzzy controller, SC stability controller.

The stabilization of the closed system ensures an appended stabilizer SC [8], without changing the fuzzy controller FC (Fig. 8).

Given is a nonlinear process with differential equation (process equation)

$$(5) \qquad x^{(n)} = f\left(x, \frac{dx}{dt}, \ldots, x^{(n-1)}\right) + g\left(x, \frac{dx}{dt}, \ldots, x^{(n-1)}\right) u ,$$

with the output variable x, the control variable u and function $g > 0$. The output variable and its derivatives can be abbreviated to the state vector $\underline{x} = (x, \frac{dx}{dt}, \ldots, x^{(n-1)})$, which should be measurable or computable. The control variable is generated by the fuzzy controller $u_f = u_f(\underline{x})$ and by the stability supervising controller $u_s = u_s(\underline{x})$.

$$(6) \qquad u = u_f(\underline{x}) + u_s(\underline{x}) .$$

The fuzzy controller is designed by a linguistic model, because the functions f and g are unknown. The only information about f and g is that $|f(\underline{x})| < f_u(\underline{x})$ and $0 < g_1(\underline{x}) < g(\underline{x})$.

The process mentioned above has the same stability behaviour like a linear process

(7) $$x^{(n)} = -\underline{k}^T \underline{x} + g(\underline{x})\, u$$

with the coefficient vector $\underline{k} = (k_n, \ldots, k_1)$ and the state-depending control amplification $g(\underline{x})$ in the case of a control function

(8) $$u^* = [-f(\underline{x}) - \underline{k}^T \underline{x}][g(\underline{x})]^{-1}, \; f(\underline{x}) = -\underline{k}^T \underline{x} - u^* g(\underline{x}) \,.$$

The polynomial $s^n + k_1 s^{n-1} + \ldots + k_n$ is a Hurwitz polynomial and has stable roots. Substituting the unknown function $f(\underline{x})$ in the process equation gives

(9) $$x^{(n)} = -\underline{k}^T \underline{x} - u^* g(\underline{x}) + g(\underline{x})[u_f(\underline{x}) + u_s(\underline{x})]$$

or written in form of state equations

(10) $$\frac{d\underline{x}}{dt} = A\underline{x} + \underline{b}(u_f + u_s - u^*)$$

with the Frobenius matrix

$$A = \begin{pmatrix} 0 & 1 & 0 & 0 & \ldots & 0 \\ 0 & 0 & 1 & 0 & \ldots & 0 \\ \ldots & \ldots & \ldots & \ldots & \ldots & \ldots \\ 0 & 0 & 0 & 0 & \ldots & 1 \\ -k_n & -k_{n-1} & \ldots & \ldots & \ldots & -k_1 \end{pmatrix}$$

and the vector

(12) $$\underline{b} = [0\; 0 \ldots 0\; g(\underline{x})]^T \,.$$

With the Lyapunov function candidate

(13) $$V = 2^{-1}\, \underline{x}^T P \underline{x}$$

and the symmetric positive definite matrix P satisfying the Lyapunov equation

(14) $$A^T P + P A = -Q$$

results the derivative

(15) $$\frac{dV}{dt} = -2^{-1}\, \underline{x}^T Q \underline{x} + \underline{x}^T P \underline{b}(u_f + u_s - u^*) < |\underline{x}^T P \underline{b}|\, (|u_f| + |u^*|) + \underline{x}^T P \underline{b}\, u_s \,.$$

To ensure $\frac{dV}{dt} \leq 0$ the supervisory controller (Fig. 8) has to implement the equation

$$u_s = -\text{sign}(\underline{x}^T P \underline{b}) \left[\frac{(f_u + |\underline{k}^T \underline{x}|)}{g_1} + |u_f| \right] \tag{16}$$

considering the bounds f_u and g_1 of the functions f and g respectively.

Usually a similar procedure is used in designing model reference adaptive control systems to get the adaptation law on the base of stability theory.

5. Conclusions. The high flexibility and variety of fuzzy units allow to design non-linear controllers performing special application objectives. Because of the non-linearity of systems with fuzzy controller the stability behaviour can not be decided easily both in the case of a mathematical process model and of a linguistic model. The discussed variants show, that often the stability behaviour of linear systems opens up a way to realize the global stability as a necessary condition for practical controls. Examples for using linear methods are describing function, large-scale linearization by compensation, linearization near the operating point, the reference to linear systems in adaptation and stability supervision. The necessary decomposition of transfer functions into linear and nonlinear parts is often a strong limitation.

The author wants to express his thanks to Prof. Dr.-Ing. Dr. rer.nat K. Reinschke for supporting his research work.

References

[1] Bindel, Th. and Mikut R., *Stabilitätssicherung durch Verwendung einer Fuzzy-Überwachungseinrichtung am Beispiel einer Durchflussregelstrecke*. In: Forschungsbericht 4. Workshop Fuzzy Control, Dortmund, Germany, (1994), pp. 185–199.
[2] Böhm, R., *Ein Ansatz zur Stabilitätsanalyse von Fuzzy-Regelungen*. In: Forschungsbericht 2. Workshop Fuzzy Control, Dortmund, Germany, (1992), pp. 24–35.
[3] Boll, M., Bornemann J. and Dörr F., *Anwendung der harmonischen Balance auf Regelkreise mit unsymmetrischen Fuzzy-Komponenten und konstanten Eingangsgrössen*. In: Forschungsbericht 4. Workshop Fuzzy Control, Dortmund, Germany, (1994), pp. 70–84.
[4] Bühler, H., *Stabilitätsuntersuchung von Fuzzy-Regelungen*. In: Forschungsbericht 3. Workshop Fuzzy Control, Dortmund, Germany, (1993), pp. 1–12.
[5] Kiendl, H., *Stabilitätsanalyse von mehrschleifigen Fuzzy-Regelungssystemen mit der Methode der Harmonischen Balance*. In: Forschungsbericht 3. Workshop Fuzzy Control, Dortmund, Germany, (1993), pp. 315–321.
[6] Strietzel, R., *Feedforward adapted control of a track bound vehicle*. In: Proceedings of the first European Control Conference (ECC '91), Grenoble, France, Vol. 2, (1991), pp. 1163–1168.
[7] Strietzel, R., *Tuning Problem of Fuzzy Controllers*. In: Proceedings of the European Seminar on Automation and Control Technology, Dresden, Germany, (1994), pp. 83–91.
[8] Wang, L.-X., *A supervisory controller for fuzzy control systems that guarantees stability*. IEEE Trans. on Automatic Control, **39**, 9, (1994), pp. 1845–1847.

Discrete Optimization Using Analog Neural Networks with Discontinuous Dynamics

M. Vidyasagar
Centre for Artificial Intelligence and Robotics, Raj Bhavan Circle
High Grounds, Bangalore 560 001, INDIA
E-Mail: sagar@cair.ernet.in

Abstract. In this paper, a new type of neural network is presented, that can be used to perform discrete optimization over a set of the form $\{0,1\}^n$. Unlike earlier neural networks, this network is characterized by the fact that the dynamics are discontinuous. It is shown that the discontinuous nature of the dynamics makes it possible to carry out quite a thorough analysis of the network trajectories. In particular, it is shown that, in the practically important case where the objective function to be maximized is a multilinear polynomial, almost all trajectories converge to a local maximum of the objective function. Moreover, it is possible to make the trajectories converge within a finite amount of time, and in fact arbitrarily quickly, to a local maximum. The results presented here open the way for the formulation of a suitable complexity theory for analog computation.

1. Introduction. The area of neural networks can by now be considered as a mature discipline. Broadly speaking, the applications of neural networks can be grouped into two categories, namely: pattern recognition, and discrete optimization. The popular back-propagation algorithm and its variations [6] represent attempts to synthesize neural networks that are capable of performing pattern recognition, whereas the use of Hopfield networks and its generalizations [5, 8] represent the use of neural networks to perform discrete optimization.

The focus of the present paper is on the latter application. It is assumed that one is given an objective function $E : \Re^n \to \Re$, and one wishes to find an element $\mathbf{x} \in \{0,1\}^n$ that maximizes $E(\mathbf{x})$ as $\mathbf{x}$ varies over the *discrete* set $\{0,1\}^n$. Initially Hopfield [3] proposed a simple algorithm to maximize[1] a quadratic function of the form

$$(1) \qquad E(\mathbf{x}) = \frac{1}{2}\mathbf{x}^t W \mathbf{x} + \mathbf{x}^t \mathbf{c},$$

where $W \in \Re^{n \times n}$ is a symmetric matrix *with zero diagonal elements* and $\mathbf{c} \in \Re^n$. To optimize such an objective function E, Hopfield proposes a neural network evolving on the discrete set $\{0,1\}^n$, whose dynamics are described by

$$(2) \qquad x_i(t+1) = \operatorname{sat}\left[\sum_{i=1}^{n} w_{ij} x_j(t) + c_i\right],$$

where the "sat" function is defined by

$$\operatorname{sat}(x) = \begin{cases} 0 & \text{if } x < 0, \\ 1 & \text{if } x \geq 0. \end{cases}$$

---
[1] Actually, Hopfield studied the *minimization* of a quadratic objective function, but this is a trivial difference.

He shows that under *asynchronous updating*, that is, if only one neuron is updated at any given time, then the network trajectory evolves in such a way that the objective function is always nondecreasing; that is,

$$E[\mathbf{x}(t+1)] \geq E[\mathbf{x}(t)], \forall t.$$

Observe that, if the weight matrix W has all of its diagonal elements equal to zero, then the neural network (2) is characterized by the *no self-interactions* property. That is, the state of neuron i at time $t+1$ depends only on the states of the *other* neurons $x_j(t), j \neq i$, and not on $x_i(t)$. In subsequent papers [5, 8], it is claimed that several combinatorial optimization problems, such as the Travelling Salesman Problem and 0-1 programming, can be solved using the above approach.

While the above algorithm appears to be intuitively appealing, it is not too difficult to show that the algorithm is merely the naive "nearest-neighbour search" procedure. Thus it is not very surprising that the algorithm is quite inefficient in general. In fact, Schäffer and Yannakakis [7] show that this algorithm can, in the worst case, take up to an exponential number of time steps before converging to a local maximum. Observe that simply enumerating the values of $E(\mathbf{x})$ at each $\mathbf{x} \in \{0, 1\}^n$ and thus finding a *global* maximum requires $O(2^n)$ operations. Thus, at present, the above approach to discrete optimization using binary neural networks cannot be considered to be viable.

In recent years, several persons have attempted to solve *discrete* optimization problems using *analog* neural networks. The basic idea is that, in order to minimize an objective function $E(\mathbf{x})$ over the *discrete* set $\{0,1\}^n$, one sets up a neural network evolving over the hypercube $[0,1]^n$, and adjusts the dynamics in such a way that the network trajectories always converge towards a *corner* of this hypercube. The hope is that the corner to which the trajectory converges will be an optimum. Hopfield [4] proposed perhaps the first such network, but more recent approaches such as tabu search represent the same philosophy. This approach is also closely related to the so-called "interior-point" methods for linear programming, and more generally, for optimization subject to a set of linear constraints; see e.g. [2].

In this paper, a new type of neural network is presented for performing discrete optimization over a set of the form $\{0,1\}^n$. Unlike earlier neural networks, this network is characterized by the fact that the dynamics are *discontinuous*. It is shown that the discontinuous nature of the dynamics makes it possible to carry out quite a thorough analysis of the network trajectories. One of the novel features of the network is the introduction of a so-called "bias" vector into the network dynamics. Many of the ideas in the present paper carry on from similar ideas proposed in [10], which studies an analog neural network with *smooth* dynamics that can be used for $0-1$ programming. It is shown that, in the practically important case where the objective function to be maximized is a multilinear polynomial, almost all trajectories converge to a local maximum of the objective function. However, this property is true only because of the introduction of the bias term, and is not true otherwise. Moreover, it is possible to make the trajectories converge within a finite amount of time, and in fact arbitrarily quickly, to a local maximum.

Although several researchers have presented analog approaches to discrete computation, a suitable *complexity theory* for solving discrete problems using analog computation still remains to be found. The results presented here perhaps open the way for the formulation of such a theory. Some suggestions are given as to how such a theory could be evolved.

2. Proposed New Neural Network.
Suppose it is desired to maximize a function $E(\mathbf{x}): \Re^n \to \Re$ as $\mathbf{x}$ varies over the discrete set $\{0,1\}^n$. In such a case, we are free to replace the given objective function E by any other function F so long as

$$E(\mathbf{x}) = F(\mathbf{x}) \; \forall \mathbf{x} \in \{0,1\}^n.$$

Now, it is not too difficult to show that, given *any* function $E: \Re^n \to \Re$, there exists a *unique multilinear polynomial* F such that the above equation holds. Thus, without loss of generality, the objective function to be maximized can be assumed to be a multilinear polynomial in the variables $x_1, \ldots, x_n$. Note that a quadratic function of the form (1) is multilinear if and only if $w_{ii} = 0$ for all i. Thus the assumption of multilinearity of the objective function is a natural generalization of the popular "no self-interactions" assumption.

Let $E(\mathbf{x})$ be the multilinear polynomial that we wish to maximize as $\mathbf{x}$ varies over the discrete set $\{0,1\}^n$. The proposed neural network is described by

(3) $$\dot{x}_i = s[x_i, \nabla E(\mathbf{x})_i] + b_i,$$

where $\nabla E(\mathbf{x})$ denotes the gradient vector of E, the number b_i is a so-called "bias" term, and $s: \Re^2 \to \Re$ is the function defined by

$$s(x,y) = \begin{cases} \max\{0,y\} & \text{if } x \leq 0, \\ y & \text{if } x \in (0,1), \\ \min\{0,y\} & \text{if } x \geq 1. \end{cases}$$

Note that the function s is *discontinuous*. Thus the neural network is described by discontinuous dynamics. In general, in order to study the trajectories of such a system, one would have to appeal to very general concepts of a "solution," such as solutions in the sense of Fillipov. However, because of the rather "mild" nature of the discontinuity in the present case, such powerful concepts are not needed. The term b_i, representing the i-th component of the so-called "bias" vector, is a novel feature of the network, and is the key to several of its properties.

3. Properties of the Proposed Neural Network.
In this section, some relevant properties of the proposed neural network are presented without proof.

Lemma 1 The hypercube $[0,1]^n$ is an invariant set of the neural network (3).

This means that, if the initial state belongs to the hypercube $[0,1]^n$, the resulting solution trajectory also belongs to the same hypercube.

Definition 1 A vector $\mathbf{x}_0 \in \Re^n$ is said to be an *equilibrium* of the network (3) if

$$x(0) = x_0 \Rightarrow x(t) = x_0 \ \forall t \geq 0.$$

Note that, in the case of a differential equation of the form

$$\dot{\mathbf{x}}(t) = \mathbf{f}[\mathbf{x}(t)],$$

where the right side is differentiable, this definition is equivalent to requiring that $\mathbf{f}(\mathbf{x}_0) = \mathbf{0}$. However, in the present case, where the right side of the differential equation is not even continuous, the present (and more general) definition is required. In the present context, equilibria correspond to those points in $[0,1]^n$ where the network trajectories go and "settle down."

Once the concept of an equilibrium is defined, it is possible to define the concepts of stability, asymptotic stability, and instability in a manner analogous to the corresponding definitions for differential equations with smooth dynamics, as in e.g. [9].

In order to analyze the dynamics of the neural network (3), we divide the equilibria of the network into three types, namely, those lying respectively in (i) the interior of $[0,1]^n$, (ii) the faces of $[0,1]^n$, and (iii) the corners of $[0,1]^n$. The next several theorems characterize these equilibria. As a prelude to this, we make precise the intuitive notions of the interior, the corners, and the faces of the hypercube $[0,1]^n$. Let $H = (0,1)^n$ denote the interior of the hypercube $[0,1]^n$. Then any equilibrium belonging to H is called an interior equilibrium. Next, the 2^n points $\{0,1\}^n \subset [0,1]^n$ are referred to as the corners of $[0,1]^n$, and any equilibrium belonging to $\{0,1\}^n$ is called a corner equilibrium. The subset of $[0,1]^n$ obtained by removing both H and $\{0,1\}^n$ is referred to as the set of faces of $[0,1]^n$. It is easy to see that the set of faces of $[0,1]^n$ is the union of vectors $\mathbf{x} \in [0,1]^n$ such that $x_i \in \{0,1\}$ for at least one, and at most $n-1$, values of the index i. To be more precise (though perhaps more obscure), let J be any nonempty and proper subset of $\{1, \ldots, n\}$, and let $I = \{1, \ldots, n\} - J$. Let $\mathbf{e} \in [0,1]^n$ be a vector such that $e_i \in \{0,1\} \ \forall i \in I$. Then the set

$$x_i \in (0,1) \ \forall i \in J, \ x_i = e_i \ \forall i \in I$$

is called a face of the hypercube $[0,1]^n$. Given a fixed subset J, there are $2^{|I|}$ different ways of choosing the binary vector $\mathbf{e}$, and the resulting $2^{|I|}$ different faces of $[0,1]^n$ are said to be parallel. If $|I| = 1$, then the face is also sometimes called an edge.

Theorem 1 For all bias vectors $\mathbf{b}$ except those belonging to a set of measure zero, the network (3) has only a finite number of equilibria in the interior of $[0,1]^n$, and these are all unstable.

Theorem 2 For all bias vectors $\mathbf{b}$ except those belonging to a set of measure zero, the network (3) has only a finite number of equilibria in the faces of $[0,1]^n$, and they are all unstable.

Before presenting the next theorem, the concept of a local maximum in the case of discrete optimization is made precise. Given a vector $\mathbf{x} \in \{0,1\}^n$, the set $N(\mathbf{x})$ consisting of all vectors $\mathbf{y} \in \{0,1\}^n$ that differ from $\mathbf{x}$ in exactly one component is called the **set of nearest neighbours** of $\mathbf{x}$. A vector $\mathbf{x} \in \{0,1\}^n$ is called a **local maximum** of E over $\{0,1\}^n$ if

$$E(\mathbf{x}) \geq E(\mathbf{y}) \ \forall \mathbf{y} \in N(\mathbf{x}).$$

In other words, $\mathbf{x}$ is a local maximum of E if the value of E cannot be increased beyond $E(\mathbf{x})$ by changing just one component of x.

Theorem 3 For all sufficiently small bias vectors $\mathbf{b}$ except those belonging to a set of measure zero, the network (3) has an equilibrium in a corner of $[0,1]^n$ if and only if that corner represents a local maximum of E over $\{0,1\}^n$. Moreover, any such equilibrium is asymptotically stable.

Theorem 4 For all sufficiently small bias vectors $\mathbf{b}$ except those belonging to a set of measure zero, and for all initial conditions $\mathbf{x}(0) \in [0,1]^n$ except those belonging to a nowhere dense set, the solution trajectory of the network (3) converges to a local maximum of E within a finite amount of time.

The above theorem can be stated roughly as: For almost all sufficiently small bias vectors, and for almost all initial conditions, the network trajectory will converge to a local maximum of the objective function.

At this point, one may ask why it is necessary to introduce the bias vector at all, if "almost all" choices will do? Why not simply choose $\mathbf{b} = \mathbf{0}$? The answer is that, while *almost* all choices of $\mathbf{b}$ will do, it may happen that the specific choice $\mathbf{b} = \mathbf{0}$ is one of the "forbidden" choices. For example, consider the problem of minimizing the function $E(x_1, x_2, x_3) = x_1 x_2 x_3$ as $\mathbf{x}$ varies over $\{0,1\}^3$. It can be shown that the statements of Theorems 1 through 3 are *false* if $\mathbf{b} = \mathbf{0}_3$. Thus, the "natural" choice $\mathbf{b} = \mathbf{0}$ may not work in general.

Note that it is possible to scale the time variable by changing t to λt, where λ is some positive number. In this way, it is possible to make the network converge to a local maximum *arbitrarily quickly*.

4. Complexity of Analog Computation. The above theorems suggest that a discrete optimization problem can be solved "arbitrarily" quickly in time by using an analog neural network. Thus it is natural to ask whether analog computation offers any advantages over discrete computation. In order to make the comparison fair, it is necessary to introduce a suitable measure of the "complexity" of analog computation, in the same way that one uses the storage requirements or the number of time steps as a measure of the complexity of discrete computation. Clearly, "time" alone is not an appropriate measure in the case of analog computation, because time can be scaled. Thus, in contrast to discrete computation, where time is quantized into time steps, the "pure" value of time is meaningless in analog computation. As a possible way of getting around this difficulty, we suggest a study of the trade-off between the accuracy of representing the data, and time needed to converge to a solution of the problem. Thus, suppose we wish to make the solution trajectory of the neural network (3) converge within time a. To what accuracy should the coefficients (for example) of the multilinear polynomial $E(\mathbf{x})$ be represented in the differential equation (3) in order to achieve this? Suppose it is necessary to represent each coefficient using at least k bits, where the largest coefficient is normalized to be no larger than unity in magnitude. Then k is actually a function of a. What happens to $k(a)$ as $a \to 0$? At what rate does the required accuracy of representation $k(a)$ approach infinity? Such questions are very easy to pose, but are very difficult to answer in general. However, in the present case, because of the particularly simple structure of the neural network, there is some hope of being able to answer such questions.

The theory of analog complexity proposed here differs substantially from earlier theories of analog computation, in which there is no notion of *dynamics* – only static computation with real-valued functions (see e.g. [1]).

5. Conclusions. In this paper, a method for minimizing an arbitrary multilinear polynomial over a discrete set of the form $\{0,1\}^n$. The results presented here generalize those of [10], in which the neural network evolves over the *open* hypercube $(0,1)^n$, and the mathematical arguments are much more technically complex.

References

[1] L. Blum, M. Shub and S. Smale, "On a theory of computation and complexity over the real numbers: NP-Completeness, recursive functions and universal machines," *Bull. Amer. Math. Soc.*, Vol. 21, pp 1–46, 1989.

[2] L. Faybusovich, "Dynamical systems which solve optimization problems with linear constraints," *IMA J. of Math. Control and Information*, Vol. 8, pp. 135–149, 1991.

[3] J. J. Hopfield, "Neural networks and physical systems with emergent collective computational capabilities," *Proc. Nat'l. Acad. Sci. (U.S.A.)*, Vol. 79, pp. 2554–2558, 1982.

[4] J. J. Hopfield, "Neurons with graded response have collective computational capabilities like those of two-state neurons," *Proc. Nat. Acad. Sci. (U.S.A.)*, Vol. 81, pp. 3088–3092, 1984.

[5] J. J. Hopfield and D. W. Tank, " 'Neural' computation of decision optimization problems," *Biological Cybernetics*, Vol. 52, pp. 141–152, 1985.

[6] D. Rumelhart and J. McClelland, *Parallel Distributed Processing*, Volumes I and II, M.I.T. Press, Cambridge, MA, 1986.

[7] A. A. Schäffer and M. Yannakakis, "Simple local search problems that are hard to solve," *SIAM J. Computing*, Vol. 20, pp. 56–87, 1991.

[8] D. W. Tank and J. J. Hopfield, "Simple 'neural' optimization networks: An A/D converter, signal decision circuit, and a linear programming circuit," *IEEE Trans. on Circ. and Sys.*, Vol. CAS-33, pp. 533–541, 1986.

[9] M. Vidyasagar, *Nonlinear Systems Analysis*, 2nd Edition, Prentice-Hall, Englewood Cliffs, NJ, 1993.

[10] M. Vidyasagar, "Minimum-Seeking Properties of Analog Neural Networks with Multilinear Objective Functions," *IEEE Transactions on Automatic Control*, **AC-40**, No. 8, 1359–1375, August 1995.

Multiplier Theory and Operator Square Roots: Application to Robust and Time-Varying Stability

Brian D.O. Anderson*
Research School of Information Sciences and Engineering
Australian National University, ACT 0200, Australia
Soura Dasgupta*
Department of Electrical and Computer Engineering,
The University of Iowa, Iowa City, IA-52242, USA.

> **Abstract.** This paper considers the extension of a number of passive multiplier theory based results, previously known only for linear time invariant scalar systems, to time varying and multivariable settings. The extensions obtained here have important applications to the stability of both adaptive systems and linear systems in general. We demonstrate in this paper that at the heart of the extensions carried out here lies the result that if a stable multivariable and/or linear time varying system is stable under all scalar constant, positive feedback gains, then it has a well defined square root. The existence of this square root is demonstrated through a constructive Newton-Raphson based algorithm. The extensions provided here (dealing with robust stability and introduction of time-varying gains) though different in form from their linear time invariant scalar counterparts, do recover these as a special case.

1. Introduction and Problem Motivation. This paper is concerned with finding time-varying, multivariable generalizations of some multiplier theory results involving Strictly Positive Real (SPR) functions.

The following is a well known result in linear systems theory [1]. Consider an asymptotically stable linear time invariant (LTI), single input single output (SISO) system with a strictly proper transfer function $H(s)$. Then the system in Fig. 1 is asymptotically stable for all

(1.1) $$0 \leq k \leq 1,$$

if, and only if, there exists a SPR scalar operator $Z(s)$, such that

(1.2) $$Z(s)(1 + H(s))$$

is SPR. The concept of a SPR operator is defined as follows.
DEFINITION 1.1. *A real, square matrix transfer function $Z(s)$ is Positive Real (PR) if:*
  1. *$Z(s)$ is analytic in the right half plane; and*
  2. *for all $Re[s] \geq 0$, $Z(s) + Z^H(s) \geq 0$ where the superscript H denotes the Hermitian transpose.*

We say $Z(s)$ is SPR if for some $\alpha > 0$, $Z(s - \alpha)$ is PR.

---
* The author wishes to acknowledge the funding of the activities of the Cooperative Research Centre for Robust and Adaptive Systems by the Australian Commonwealth Government under the Cooperative Research Centres Program.
* He was visiting the Department of Systems Engineering, Australian National University, ACT 0200, Australia, when this work was completed. Supported in part by NSF grants ECS-9211593 and ECS-9350346.

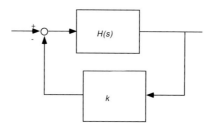

Fig. 1. *A closed-loop configuration*

We note that strictly speaking this result as presented in [1] allows for the presence of simple purely imaginary poles in both $H(s)$ and the closed loop above. In such a case all appearances of SPR operators in the statement of the result of [1], must be replaced by PR operators. For the purposes of this paper, however, only the case concerning asymptotic stability is relevant, and imaginary axis poles are precluded.

From this result spring a number of other important results of which two are cited below: The first states that two scalar polynomials of equal degree $p_1(s)$ and $p_2(s)$, have the property that $p_1(s) + kp_2(s)$ is Hurwitz (i.e. has roots in the open left half plane) for all k as in (1.1) iff there exists an asymptotically stable minimum phase $G(s)$, such that $G(s)(p_1(s) + kp_2(s))$ is Strictly Positive Real (SPR) for all k as in (1.1); in turn, this holds iff there exists an asymptotically stable minimum phase $G(s)$ such that $G(s)p_1(s)$ and $G(s)(p_1(s)+p_2(s))$ are SPR. As will be evident in a later section of this paper, this has an important application in certain adaptive systems problems involving a single unknown parameter. An alternative way of viewing this result is that every convex combination of two monic polynomials with the same degree is Hurwitz iff there exists a *single* stable minimum phase operator whose product with every such convex combination is SPR (see [2] which in fact considers the more general case of convex combinations of more than two polynomials).

The second result concerns the stability of a class of linear time varying (LTV) systems. Specifically, suppose that the configuration in Fig. 1 is stable with *a degree of stability* α for all k as in (1.1). Now consider the LTV systems obtained in Fig. 1, when the feedback gain $k(t)$ is allowed to be time varying while obeying

$$(1.3) \qquad 0 < k(t) < 1.$$

Then it has been shown in [3, 4] that the closed loop retains stability whenever, there exist T and $\delta \in (0, \alpha)$ for which

$$(1.4) \qquad \sup_{t \geq 0} \frac{1}{T} \int_t^{t+T} \left[\frac{d}{d\tau} \ln \frac{k(\tau)}{1 - k(\tau)} \right]^+ d\tau < 2(\alpha - \delta)$$

where

$$[a]^+ = \begin{cases} a; & a \geq 0 \\ 0; & a < 0 \end{cases}$$

See [5] for an association between the result of [1] and that of [3, 4], using tools that include the Popov-Kalman-Yakubovic (PKY) Lemma.

The question addressed in this paper is: *to what extent do these results extend to systems that are MIMO LTI or for that matter LTV*. The ability to answer this question depends critically on the existence of the square root of certain MIMO and/or LTV systems. This can be understood by noting that the result of [1] can itself be viewed in the following terms. The stability of the closed loop of Fig. 1 for all $k \in [0, 1]$ is equivalent to the transfer function $1 + H(s)$ having a phase function that lies in $(-\pi, \pi)$, [6]. Accordingly, there is a well defined square root of $1 + H(s)$. Then a $Z(s)$ chosen as a suitable approximant of the inverse of this square root will be SPR, as indeed will be the product in (1.2).

Having dispensed with some preliminaries in Section 2, the first question we ask concerns square, MIMO and/or LTV, strictly causal continuous operators H such that both H and $[I + kH]^{-1}$ are stable for all k as in (1.1). Observe, that this corresponds to a stable closed loop of the form of Fig. 1, with the MIMO or LTV operator H occupying the position of $H(s)$. Then, using a Newton-Raphson technique, we explain in Section 3, that in such a case $I + H$ does indeed have a square root. Of course H is presumed to be the operator relating inputs and outputs of a strictly causal system. Stability corresponds to the boundedness of the operator given suitable input and output norms.

Sections 4 and 5 respectively provide the analogs of the result of [1] and its first consequence mentioned in the foregoing. Both these results assume that H is finite dimensional, i.e. has a finite dimensional state variable description. Section 5 also discusses the application of this latter result to certain types of adaptive identification algorithms involving MIMO, LTI systems. The results of this Section also resolve an open problem presented in [7]. Section 6 states the analog of the [3, 4] result. Each of the results in Sections 4 through 6, though different from their SISO, LTI counterparts, capture these as special cases. Section 7 contains conclusions.

2. Preliminaries. In this Section we make precise the general framework of this paper by presenting some definitions and assumptions. All systems in this paper will be represented by square, possibly MIMO or LTV, real, continuous operators mapping L_2 to L_2. Consider such an operator G. Then possibly MIMO or G^a will denote the *Adjoint* of G, i.e. if G has impulse response $g(t, \tau)$ then G^a has the impulse response $g'(\tau, t)$. For an input signal $x(t)$, Gx will denote the corresponding output, i.e. if $g(t, \tau)$ is the impulse response of G then

$$(2.1) \qquad [Gx](t) = \int_{-\infty}^{\infty} g(t, \tau) x(\tau) d\tau.$$

This operator is *causal* if $g(t,\tau) = 0 \ \forall t < \tau$. In this case the upper limit in the integral of (2.1) can be replaced by t. The *inner product* between two signals $x(t)$ and $y(t)$ will be

$$(2.2) \qquad \langle x, y \rangle = \int_{-\infty}^{\infty} x'(t) y(t) dt$$

and the *norm* of a signal $x(t)$ will denote the L_2-norm

$$(2.3) \qquad \|x\| = \sqrt{\langle x, x \rangle}.$$

The norm of G will be the induced L_2 operator norm

$$(2.4) \qquad \|G\| = \sup_{\|x\|=1} Gx.$$

In the sequel we will use the terms bounded and stable interchangeably to signify operators that have a finite norm. Moreover, the operator G^n for a positive integer n will designate the combined operator obtained by a cascade of n operators G. A bounded operator $R : L_2 \to L_2$ will be called the inverse of G if $GR = RG = I$. In such a case we denote $R = G^{-1}$ and note that the existence of G^{-1} automatically signifies its stability. Further, G will be called *symmetric* or *self adjoint* if

$$(2.5) \qquad G^a = G.$$

Every symmetric operator G can in turn be expressed as:

$$(2.6) \qquad G = G_c + G_{ac}$$

where G_c is causal and called the *causal part* of G; G_{ac} is anticausal and called the *anticausal part* of G; and together they obey

$$(2.7) \qquad G_c^a = G_{ac}.$$

If a term such as αI appears in G then it will be shared equally between G_c and G_{ac}; i.e. each of G_c and G_{ac} will get $0.5\alpha I$.

While the results of Section 3 rely only on an operator based description of the underlying systems, those of the subsequent Sections do require that some of these operators have a State Variable Description. In either case certain stability and existence assumptions are made. The first is on the operator based description of the open loop system H mentioned in the Introduction.

ASSUMPTION 2.1. *The operator $H : L_2 \to L_2$ is causal and $[I + kH]^{-1} : L_2 \to L_2$ is invertible and causal for all $k \in [0, 1]$. Further, there exist numbers M_1 and M_2 such that:*

$$(2.8) \qquad \|[I + kH]^{-1}\| \leq M_1 \quad \forall k \in [0, 1];$$

and

(2.9) $$\|H\| \leq M_2.$$

Moreover, the impulse response $h(t,\tau)$ of H is finite for all finite t and τ.

REMARK 2.2. *The boundedness assumption on $h(t,\tau)$ precludes the presence of impulse functions in $h(t,\tau)$, much like the strict properness assumption in the LTI case. In actual fact all the results derived here remain valid, if one permits the presence of terms like $\delta(t-\tau)$ in $h(t,\tau)$, as long as $[I+kH]^{-1}$ exists and is causal $\forall k \in [0,1]$. This boundedness assumption is made as it considerably simplifies the notation in Section 6.*

REMARK 2.3. *Under these conditions both H and the feedback loop of Fig. 1 are Bounded Input, Bounded Output (BIBO) stable.*

To provide the assumption on the state variable realization (SVR) of H we introduce the following notation. For a given continuous square matrix function $A(t)$ we designate

(2.10) $$A_\alpha(t) = \alpha I + A(t).$$

We will be concerned with the notion of degree of stability of operators such as H. To this end we introduce H_α having SVR:

(2.11) $$\{A_\alpha(t), B(t), C(t)\},$$

where each of $A_\alpha(t), B(t), C(t)$ is a bounded, continuous function of time. We also make the following definition.

DEFINITION 2.4. *The matrix $A(t)$ is exponentially asymptotically stable with degree of stability $\alpha > 0$ (α-eas) if for the LTV system*

(2.12) $$\dot{x}(t) = A(t)x(t)$$

$\exists c, \gamma > 0$ *such that for all $x(t_0)$ and $t \geq t_0$,*

(2.13) $$\|x(t)\|e^{\alpha(t-t_0)} \leq c\|x(t_0)\|e^{-\gamma(t-t_0)}.$$

If $\alpha = 0$, we simply say that $A(t)$ is eas. Further, we will call a system with SVR, $\{A(t), B(t), C(t), D(t)\}$, all matrices bounded and continuous, α-eas (resp. eas) if $A(t)$ is α-eas (resp. eas).

REMARK 2.5. *In (2.10) $A_\alpha(t)$ is eas iff $A(t)$ is α-eas.*

Then we have the following assumption.

ASSUMPTION 2.6. *The system H_α has an SVR*

$$\{A_\alpha(t), B(t), C(t)\},$$

such that $[A_\alpha(t), B(t)]$ is uniformly completely controllable (u.c.c.), [8], $[A_\alpha(t), C(t)]$ is uniformly completely observable (u.c.o.), [8], and both the systems H_α and

$$\{A_\alpha(t) - kB(t)C'(t), B(t), -kC(t), I\}$$

are eas, for all $k \in [0,1]$.

REMARK 2.7. *It should be noted that any H_α that satisfies Assumption 2.6 also satisfies Assumption 2.1 as $\{A_\alpha(t) - kB(t)C'(t), B(t), -kC(t), I\}$ is precisely the SVR of $[I + kH_\alpha]^{-1}$. Further under the given uco and ucc conditions, BIBO stability, as exemplified by (2.8,2.9), is equivalent to eas, [8].*

REMARK 2.8. *Observe that a system having an SVR is necessarily causal. Since in this paper eas as a property has been defined in terms of SVR's, in any statement to the effect that a given system is eas will implicitly be indicating the causality of that system.*

REMARK 2.9. *Part of the utility of working with H_α stems from the easily proved facts that $\|H_\alpha\| > M_2$ implies $\|H\| < M_2$, and $\|[I + kH_\alpha]^{-1}\| < M_1$, implies $\|[I+kH]^{-1}\| < M_1$.* As is clear from the Introduction, the motivating result for this paper, namely that of [1], concerns SPR operators. As such, SPR as a concept does not apply to LTV systems. Instead the more appropriate concept is that of Strict Passivity [4]. Note that for LTI systems SPR and Strict Passivity are equivalent properties. In operator theoretic terms, Strict Passivity in turn is equivalent to the concept of a positive operator defined in Definition 2.10 below. Thus it is this concept of positivity that will replace the SPR property that underlies the LTI, SISO results.

DEFINITION 2.10. *An operator $P : L_2 \to L_2$ is called Strictly Positive ($P \geq \epsilon I > 0$) if for all x in L_2*

$$(2.14) \qquad \langle x, Px \rangle \geq \epsilon \langle x, x \rangle.$$

In Section 4 we will need the concept of *Spectrum* of a LTV operator.

DEFINITION 2.11. *The resolvent set $\rho(H)$ of an operator $H : L_2 \to L_2$ is the set of all complex numbers λ such that $[\lambda I - H]^{-1} : L_2 \to L_2$ exists. The complement of all $\rho(H)$ in the complex plane is called the spectrum of H and is denoted $\sigma(H)$.*

REMARK 2.12. *The set $\sigma(H)$ is a bounded closed set, [9]. Further, [10] its elements vary continuously with continuous variations in H.*

3. Existence of the Square Root. The principal contribution of this Section is:
  1. to argue that subject to Assumption 2.1, I+H has a square root and
  2. to give an algorithm for constructing this square root.

In the sequel, we say that $G : L_2 \to L_2$ is the square root of $I + F$ with $F : L_2 \to L_2$ if

$$(3.1) \qquad G^2 = I + F.$$

To compute the square root we propose a Newton-Raphson algorithm obtained as follows. Suppose the current estimate of the square root, should of course the square root exist, is G_i and that the true square root is $G_i + \Delta G$. Then

$$[G_i + \Delta G]^2 = I + F,$$

whence neglecting the second order term, and assuming that G_i is invertible and that G_i and ΔG commute, one obtains

$$\Delta G = \frac{1}{2}[(I + F)G_i^{-1} - G_i]$$

or

(3.2) $$G_{i+1} = \frac{1}{2}[(I+F)G_i^{-1} + G_i].$$

It is possible to establish the following result.

THEOREM 3.1. *Let $F : L_2 \to L_2$ be bounded and causal, $[I + kF]^{-1}$ exist and be causal for all $k \in [0,1]$ and let G_i, $i = 0, 1, \cdots$, be the sequence of operators defined by (3.2). Then, there exists an $\epsilon > 0$ such that whenever $\|F\| < \epsilon$ holds, so do the following for all $k \in [0,1]$: (i) there exists bounded $G(kF) = \lim_{i \to \infty} G_i(kF) : L_2 \to L_2$; (ii) $G(kF)^{-1} : L_2 \to L_2$ exists and both $G(kF)$ and $G(kF)^{-1}$ are causal; (iii) $G(kF)$ and $G(kF)^{-1}$ commute with any operator that commutes with F; (iv) $G(0) = I$; (v) $G(kF)$ varies continuously with k; and (vi) $G^2(kF) = I + kF$.*

This above theorem has a serious restriction, viz that $\|F\| < \epsilon$. The operator H may not have this property. It is possible to adopt a nested strategy for determining the square root. Let $NR(I+F, I)$ describe the quantity obtained via (3.1) and (3.2), with $G_0 = I$. With M_1 and M_2 as defined in Assumption 2.1, we choose δ_1, δ_2 such that

(3.3) $\quad \delta_1 M_2 \leq \min\{\epsilon, 1\} \quad \delta_2 M_1 M_2 \leq \min\{\epsilon, 1\} \quad N = \dfrac{1 - \delta_1}{\delta_2}$ is integer

Then a finite number of square root operators are defined by

(3.4a) $\quad U_o = NR(I + \delta_1 H, I)$
(3.4b) $\quad V_m = NR(I + \delta_2 U_{m-1}^{-1} H U_{m-1}^{-1}, I)$
(3.4c) $\quad U_m = U_{m-1} V_m$

and there results

(3.5) $$U_N^2 = I + H$$

In this procedure, $\|\delta_2 U_{m-1}^{-1} H U_{m-1}^{-1}\| \leq \epsilon$ holds at each step.

A number of further points should be made:
- each U_i commutes with H
- an arbitrarily accurate approximation to U_N can be obtained when the Newton-Raphson iteration sequences are truncated; such approximations are rational in H
- there exists a square root $X(kH)$ of $I + kH$ for $k \in [0,1]$ which varies continuously with k, $X(H) = U_N$
- if H has a state variable realization with degree of stability α, and $\widehat{X}$ is an approximation to $X(H)$ obtained by truncation of the Newton-Raphson sequence, then $\widehat{X}$ has degree of stability α.

4. Existence of Passive Multipliers.

Having demonstrated the existence of the square root of $I + H$, we now generalize the result of [1] and its first implication discussed in the Introduction. Instead of focussing on PR type properties, we will consider SPR type (or strict passivity) properties. This is simply a matter of minor technicality in an attempt to avoid having to deal with singular situations.

In the spirit of [1] the principal result we derive takes the following form: Under Assumption 2.6 there exist operators, $X_{1\alpha}$ and $X_{2\alpha}$, both eas and having eas inverses, for which:

1. $X_{1\alpha}X_{2\alpha}$ is Strictly Positive.
2. $X_{1\alpha}[I + H_\alpha]X_{2\alpha}$ is Strictly Positive.

Observe that, since in the LTI, SISO case all operators in question are mutually commutative, this directly reduces to one direction of the Brockett and Willems result. The other direction will be discussed later. In keeping with the requirements of the next Section, in our discussion here, we will pay special attention to degree of stability considerations.

In view of the results of Section 3, the starting point of our development here will be that there exists an X_α which is causal stable and has a causal stable inverse such that

$$(4.1) \qquad X_\alpha^2 = I + H_\alpha.$$

Hence,

$$(4.2) \qquad X_\alpha = [I + H_\alpha]X_\alpha^{-1} = X_\alpha^{-1}[I + H_\alpha].$$

Because $[I + kH_\alpha]^{-1}$ exists for all $k \in [0, 1]$, it turns out that the spectrum of $I + H_\alpha$ avoids the negative real axis, and the spectrum of X_α is then necessarily confined to $Re(\lambda) > \epsilon$ for some $\epsilon > 0$. From this follows the existence of a self adjoint operator P guaranteeing that $X_\alpha^a P + P X_\alpha > 0$. Moreover P is positive definite and has a factorization $W^a W$ for a causal and causally invertible W.

We then have the following Theorem that captures one direction of the results of [1]

THEOREM 4.1. *Under Assumption 2.6, there exist finite dimensional eas and eas invertible operators X_α^{-1} and W, such that*

$$(4.3) \qquad [W(I + H_\alpha)X_\alpha^{-1}W^{-1}]^a + W(I + H_\alpha)X_\alpha^{-1}W^{-1} > 0.$$

and

$$(4.4) \qquad [WX_\alpha^{-1}W^{-1}]^a + [WX_\alpha^{-1}W^{-1}] > 0.$$

In the Theorem statement, X_α and W can be taken to be finite dimensional approximations of the operators X_α, W described in the preamble to the theorem.

REMARK 4.2. *As an alternative to the interpretation provided at the beginning of this Section, this theorem also says that there is a causal operator* $WX_\alpha^{-1}W^{-1}$ *that is strictly positive (i.e. (4.4) holds) and such that the product of this operator with* $W(I+H_\alpha)W^{-1}$ *is also positive (i.e. 4.3) holds). Notice that Assumption 2.1 holds with H replaced by* WHW^{-1}. *So the difference with the original time-invariant scalar result lies in the introduction of W. Because in the time-invariant scalar case the various operators commute, W drops out of the picture. This difference reappears in the next section when we generalize the [3] result.*

Before discussing the second direction of the [1] result we turn now to the following Corollary.

COROLLARY 4.3. *Under Assumption 2.6, there exist finite dimensional eas and eas invertible operators* X_α^{-1} *and W, such that for all* $k \in [0,1]$

$$[W(I+kH_\alpha)X_\alpha^{-1}W^{-1}]^a + W(I+kH_\alpha)X_\alpha^{-1}W^{-1} > 0.$$

Proof. Follows from the fact that the above equation holds for $k=0$ and $k=1$ and the fact that positivity is a convex property. □

Observe that eas Strictly Positive operators have an inverse that is eas. Thus as long as H_α is eas and one can find eas and eas invertible operators W, X_α such that (4.3) and (4.4) hold, then the operator $(I+H_\alpha)^{-1}$ must be eas for all $k \in [0,1]$. Thus the analog of the reverse direction of the [1] result also holds.

5. Solution to a Problem Posed in [7].

Motivated by adaptive systems problems, [7] had posed the following question: Suppose the following set of square Matrix Polynomials:

(5.1) $$\{A_1(s) + kA_2(s) | k \in [0,1]\}$$

has all its members Hurwitz (i.e. the determinant is Hurwitz). Does there exist a single LTI operator $Z(s)$ such that all members of the set

$$\{[A_1(s) + kA_2(s)]Z(s) | k \in [0,1]\}$$

are SPR? The next Theorem shows that such construction of SPR products is possible provided one allows multiplication from both sides.

THEOREM 5.1. *There exist, square, stable minimum phase matrix transfer functions* $Z_1(s)$ *and* $Z_2(s)$ *with the former strictly proper and the latter biproper, such that with* $A_1(s)$ *and* $A_2(s)$ *two square matrix polynomials, and* $A_1^{-1}A_2$ *strictly proper, all members of the set*

$$\{Z_1(s)[A_1(s) + kA_2(s)]Z_2(s) | k \in [0,1]\}$$

are biproper and SPR, iff all members of the set (5.1) are Hurwitz.

Proof. For necessity, note (i) that the inverse of an SPR matrix is SPR, and thus stable and (ii) the stability of Z_i prevents unstable zeros of $A_1 + kA_2$ from being cancelled in forming the product $Z_1[A_1 + kA_2]Z_2$. For sufficiency, note that the operator corresponding to the biproper transfer function

$$I + k[A_1(s)]^{-1}A_2(s) \tag{5.2}$$

satisfies Assumption 2.6. Thus, from Corollary 4.3, there exist biproper (as can be seen from their SVR), operators $W(s)$, and $X(s)$, each eas and having eas inverse, such that:

$$W(s)[I + k[A_1(s)]^{-1}A_2(s)]X^{-1}(s)W^{-1}(s)$$

is SPR for all $k \in [0,1]$. Then, choosing $Z_1(s) = W(s)[A_1(s)]^{-1}$ and $Z_2(s) = X(s)W^{-1}(s)$, the result follows. □

The main application of this result is in output error adaptive identification [11]. Consider the identification of the proper MIMO plant:

$$[A_1(s) + kA_2(s)]Y(s) = [B_1(s) + kB_2(s)]U(s) \tag{5.3}$$

with k a scalar unknown parameter and $u(t)$ and $y(t)$, the input and output of the plant. To identify the plant generally, one performs state variable filtering to avoid explicit differentiation of the various signals. This requires rewriting of the model as

$$Z_1(s)[A_1(s) + kA_2(s)]Y(s) = Z_1(s)[B_1(s) + kB_2(s)]U(s) \tag{5.4}$$

such that $Z_1(s)[A_1(s) + kA_2(s)]$ is biproper. Then, for exponential convergence of the underlying identification algorithm, one requires that $Z_1(s)[A_1(s) + kA_2(s)]$ be SPR. This can be seen readily from the result of [12] which treats the SISO case. As k is unknown the underlying SPR condition is difficult to ensure. However, suppose *a priori* bounds are available for k. In fact without sacrificing generality, assume that $k \in [0,1]$. Then as long as $[A_1(s) + kA_2(s)]$ is Hurwitz for all $k \in [0,1]$, one can choose square, stable, minimum phase matrix transfer functions $Z_1(s)$ and $Z_2(s)$ such that the requirements of Theorem 5.1 are satisfied. Then, noting that $Z_2(s)$ is biproper, one can rexpress the plant as

$$Z_1(s)[A_1(s) + kA_2(s)]Z_2(s)\bar{Y}(s) = Z_1(s)[B_1(s) + kB_2(s)]U(s)$$

where

$$\bar{Y}(s) = Z_2^{-1}(s)Y(s)$$

acts as the converted output. Observe it can be constructed from $Y(s)$ without any explicit differentiation. Further, as $Z_1(s)[A_1(s) + kA_2(s)]Z_2(s)$ is SPR, the output error identification algorithm for this redefined system will be exponentially convergent.

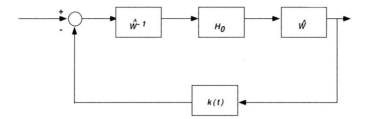

FIG. 2. *Closed-loop under time varying feedback.*

6. Generalization of the Freedman Zames Result. In this Section we consider the second consequence of the result of [1] namely that of [3]. To this end the principal result to be derived is as follows:

THEOREM 6.1. *Suppose Assumption 2.6 holds. Then there exists an eas and eas invertible operator $\hat{W}$ (independent of k) such that the operator $[I+k(t)\hat{W}H_0\hat{W}^{-1}]^{-1}$ is eas provided (1.3) and (1.4) hold.*

Essentially, this result states that provided the closed loop of Fig. 1 (with $H(s)$ replaced by H_0) is α-eas for all time invariant feedback gains in the open interval [0,1], then under a logarithmic time variation bound as in [3], by suitable pre and post filtering of H_0, the closed loop in Fig. 2 is also stable. Observe, if $H(s)$ is scalar LTI, then the underlying commutativity recovers the result of [3]. Moreover, the fact that the pre and post filters $\hat{W}^{-1}$ and $\hat{W}$ are independent of the particular trajectory that the time-varying feedback gain follows simplifies their selection.

7. Conclusion. In this paper we have a generalized a number of results concerning passive multiplier theory to MIMO LTV systems. Between them these results cover important issues spanning a number of areas. Though the focus here is on continuous time systems extension to discrete time systems is straightforward. To what extent generalizations to nonlinear time varying systems problems are possible remains however an open problem. It would also be interesting to establish a result on the stability of the determinant of convex combinations of more than two matrix polynominals.

References

[1] R. W. Brockett and J. L. Willems, "Frequency domain stability criteria–part I," *IEEE Trans. Auto. Contr.*, vol. 10, pp.255-271, 1965.
[2] B.D.O. Anderson, S. Dasgupta, P.P. Khargonekar, F.J. Kraus and M. Mansour, "Robust strict positive realness: characterization and construction", *IEEE Transac. on Circuits and Systems,* vol. 37, pp. 869-876, 1990.
[3] M. Freedman and G. Zames,"Logarithmic variation criteria for the stability of systems with time varying gains", *J. SIAM Control,* vol. 6(3), pp. 487-507, 1968.
[4] C. A. Desoer and M. Vidyasagar, *Feedback Systems: Input-Output Properties* , Academic Press, 1975.

[5] S. Dasgupta, G. Chockalingam, B.D.O. Anderson and M. Fu,"Parameterized Lyapunov functions with applications to the stability of linear time varying systems", *IEEE Transactions on Circuits and Systems I: Fundamental Theory*, vol. 41, pp 93-106, February 1994.

[6] S. Dasgupta, P.J. Parker, B.D.O. Anderson, F.J. Kraus, and M. Mansour, "Frequency domain conditions for the robust stability of linear and nonlinear dynamic systems," *IEEE Transactions on Circuits and Systems*, vol. 38, pp 389-397, 1991.

[7] S. Dasgupta, "Strictly positive realness of matrix products", Contribution to "Open Problems", in *Robustness of Dynamic Systems with Parameter Uncertainties,*, M. Mansour, W.Truol and S. Balemi, Ed.s, p 307, Birkhauser, 1992.

[8] B.D.O. Anderson and J.B. Moore, "New results in linear systems stability", *SIAM Journal of Control*, vol. 7, pp 398-414, 1969.

[9] I. Gohberg, S. Goldberg and M.A. Kaashoek, *Classes of Linear Operators Vol. 1*, Birkhauser Verlag, 1990.

[10] F. Riesz and B. Sz.-Nagy, *Functional Analysis*, Frederic Ungar Publishing Company, New York, 1955.

[11] B.D.O. Anderson, R.R. Bitmead, C.R. Johnson, Jr., P.V. Kokotovic, R.L. Kosut, I.M.Y. Mareels, L. Praly and B. Riedle, *Stability of Adaptive systems,* MIT press, 1986.

[12] S. Dasgupta, B.D.O. Anderson and R.J. Kaye, "Output-error identification methods for partially known systems", *International Journal of Control*, vol. 43, pp 177-191, Jan. 1986.

Adaptive control of non-minimum phase systems subject to unknown bounded disturbances

Rogelio Lozano and Dionisio A. Suárez
Université de Technologie de Compiègne, Compiègne, France

1. Introduction. Robustness is currently an active area of research in adaptive control. In the so-called ideal case, global stability of adaptive control systems can be established under certain ideal conditions, i.e. in noise-free processes without unmodeled dynamics, in spite of parametric uncertainty [2, 4, 11, 13]. However, it has been shown that these algorithms may become unstable in non ideal conditions [16].

Several modifications to the standard identification algorithms have been proposed to retain stability in spite of bounded disturbance. Among them we find σ-modifications, constrained estimation and dead-zones [14].

The introduction of a dead-zone in an adaptive controller is standard practice in which the algorithm is turned off if a suitable defined error is smaller than the disturbance bound. This idea was introduced in [2] and subsequently developed in [1, 6, 15, 17]. An upper bound for the disturbance must be explicitly known to implement these algorithms.

In recent work [3], it has been shown that the above knowledge is not necessary to adaptively control minimum phase discrete-time systems. In this approach the disturbance upper bound is estimated and then the dead-zone used in [1, 6] is employed with a projection algorithm. It is worth pointing out that a lower bound on the high frequency gain must be known.

In this work we introduce a robust adaptive control law for possibly non-minimum phase continuous-time systems. The only required *a priori* knowledge is the process order. A standard least squares algorithm with a dead-zone is used. The disturbance upper bound is estimated as in [3], however, the dead-zone definition has been slightly modified. We exploit the least squares properties to define a parameter modification which allows us to cope with the singularities appearing in the control law.

2. Problem statement. Let us consider the following Linear Time Invariant (LTI) plant with unknown parameters

(2.1) $$A(D)^* y = B(D)^* u + v^*$$

where

(2.2) $$A(D)^* = D^n + a_1^* D^{n-1} + ... + a_n^*$$

(2.3) $$B(D)^* = b_0^* D^{n-1} + b_1^* D^{n-2} + ... + b_{n-1}^*$$

with $D = \frac{d}{dt}$ and v^* a bounded disturbance. We make the following assumptions on the system (2.1)

A.1 The plant order n is known.

A.2 The disturbance v^* satisfies $|v^*| \leq v_0^*$ where v_0^* is an unknown constant.

Under the above assumptions the aim of this paper is to show that the proposed adaptive control law ensures boundedness of all the closed loop system signals and the closed loop poles are asymptotically assigned to arbitrary values.

3. System parameter identification. As has been shown in [8], the introduction of a stable filter applied to the plant input and output produce a filtered system having the same structure that the plant except for an exponentially decaying term ϵ, so as in [8] we have

$$A^* y_f = B^* u_f + \epsilon + v_f^* \tag{3.1}$$

where v_f^* is bounded and all the time derivatives of u_f and y_f in (3.1) are measurable signals which is not the case in (2.1). We can also rewrite (3.1) as

$$y_f^{(n)} = \phi^T \theta^* + \epsilon + v_f^* \tag{3.2}$$

where we have defined the regressor vector

$$\phi = [u_f^{(n-1)}, \ldots, u_f, -y_f^{(n-1)}, \ldots, -y_f]^T \tag{3.3}$$

and the parameter vector

$$\theta^* = [b_0^*, \ldots, b_{n-1}^*, a_1^*, \ldots, a_n^*]^T \tag{3.4}$$

where $u_f^{(i)}$ and $y_f^{(i)}$ are time derivatives of u_f and y_f. Then, θ^* can be estimated by using the following Least Squares (LS) algorithm with dead-zone

$$e = y_f^{(n)} - \phi^T \theta \tag{3.5}$$

$$\dot{\theta} = \lambda P \phi e \tag{3.6}$$

$$\dot{P} = -\lambda P \phi \phi^T P \qquad P(0) > 0 \tag{3.7}$$

$$\dot{\mu} = \lambda \qquad \mu(0) \geq 0 \tag{3.8}$$

where e is the prediction error, μ is the estimate of $\mu^* \triangleq v_{f_0}^{*2}$, i.e., the square of the disturbance upper bound and θ is the estimate of θ^*

$$\theta = [b_0, \ldots, b_{n-1}, a_1, \ldots, a_n]^T \tag{3.9}$$

The parametric errors $\tilde{\theta}$ and $\tilde{\mu}$ are defined as

$$\tilde{\theta} = \theta - \theta^* \tag{3.10}$$

and

(3.11) $$\tilde{\mu} = \mu - \mu^*.$$

Let us now define the following scheduling variable e_λ for the dead-zone

(3.12) $$e_\lambda = (e^2 + \phi^T P^2 \phi)^{1/2}$$

then, λ is defined by the following relative dead-zone with hysteresis width γ_2.
- λ switches from $\lambda = 1$ to 0 when $e_\lambda^2 - 2\mu$ decreases and reaches the value $\gamma_1 e_\lambda^2$
- λ switches from $\lambda = 0$ to 1 when $e_\lambda^2 - 2\mu$ increases and reaches the value $\gamma_1 e_\lambda^2 + \gamma_2$

The following lemma gives the main properties of the LS algorithm with dead-zone.

Lemma 1: Along the solutions of (2.1) and (3.5)-(3.8) the following results hold
1. $\tilde{\theta}$, P and μ are bounded
2. $\lambda e_\lambda \in \mathcal{L}_2$ and $e_\lambda^2 \leq \dfrac{2\mu + \gamma_2}{1 - \gamma_1}$ for $\lambda = 0$
3. $0 < P \leq P(0)$
4. $\tilde{\theta}$, P and μ converge, i.e. $\int_0^t \lambda\, dt$ converges.

Proof: See Appendix A in [8].

4. Adaptive control scheme. In this section we present an adaptive pole placement algorithm that is free from singularities in the sense that the control law is always well defined; i.e., the parameters used to synthesize the control law always form a controllable model. In order to obtain this sort of model we will use the particular parameter estimates modification procedure proposed in [9].
Let us define the following parameter modification

(4.1) $$\bar{\theta} = \theta + P\beta$$

where P is the covariance matrix inverse, $\bar{\theta}$ the modified parameter vector

(4.2) $$\bar{\theta} = [\bar{b}_0,\ \ldots,\ \bar{b}_{n-1},\ \bar{a}_1,\ \ldots,\ \bar{a}_n]^T$$

and β will be defined later.
Introducing (4.1) into (3.5), yields

(4.3) $$e = y_f^{(n)} - \bar{\theta}^T \phi + \beta^T P \phi$$

which can also be rewritten as

(4.4) $$\bar{A}(D)y_f = \bar{B}(D)u_f + w$$

where

(4.5) $$\bar{A}(D) = D^n + \bar{a}_1 D^{n-1} + \ldots + \bar{a}_n$$

(4.6) $$\bar{B}(D) = \bar{b}_0 D^{n-1} + \bar{b}_1 D^{n-2} + \ldots + \bar{b}_{n-1}$$

(4.7) $$w = e - \beta^T P\phi.$$

Equation (4.4) can be viewed as a modified estimated filtered system having u_f and y_f as the input and output respectively with the term w playing the role of a disturbance.

The following adaptive pole placement control scheme can be used to meet the control objectives defined in section 2

(4.8) $$S(D)u_f = -R(D)y_f + r^M$$

where

(4.9) $$S(D) = D^n + s_1 D^{n-1} + \ldots + s_n$$

(4.10) $$R(D) = r_0 D^{n-1} + r_1 D^{n-2} + \ldots + r_n$$

are the solutions of the following Bezout equation

(4.11) $$\bar{A}(\kappa)S(\kappa) + \bar{B}(\kappa)R(\kappa) = C(\kappa)F(\kappa)$$

where we have replaced the operator D by the dummy variable κ to denote that the Bezout equation is an algebraic relation of the coefficients on the involved polynomials where the time varying nature of A and B plays no role.

The $C(D)$ polynomial is a Hurwitz polynomial whose zeros represent the desired closed loop poles

(4.12) $$C(D) = D^n + c_1 D^{n-1} + \ldots + c_n$$

Define the vector containing the coefficients of the product $C(D)F(D)$ as

(4.13) $$p = [1, \ p_1, \ \ldots, \ p_{2n}]^T$$

and the controller parameter vector for $S(D)$ and $R(D)$ as

(4.14) $$q = [1, \ s_1, \ \ldots, \ s_n, \ r_1, \ \ldots, \ r_n]$$

Equating the coefficients of the same power of κ in (4.11), we obtain the following matrix equation

(4.15) $$\bar{A}(\kappa)S(\kappa) + \bar{B}(\kappa)R(\kappa) - C(\kappa)F(\kappa) = [\kappa^{2n}, \ \kappa^{2n-1}, \ \ldots, \ \kappa, \ 1][M(\bar{\theta})q - p]$$

where $M(\bar{\theta})$ is the Sylvester resultant matrix associated with the modified parameter estimates $\bar{\theta}$ (see [8]).

Let us now use the equation (4.8) and the definition of $F(D)$ to obtain the following expression for u

(4.16) $$u = (F(D) - S(D))u_f - R(D)y_f + r^M$$

Note that $y_f^{(n)}$ and $u_f^{(n)}$ are not required in (4.16).

5. Estimates modification procedure.

A careful selection of the vector β in (4.1) should ensure that the two following conditions are satisfied

C.1 β should converge.

C.2 β should be such that the absolute value of the determinant of the Sylvester resultant matrix associated with $\bar{\theta}$, $|\det M(\bar{\theta})|$, is uniformly bounded from below.

In order to satisfy condition C.1, we will introduce a switching function with an hysteresis width $\gamma > 0$ ($\gamma \ll 1$) in the commutations of β. See [7, 10, 12]. On the other hand, $|\det M(\bar{\theta})|$ in condition C.2 is a measure of the coprimeness of the polynomials $\bar{A}(D)$ and $\bar{B}(D)$ associated with modified parameter estimate which implies the controllability of system (4.4).

The parameter estimates modification procedure follows the reasoning in [9]. This is carried out by selecting a value σ within a set **D** containing a finite number of constants. The vector β in (4.1) is a function of the value σ which will switch from a value to another in **D** so that β satisfies condition C.2. β will be chosen to have the following structure

$$(5.1) \qquad \beta = [\sigma(t), \sigma(t)^m, \sigma(t)^{m^2}, \ldots, \sigma(t)^{l'}]^T$$

with

$$(5.2) \qquad m = 2n \qquad \text{and} \qquad l' = m^{m-1}$$

$\sigma(t)$ in (5.1) takes values in the set **D**

$$(5.3) \qquad \mathbf{D} : \{\sigma_1, \sigma_2, \ldots, \sigma_l\}$$

with

$$(5.4) \qquad \sigma_i \in \Re, \quad \sigma_i \geq \sigma_{i-1} + 1, \quad i = 1, \ldots, l$$

and

$$(5.5) \qquad l = m^m.$$

For simplicity we define

$$(5.6) \qquad Z(\alpha) = |\det M(\theta + P\,[\alpha, \alpha^m, \alpha^{m^2}, \ldots, \alpha^{l'}]^T)|$$

where α is a dummy variable. Note that $Z(\alpha)$ above is equal to $|\det M(\bar{\theta})|$ when β is chosen as in (5.1) with $\sigma(t) = \alpha$.

Assume that for some σ_k and $\epsilon > 0$, $\sigma(t) = \sigma_k$ for $t - \epsilon \leq \tau < t$. The hysteresis switching function defining $\sigma(t)$ (and therefore β in (5.1)) is generated as follows

$$(5.7)\ \sigma(t) = \begin{cases} \sigma_k & \text{if } Z(\sigma_j) < (1+\gamma) Z(\sigma_k) \text{ for all } \sigma_j \in \mathbf{D} \\ \sigma_j & \text{if } j \text{ is the smallest integer such that} \\ & Z(\sigma_j) \geq (1+\gamma)\, Z(\sigma_k) \text{ and } Z(\sigma_j) \geq Z(\sigma_i)\ \forall\ \sigma_i \in \mathbf{D}. \end{cases}$$

Since the proposed estimates modification procedure introduces commutations in the closed loop system, we have to discuss the issue of existence of solutions of the differential equations involved in the system before we proceed to convergence analysis. The existence of solutions is proved by using arguments similar to those in [5]. Briefly, solutions obtained by concatenating the solutions defined between the time instants at which commutations occur. On the other hand, since $|\det M(\bar{\theta})| \geq \epsilon > 0$, the controller parameters in $S(D)$ and $R(D)$ are bounded. Therefore the closed loop system is linear and the plant output and input can not grow faster than exponentially. Furthermore, since θ converges (see lemma 1) and in view of the hysteresis defined in (5.7), we conclude that β converges. The hysteresis also prevents the commutations from occurring an infinite number of times or infinitely often. We can then state the! following lemma whose proof follows the ideas in [5].

Lemma 2: The plant input u and output y do not grow faster than exponentially.

6. Convergence and stability analysis. Combining (4.4) and (4.8) we have an adaptive closed-loop system as follows

$$(6.1) \qquad \dot{x} = A_{cl}x + B_1 w + B_2 r^M$$

$$(6.2) \qquad y_f = C_{cl}^T x$$

where A_{cl} is a $(2n) \times (2n)$ matrix whose entries are only the modified parameter estimates and the controller parameters, $x = [y_f^{(n-1)}, \ldots, y_f, u_f^{(n-1)}, \ldots, u_f]^T$, and B_1, B_2 and C_{cl} are constant vectors (see [8]).

Since $\bar{\theta}$ converges, we have

$$(6.3) \qquad A_{cl} \to A_s$$

or

$$(6.4) \qquad A_{cl} = A_s + \Delta$$

with

$$(6.5) \qquad \Delta \to 0$$

therefore (6.1) can be rewritten as

$$(6.6) \qquad \dot{x} = (A_s + \Delta)x + B_1 w + B_2 r^M.$$

From (4.7), defining $\beta_{\max} = \max(1, \|\beta\|)$ and using inequality $(a+b)^2 \leq 2(a^2+b^2)$ and (3.12) we can now show that $w \in \mathcal{L}_\infty \cup \mathcal{L}_2$ and

$$(6.7) \qquad |w| \leq \sqrt{2}\beta_{\max} e_\lambda$$

From λ of Lemma 1, it follows that $e_\lambda \in \mathcal{L}_\infty \cup \mathcal{L}_2$ since for $\lambda = 1$, $e_\lambda = \lambda e_\lambda$ which is $\mathcal{L}_2$, and otherwise, for $\lambda = 0$, e_λ is bounded.

Note that if the coefficients of $\bar{A}(D)$ and $\bar{B}(D)$ in (4.4) were constant, then the control law (4.8) would be such that the closed loop poles would be located exactly at their desired values given by the zeros of $C(D)F(D)$ in (4.11). Therefore the eigenvalues of the matrix A_{cl} in (6.1) would also be equal to the zeros of $C(D)F(D)$. This argument leads us to conclude that, even though A_{cl} is time varying, the eigenvalues of A_{cl} are equal to the zeros of $C(D)F(D)$ at every fixed time. This does not allow us to conclude directly the stability of system (6.6) but insures us that A_s in (6.6) is a stable matrix.

Since $\Delta \to 0$ and $w \in \mathcal{L}_\infty \cup \mathcal{L}_2$ in (6.6), we can expect x to converge to a certain bounded signal depending only on r^M and w. Note that A_s depends only on the final values of $\bar{A}(D)$ and $\bar{B}(D)$ in (4.4). Searching for the function describing x in the limit, we were lead to use the artifice of defining an auxiliary system described as follows

$$(6.8) \qquad A_\infty(D)y^* = B_\infty(D)u^* + F(D)w$$

where A_∞ and B_∞ denote the final estimated parameter values for $\bar{A}(D)$ and $\bar{B}(D)$ respectively, and u^* and y^* are the input and output of the artificial system respectively.

Filtering the input and output of the artificial system (6.8) with a stable filter, as has been done for the plant (2.1), we obtain

$$(6.9) \qquad A_\infty y_f^* = B_\infty u_f^* + w' + w$$

where w' is a exponentially decaying term depending of the filter parameters and starting with unknown initial conditions.

Define the controller

$$(6.10) \qquad S_\infty u_f^* = -R_\infty y_f^* + r^M$$

with

$$(6.11) \qquad A_\infty S_\infty + B_\infty R_\infty = CF$$

where S_∞, R_∞ denote the final controller polynomials for S and R in (4.11) respectively.

We can show that the input and output signals of the artificial system (6.8) satisfy (see [8] for more details)

$$(6.12) \qquad Cy^* = B_\infty r^M + S_\infty(w' + w)$$

$$(6.13) \qquad Cu^* = A_\infty r^M - R_\infty(w' + w).$$

Therefore the closed loop poles are located at the positions of the zeros of C and the control input is bounded.

We can also write the closed loop system (6.9) and (6.10) as

$$\dot{x}^* = A_s x^* + B_1(w' + w) + B_2 r^M \tag{6.14}$$

$$y_f^* = C_{cl}^T x^* \tag{6.15}$$

where

$$x^* = \begin{bmatrix} y_f^{*(n-1)}, & y_f^{*(n-2)}, & \ldots, & y_f^*, u_f^{*(n-1)}, & u_f^{*(n-2)}, & \ldots, & u_f^* \end{bmatrix}^T \tag{6.16}$$

and A_s, B_1, B_2 and C_{cl}^T are given in (6.1)-(6.5).
Define

$$x_e = x - x^* \tag{6.17}$$

from (6.6) and (6.14), we have

$$\dot{x}_e = (A_s + \Delta)x_e + \vartheta_0 \tag{6.18}$$

where

$$\vartheta_0 = \Delta x^* - B_1 w' \tag{6.19}$$

it is clearly that $\vartheta_0 \in \mathcal{L}_\infty$ and $\vartheta_0 \to 0$.

We can now state our final result.

Theorem 1: Consider system (2.1) in the closed loop with the adaptive controller defined in sections 3, 4 and 5, then x_e, y and u are bounded by the noise upper bound v^* and the initial values of $\tilde{\theta}$, P, $y^{(i)}$, $u^{(i)}$, $y_f^{(i)}$, $u_f^{(i)}$. Furthermore x_e, $(y-y^*)$ and $(u - u^*)$ all converge to zero with u^* and y^* as in (6.8), (6.12) and (6.13).

Proof: The proof is similar to the one in appendix A in [9].

7. Conclusion. In this paper we have presented a globally convergent adaptive control of non minimum phase systems subject to unknown bounded disturbances. The singularities that may appear in standard adaptive controllers have been circumvented here by introducing an ad hoc estimates modification based upon the properties of the least squares covariance matrix. The system order is the only a-priori knowledge required on the plant. The disturbance upper bound is estimated which makes unnecessary its explicit knowledge.

References

[1] A.L. Bunich. Rapidly converging algorithms for the identification of a linear system with limited nois. *Automn. Remote Control*, 44:1047–1054, 1983.

[2] B. Egardt. Stability of adaptive controllers. *Lectures notes in control and information sciences*, 20, 1979. Springer Berlin.

[3] G. Feng. A robust discrete-time direct adaptive control algorithm. Technical report, University of New South Wales. Dept. of Systems and Control, 1993.

[4] G.C. Goodwin, P.J. Ramadge, and P.E. Caines. Discrete time multivariable adaptive control. *IEEE Trans. Aut. Control*, 25:449–456, 1980.

[5] R. Lozano and B. Brogliato. Adaptive control of a simple nonlinear system without a priori information on the parameters. *IEEE Trans. Aut. Control*, 37:30–37, 1992.

[6] R. Lozano and R. Ortega. Reformulation of the parameter identification problem for systems with bounded disturbances. *Automatica*, 23:247–251, 1987.

[7] R. Lozano, A. Osorio, and J. Torres. Adaptive stabilization of nonminimal phase 1-th order continuous time systems. In *ACC Chicago*, June 1992.

[8] R. Lozano and D.A. Suárez. Adaptive control of non-minimum phase systems subject to unknown bounded disturbances. Technical report, Université de Technonolie de Compiègne. HEUDIASYC URA CNRS 817, 1995.

[9] R. Lozano and X. Zhao. Adaptive pole placement without excitation probing signals. *IEEE Trans. Aut. Control*, 39(1), 1994.

[10] R.H. Middleton, G.C. Goodwin, D.J. Hill, and D.Q. Mayne. Design issues in adaptive control. *IEEE Trans. Aut. Control*, 33:50–58, Jan 1988.

[11] A.S. Morse. Global stability of parameter adaptive control systems. *IEEE Trans. Aut. Control*, 25:433–439, 1980.

[12] A.S. Morse, D.Q. Mayne, and G.C. Goodwin. Applications of hysteresis switching in parameter adaptive control. Technical report, Yale University, 1991. Report No. 9104.

[13] K.S. Narendra, Y.H. Lin, and L.S. Valavani. Stable adaptive controller design, part II: Proof of stability. *IEEE Trans. Aut. Control*, 25:440–448, 1980.

[14] R. Ortega and Y. Tang. Robustness of adaptive controllers-a suvey. *Automatica*, 25:651–677, 1989.

[15] B.B. Peterson and K.S. Narendra. Bounded error adaptive control. *IEEE Trans. Aut. Control*, 27:1161–1168, 1982.

[16] C.E. Rohrs, M. Athans, and G. Stein. Robustness of continuous-time adaptive control algorithms in the presence of unmodeled dynamics. *IEEE Trans. Aut. Control*, 30:881–889, 1985.

[17] C. Samson. Stability analysis of adaptively controlled system subject to bounded disturbances. *Automatica*, 19:81–86, 1983.

On the Robust Stability of Time-Varying Linear Systems

Mohamed Mansour*
Institut für Automatik, Swiss Federal Institute of Technology
CH-8092 Zürich, Switzerland.

Brian D.O. Anderson*
Research School of Information Sciences and Engineering and
Cooperative Research Centre for Robust and Adaptive Systems
Australian National University, ACT 0200, Australia.

Abstract. This paper considers the robust stability of time-varying linear systems described by a linear differential equation whose coefficients vary inside given intervals and with restricted magnitudes of the rates of change of the coefficients. This problem can be considered as a generalization of the Kharitonov problem, which is in turn a generalization of the Hurwitz problem, and it was formulated as an open problem in [7].
To solve this problem Lyapunov theory is used where a Lyapunov function is obtained (using characteristics of positive real functions [2], [3]) which is multiaffine in the polynomial coefficients. With this Lyapunov function extreme point results are obtained. The structure of the Lyapunov matrix as well as the structure of the conditions for the solution of a robust positive real function problem are characterized. A second approach based on the critical stability conditions is also suggested but the Lyapunov matrix thus obtained is no longer in general multiaffine in the parameters. Examples of low order systems are given. The resulting stability conditions are only sufficient.

1. Introduction. Robust stability of time-invariant systems with parameter uncertainty has received much attention since the seminal paper of Kharitonov [7]. Also, robust stability of time-varying systems was investigated in several publications e.g. [2-6]. In [6] the bounds on the variation of the parameters of discrete time systems were determined which depend on predetermined parameter regions in the parameter space. In [7] the following problem was formulated as an open problem:

Given a time-varying differential equation

(1.1) $$y^{(n)} + a_1 y^{(n-1)} + \cdots + a_n y = 0 \text{ with}$$

(1.2) $$\underline{a}_i \leq a_i(t) \leq \bar{a}_i$$

(1.3) $$-\alpha_i \leq \frac{da_i}{dt} \leq \alpha_i$$

find conditions for exponential asymptotic stability of (1.1) subject to (1.2) and (1.3). A discrete version of this problem with no restriction on the time-variation

---

* The author wishes to acknowledge the support of the Australian National University.
* The author wishes to acknowledge the funding of the activities of the Cooperative Research Centre for Robust and Adaptive Systems by the Australian Commonwealth Government under the Cooperative Research Centres Program.

rate is solved in [4]. This problem is a generalization of the time-invariant problem of Kharitonov, which is in turn a generalization of the Hurwitz problem. In order to derive sufficient conditions for stability of the time-variable system it is natural to use Lyapunov theory and indeed quadratic Lyapunov functions. Equations (1.4) through (1.7) define the state variable representation of (1.1), Lyapunov function and its derivative:

(1.4) $$\dot{\underline{x}}(t) = A(t)\underline{x}(t)$$

(1.5) $$V(\underline{x},t) = \underline{x}^T P(t)\underline{x}(t)$$

(1.6) $$\dot{V}(\underline{x}) = -\underline{x}^T(Q - \dot{P})\underline{x}$$

(1.7) $$A^T P + PA = -Q.$$

It is well known that if $V(\underline{x},t)$ is a positive definite decrescent function with negative definite derivative then the equilibrium is exponentially stable, [8].

Now if P and Q as solutions of the Lyapunov equation (1.7) are multiaffine in the uncertain coefficients a_i then convexity implies the positive definiteness of $Q - \dot{P}$ is determined by the positive definiteness at the extreme values the coefficients (the vertices), which can be checked easily.

The multiaffine Lyapunov function and derivative are obtained using positive real functions and the Kalman Yakubovic-Popov Lemma [9], [10].

In Section 2 some preliminary results are given. In Section 3 we show how to construct the Lyapunov function using the characteristics of positive real functions and how a modified Hurwitz matrix appears in the procedure. Solutions for low order systems are given.

In Section 4 the solution of the Lyapunov equation is given using the determinant of the Hurwitz matrix. In this latter case P and Q are no longer multiaffine in the parameters and therefore no vertex results are obtained except for $n = 2$.

Section 5 gives examples solved by different Lyapunov functions, and Section 6 gives the conclusions.

2. Preliminaries.

2.1. Let $p(s) = s^n + a_1 s^{n-1} + \cdots + a_n$ where $\underline{a}_i \leq a_i \leq \bar{a}_i$ i.e. $p(s) \in \mathcal{P}$ where $\mathcal{P}$ is a box in the coefficient space. If there exists $B(s)$ such that $\frac{p_i(s)}{B(s)}, i = 1, 2, 3, 4$ is positive real (SPR) where $p_i(s)$ are the four Kharitonov polynomials, then $\frac{p(s)}{B(s)}$ is SPR for all $p \in \mathcal{P}$ [11].

2.2. In [9] the following result was proved (Corollary 3.1): Consider a convex polytope $\mathcal{P}$ of $\ell-th$ degree polynomials in s. There exists a non-negative integer M and a polynomial $B(s)$ of degree $\ell + M$ such that $\frac{p(s)(1+s)^M}{B(s)}$ is strict positive real (SPR) for all $p(s) \in \mathcal{P}$ if and only if $\mathcal{P}$ is Hurwitz invariant, i.e. for all $p(s) \in \mathcal{P}, p(s_0) = 0$ implies $Re[s_0] < 0$.

2.3. Given a scalar transfer function $T(s) = 1 + \underline{c}^T(sI - A)^{-1}\underline{b}$ with $[A, \underline{c}]$ completely observable and $[A, \underline{b}]$ completely reachable, $T(s)$ is strictly positive real (i.e. poles are in $Res < 0$ and $ReT(j\omega) > 0$ $\forall \omega$ if and only if there exist symmetric positive definite matrices P and Q and a vector $\underline{q}$ such that

(2.1) $$A^T P + PA + \underline{q}\underline{q}^T = -Q \text{ and}$$

(2.2) $$P\underline{b} = \underline{c} + \sqrt{2}\underline{q}.$$

This is a version of the celebrated Kalman-Yakubovic-Popov Lemma.

2.4. Let

(2.3) $$K = \{\underline{k} = [k_1 \cdots k_m]^T : \underline{k}_i \leq k_i \leq \overline{k}_i\},$$

let $h(k) \in R^n$ be affine in the elements of k, let $g \in R^n$ and $F \in R^{n \times n}$. Define

(2.4) $$\Omega = \{A(k) = F + \underline{g}h'(k) \in R^{n \times n} : k \in K\}.$$

It is shown in [10] that Ω is Hurwitz invariant if and only if there exists a square matrix with stable characteristic polynomial, a compatibly dimensioned vector $\underline{\omega}$ and a Lyapunov pair $P(k), Q(k)$ depending multiaffinely on the elements of $\underline{k}$ which satisfies the Lyapunov equation (2.5) for all $k \in K$:

(2.5) $$\Pi^T(k)P(k) + P(k)\Pi(k) < -Q(k)$$

where

(2.6) $$\Pi(k) = \begin{bmatrix} \triangle & \underline{\omega}\, \underline{h}^T(k) \\ \bigcirc & A(k) \end{bmatrix}.$$

2.5. We prove the following lemma:

LEMMA 2.1. *Let $z(s)$ satisfy*

1. *all poles of $z(s)$ lie in $Re[s] \leq -\sigma < 0$*
2. $Rez(j\omega) \geq \delta > 0$ $\forall \omega$
3. $\lim_{\omega \to \infty} z(j\omega) = 1.$

Then for suitably large a, there holds

(a) *all poles of $\frac{z(s)}{s+a}$ lie in $Re[s] \leq -\sigma < 0$*
(b) $Re\frac{z(j\omega)}{j\omega + a} > 0$ $\forall \omega < \infty.$

Proof. The first condition holds if $a > \sigma$. For the second condition

$$Re\frac{z(j\omega)}{j\omega + a} = \frac{z(j\omega)}{j\omega + a} + \frac{z(-j\omega)}{-j\omega + a} = 2\frac{aRez(j\omega) - \omega Imz(j\omega)}{\omega^2 + a^2}.$$

Because $\lim_{\omega \to \infty} Imz(j\omega) = 0$ and $Imz(j\omega)$ is rational in ω

$$\left|\lim_{\omega \to \infty} \omega \cdot Imz(j\omega)\right| < \infty.$$

Also, $Imz(j\omega)$ is bounded in ω, so that for some K,
$$|\omega Imz(j\omega)| < K \quad \forall \text{ real } \omega.$$
Now choose $a > \frac{K}{\delta}$ to secure $Re\frac{z(j\omega)}{j\omega+a} > 0$.   □

COROLLARY 2.2. *Let $\mathcal{P}$ be a convex family of stable monic polynomials*
(2.7) $$p(s) = s^n + a_1 s^{n-1} + \cdots a_n$$
where $[a_1 a_2 \cdots a_n] \in \ell$, a convex stable set.

Then there exists a transfer function $w(s)$ such that
(i) $w(s)p^{-1}(s)$ is stable $\quad \forall p \in \mathcal{P}$
(ii) $Re w(j\omega)p^{-1}(j\omega) > 0 \quad \forall$ finite real ω
(iii) $\lim_{s\to\infty} w(s)p^{-1}(s) = 0$.

An application of this result to low degree polynomials is given in Section 3.

3. The Affine Lyapunov Function. As mentioned in the introduction, solutions P and Q of the Lyapunov equation (1.7) which are affine or multiaffine in the coefficients of the differential equation (1.1), yields a vertex-type result for stability of the time variable system. The Lyapunov pair can be determined using SPR functions results of Section 2 together with the Kalman-Yakubovic-Popov Lemma (KYP). The following subsection illustrates the approach.

3.1. $n = 2$:
(3.1) $$p(s) = s^2 + a_1 s + a_2, \quad \underline{a}_i \le a_i \le \overline{a}_i .$$
For stability of $p(s)$
(3.2) $$\underline{a}_i > 0, \quad i = 1, 2.$$
For $z(s) = \frac{p(s)}{B(s)}$ to be SPR we choose $B(s)$ to be of degree 1 or 2.

Case (a):
(3.3) $$B(s) = b_1 s + b_2, \quad b_1 \text{ and } b_2 > 0,$$
b_1 or b_2 can be chosen arbitrarily. Then
(3.4) $$Rez(j\omega) = \frac{(b_1 a_1 - b_2)\omega^2 + b_2 a_2}{\omega^2 b_1^2 + b_2^2}$$
is positive if
(3.5) $$b_1 \underline{a}_1 - b_2 > 0$$
which can always be satisfied by appropriate choice of b_1 and b_2.

A realization of $\frac{1}{z(s)} = \underline{c}^T(sI - A)^{-1}\underline{b}$ in controllable normal form gives
(3.6) $$A = \begin{bmatrix} 0 & 1 \\ -a_2 & -a_1 \end{bmatrix}, \quad \underline{b} = \begin{bmatrix} 0 \\ 1 \end{bmatrix}, \quad \underline{c}^T = [b_2 \ b_1].$$

Then the KYP lemma for strictly proper transfer functions has

(3.7) $$A^T P_1 + P_1 A = -Q_1, \quad P_1 \underline{b} = \underline{c}.$$

Hence,

(3.8) $$P_1 = \begin{bmatrix} b_1 a_2 + b_2 a_1 & b_2 \\ b_2 & b_1 \end{bmatrix} = \begin{bmatrix} \begin{vmatrix} b_1 & b_2 \\ -a_1 & a_2 \end{vmatrix} & b_2 \\ b_2 & b_1 \end{bmatrix}$$

and

$$Q_1 = 2 \begin{bmatrix} b_2 a_2 & 0 \\ 0 & b_1 a_1 - b_2 \end{bmatrix}.$$

The result in (3.8) can also be obtained if we solve the Lyapunov equation for P_1 assuming Q_1 diagonal, and assuming the last column of P_1 to be independent of a_1, a_2; this ensures P_1 and Q_1 are affine in the parameters a_1 and a_2.

We come now to Case (b):

(3.9) $$B(s) = s^2 + b_1 s + b_2, \quad b_1 \text{ and } b_2 > 0.$$

Then

(3.10) $$Re z(j\omega) = \frac{\omega^4 + (b_1 a_1 - b_2 - a_2)\omega^2 + a_2 b_2}{(-\omega^2 + b_2)^2 + b_1^2}$$

is positive if

(3.11) $$b_1 \underline{a}_1 - b_2 - \overline{a}_2 > 0$$

which can always be satisfied by appropriate choice of b_1 and b_2.

Now writing $z^{-1}(s) = 1 + \underline{c}^T (sI - A)^{-1} \underline{b}$ with

(3.12) $$A = \begin{bmatrix} 0 & 1 \\ -a_2 & -a_1 \end{bmatrix}, \quad \underline{b} = \begin{bmatrix} 0 \\ 1 \end{bmatrix}, \quad \underline{c}^T = [b_2 - a_2 \quad b_1 - a_1]$$

and using the KYP lemma, we get

(3.13) $$\begin{aligned} P_2 &= \begin{bmatrix} b_1 a_2 + b_2 a_1 & b_2 + a_2 \\ b_2 + a_2 & b_1 + a_1 \end{bmatrix} = \begin{bmatrix} \begin{vmatrix} b_1 & b_2 \\ -a_1 & a_2 \end{vmatrix} & b_2 + a_2 \\ b_2 + a_2 & b_1 + a_1 \end{bmatrix} \\ &= P_1 + \begin{bmatrix} 0 & a_2 \\ a_2 & a_1 \end{bmatrix}, \end{aligned}$$

$$Q_2 = 2 \begin{bmatrix} b_2 a_2 & 0 \\ 0 & b_1 a_1 - b_2 - a_2 \end{bmatrix} \quad \text{and} \quad \underline{q} = \sqrt{2} \begin{bmatrix} a_2 \\ a_1 \end{bmatrix}.$$

3.2. $n = 3$:

(3.14) $$p(s) = s^3 + a_1 s^2 + a_2 s + a_3, \quad \underline{a}_i \leq a_i \leq \overline{a}_i, \quad a_i > 0.$$

For stability of all $p(s)$ we require

(3.15) $$\underline{a}_1 \underline{a}_2 - \overline{a}_3 > 0.$$

For $z(s) = \frac{p(s)}{B(s)}$ to be SPR, we choose $B(s)$ to be of degree 2 or 3.

(3.16) $\qquad$ (a) $\quad B(s) = b_1 s^2 + b_2 s + b_3, \quad b_i > 0$.

Any one of b_1 or b_2 or b_3 can be chosen arbitrarily. It follows that $Rez(j\omega)$ is positive if its numerator coefficients are positive, i.e. if

(3.17) $\qquad b_2 \underline{a}_2 - b_1 \bar{a}_3 - b_3 \bar{a}_1 > 0, \quad b_1 \underline{a}_1 - b_2 > 0$.

A realization of $\frac{1}{z(s)} = \underline{c}^T (sI - A)^{-1} \underline{b}$ in controllable normal form gives

(3.18) $\quad A = \begin{bmatrix} 0 & 1 & 0 \\ 0 & 0 & 1 \\ -a_3 & -a_2 & -a_1 \end{bmatrix}, \quad \underline{b} = \begin{bmatrix} 0 \\ 0 \\ 1 \end{bmatrix}, \quad \underline{c}^T = [b_3 \quad b_2 \quad b_1]$.

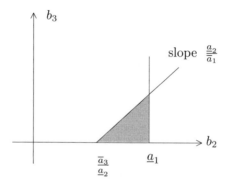

Fig. 1: Graphical interpretation of stability conditions

Then (3.7) gives

(3.19)
$$P_1 = \begin{bmatrix} \begin{vmatrix} b_2 & b_3 \\ -a_2 & a_3 \end{vmatrix} & \begin{vmatrix} b_1 & b_3 \\ -a_1 & a_3 \end{vmatrix} & b_3 \\ \begin{vmatrix} b_1 & b_3 \\ -a_1 & a_3 \end{vmatrix} & \begin{vmatrix} b_1 & b_2 \\ -a_1 & a_2 \end{vmatrix} - b_3 & b_2 \\ b_3 & b_2 & b_1 \end{bmatrix}$$

and $Q_1 = 2 \begin{bmatrix} b_3 a_3 & 0 & 0 \\ 0 & b_2 a_2 - b_1 a_3 - b_3 a_1 & 0 \\ 0 & 0 & b_1 a_1 - b_2 \end{bmatrix}$.

The same P_1 and Q_1 follow if one simply solves the Lypunov equation assuming affine P and Q, diagonal Q and last column of P constant.

If we choose $b_1 = 1$, the two conditions in (3.17) which guarantee the positive definiteness of Q_1 can always be satisfied due to the stability condition (3.15). See Figure 1, where the shaded region corresponds to (3.17).

(3.20) $\qquad$ (b) $\quad B(s) = s^3 + b_1 s^2 + b_2 s + b_3, \quad b_i > 0$.

Then $Rez(j\omega)$ is positive if its numerator coefficients are positive, i.e. if

(3.21) $\quad\quad\quad b_2\underline{a_2} - b_1\bar{a}_3 - b_3\bar{a}_1, > 0, \quad b_1\underline{a_1} - b_2 - \bar{a}_2 > 0.$

Let $\frac{1}{z(s)} = 1 + \underline{c}^T(sI - A)^{-1}\underline{b}$ with

(3.22) $A = \begin{bmatrix} 0 & 1 & 0 \\ 0 & 0 & 1 \\ -a_3 & -a_2 & -a_1 \end{bmatrix}, \underline{b} = \begin{bmatrix} 0 \\ 0 \\ 1 \end{bmatrix}, \underline{c}^T = [b_3 - a_3 \;\; b_2 - a_2 \;\; b_1 - a_1].$

Then using the KYP lemma we get

$$P_2 = \begin{bmatrix} \begin{vmatrix} b_2 & b_3 \\ -a_2 & a_3 \end{vmatrix} & \begin{vmatrix} b_1 & b_3 \\ -a_1 & a_3 \end{vmatrix} & b_3 + a_3 \\ \begin{vmatrix} b_1 & b_3 \\ -a_1 & a_3 \end{vmatrix} & \begin{vmatrix} b_1 & b_2 \\ -a_1 & a_2 \end{vmatrix} - b_3 & b_2 + a_2 \\ b_3 + a_3 & b_2 + a_2 & b_1 + a_1 \end{bmatrix}$$

(3.23)
$$= P_1 + \begin{bmatrix} 0 & 0 & a_3 \\ 0 & 0 & a_2 \\ a_3 & a_2 & a_1 \end{bmatrix},$$

$$Q_2 = \begin{bmatrix} b_3 a_3 & 0 & 0 \\ 0 & b_2 a_2 - b_1 \bar{a}_3 - b_3 \bar{a}_1 & 0 \\ 0 & 0 & b_1 \underline{a}_1 - b_2 - \bar{a}_2 \end{bmatrix}, \underline{q} = \sqrt{2}\begin{bmatrix} a_3 \\ a_2 \\ a_1 \end{bmatrix}.$$

It is shown in [9] that (3.15) ensures it is always possible to find b_i, such that (3.21) holds.

3.3. $n = 4$:

(3.24) $\quad\quad P(s) = s^4 + a_1 s^3 + a_2 s^2 + a_3 s + a_4, \quad \underline{a}_i \leq a_i \leq \bar{a}_i, a_i > 0.$

For stability of $p(s)$ we require [12]

(3.25)
$$\underline{a}_1\underline{a}_2 - \bar{a}_3 > 0$$
$$\underline{a}_1\underline{a}_2\bar{a}_3 - \underline{a}_1^2\bar{a}_4 - \bar{a}_3^2 > 0$$
$$\bar{a}_1\underline{a}_2\underline{a}_3 - \bar{a}_1^2\bar{a}_4 - \underline{a}_3^2 > 0.$$

For $z(s) = \frac{p(s)}{B(s)}$ to be SPR, we choose $B(s)$ to be of degree 3 or 4.

(3.26) $\quad\quad$ (a) $\;\; B(s) = b_1 s^3 + b_2 s^2 + b_3 s + b_4, \quad b_i > 0.$

Any one of $b_1 \cdots b_4$ can be chosen arbitrarily. Also $Rez(j\omega)$ is positive if its numerator coefficients are positive, i.e. if

(3.27)
$$b_1\underline{a}_1 - b_2 > 0, \quad b_2\underline{a}_2 - b_1\bar{a}_3 - b_3\bar{a}_1 + b_4 > 0 \text{ and}$$
$$b_3\underline{a}_3 - b_2\bar{a}_4 - b_4\bar{a}_2 > 0.$$

A realization of $\frac{1}{z(s)} = \underline{c}^T(sI - A)^{-1}\underline{b}$ in controllable normal form gives

(3.28) $A = \begin{bmatrix} 0 & 1 & 0 & 0 \\ 0 & 0 & 1 & 0 \\ 0 & 0 & 0 & 1 \\ -a_4 & -a_3 & -a_2 & -a_1 \end{bmatrix}$, $\underline{b} = \begin{bmatrix} 0 \\ 0 \\ 0 \\ 1 \end{bmatrix}$, $\underline{c}^T = [b_4 \ b_3 \ b_2 \ b_1]$.

Then (3.7) gives

$$P_1 = \begin{bmatrix} \begin{vmatrix} b_3 & b_4 \\ -a_3 & a_4 \end{vmatrix} & \begin{vmatrix} b_2 & b_4 \\ -a_2 & a_4 \end{vmatrix} & & \begin{vmatrix} b_1 & b_4 \\ -a_1 & a_4 \end{vmatrix} & b_4 \\ \begin{vmatrix} b_2 & b_4 \\ -a_2 & a_4 \end{vmatrix} & \begin{vmatrix} b_2 & b_3 \\ -a_2 & a_3 \end{vmatrix} - \begin{vmatrix} b_1 & b_4 \\ -a_1 & a_4 \end{vmatrix} & \begin{vmatrix} b_1 & b_3 \\ -a_1 & a_3 \end{vmatrix} - b_4 & b_3 \\ \begin{vmatrix} b_1 & b_4 \\ -a_1 & a_4 \end{vmatrix} & \begin{vmatrix} b_1 & b_3 \\ -a_1 & a_3 \end{vmatrix} - b_4 & \begin{vmatrix} b_1 & b_2 \\ -a_1 & a_2 \end{vmatrix} - b_3 & b_2 \\ b_4 & b_3 & b_2 & b_1 \end{bmatrix}$$

(3.29)

$$Q_1 = 2 \begin{bmatrix} b_4 a_4 & & & \\ & -b_4 a_2 + b_3 a_3 - b_2 a_4 & & \\ & & b_4 - b_3 a_1 + b_2 a_2 - b_1 a_3 & \\ & & & b_1 a_1 - b_2 \end{bmatrix}.$$

The same P_1 and Q_1 can be obtained by solving the Lyapunov equation assuming affine P_1 and Q_1 with diagonal Q_1 and the last column of P_1 independent of a_i.

Without loss of generality let $b_2 = 1$. Then from (3.27) $b_1 > \frac{1}{\underline{a}_1}$ which means $b_1 = \frac{\alpha}{\underline{a}_1}$ where $\alpha > 1$.

The other two conditions of (3.27) can be rewritten as

(3.30) $$\underline{a}_2 - \alpha \frac{\bar{a}_3}{\underline{a}_1} - b_3 \bar{a}_1 + b_4 > 0$$
$$b_3 \underline{a}_3 - \bar{a}_4 - b_4 \bar{a}_2 > 0.$$

Figure 2 shows that a solution exists if and only if

(3.31) $$\underline{a}_1 \underline{a}_2 \underline{a}_3 - \bar{a}_4 \underline{a}_1 \bar{a}_1 - \alpha \underline{a}_3 \bar{a}_3 > 0$$

(3.32) (b) $B(s) = s^4 + b_1 s^3 + b_2 s^2 + b_3 s + b_4, b_i > 0$.

Then $Rez(j\omega)$ is positive if its numerator coefficients are positive, or if

(3.33) $$b_1 \underline{a}_1 - b_2 - \bar{a}_2 > 0$$
$$b_4 - b_3 \bar{a}_1 + b_2 \underline{a}_2 - b_1 \bar{a}_3 + \underline{a}_4 > 0$$
$$-b_4 \bar{a}_2 + b_3 \underline{a}_3 - b_2 \bar{a}_4 > 0.$$

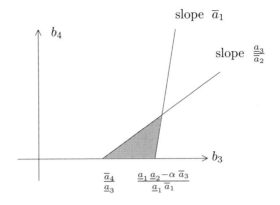

Fig. 2: Graphical interpretation of the stability conditions

Let $\frac{1}{z(s)} = 1 + \underline{c}^T(sI - A)^{-1}\underline{b}$ with

(3.34)
$$A = \begin{bmatrix} 0 & 1 & 0 & 0 \\ 0 & 0 & 1 & 0 \\ 0 & 0 & 0 & 1 \\ -a_4 & -a_3 & -a_2 & -a_1 \end{bmatrix}, \quad \underline{b} = \begin{bmatrix} 0 \\ 0 \\ 0 \\ 1 \end{bmatrix},$$

$$\underline{c}^T = [b_4 - a_4 \quad b_3 - a_3 \quad b_2 - a_2 \quad b_1 - a_1].$$

Then using the KYP lemma we get

$$P_2 = \begin{bmatrix} \begin{vmatrix} b_3 & b_4 \\ -a_3 & a_4 \end{vmatrix} & \begin{vmatrix} b_2 & b_4 \\ -a_2 & a_4 \end{vmatrix} & \begin{vmatrix} b_1 & b_4 \\ -a_1 & a_4 \end{vmatrix} & b_4 + a_4 \\ \begin{vmatrix} b_2 & b_4 \\ -a_2 & a_4 \end{vmatrix} & \begin{vmatrix} b_2 & b_3 \\ -a_2 & a_3 \end{vmatrix} - \begin{vmatrix} b_1 & b_4 \\ -a_1 & a_4 \end{vmatrix} & \begin{vmatrix} b_1 & b_3 \\ -a_1 & a_3 \end{vmatrix} - b_4 & b_3 + a_3 \\ \begin{vmatrix} b_1 & b_4 \\ -a_1 & a_4 \end{vmatrix} & \begin{vmatrix} b_1 & b_3 \\ -a_1 & a_3 \end{vmatrix} - b_4 & \begin{vmatrix} b_1 & b_2 \\ -a_1 & a_2 \end{vmatrix} - b_3 & b_2 + a_2 \\ b_4 + a_4 & b_3 + a_3 & b_2 + a_2 & b_1 + a_1 \end{bmatrix}$$

$$= P_1 + \begin{bmatrix} 0 & 0 & 0 & a_4 \\ 0 & 0 & 0 & a_3 \\ 0 & 0 & 0 & a_2 \\ a_4 & a_3 & a_2 & a_1 \end{bmatrix},$$

(3.35)

(3.35 cont.)

$$Q_2 = \begin{bmatrix} b_4 a_4 & & & \\ & -b_4 a_2 + b_3 a_3 - b_2 a_4 & & \\ & & b_4 - b_3 a_1 + b_2 a_2 - b_1 a_3 + a_4 & \\ & & & b_1 a_1 - b_2 - a_2 \end{bmatrix},$$

$$\underline{q} = \sqrt{2} \begin{bmatrix} a_4 \\ a_3 \\ a_2 \\ a_1 \end{bmatrix}.$$

In [9] it was wrongly asserted using an argument involving a dual linear programming problem that $b_1 \cdots b_4$ always exist to satisfy the SPR conditions (3.35).

REMARK 3.1. *The SPR sufficiency conditions (3.27) can be written as*

$$(3.36) \quad \begin{bmatrix} \underline{a_1} & -1 & 0 & 0 \\ -\bar{a}_3 & \underline{a_2} & -\bar{a}_1 & 1 \\ 0 & -\bar{a}_4 & \underline{a_3} & -\bar{a}_2 \\ 0 & 0 & 0 & \underline{a_4} \end{bmatrix} \begin{bmatrix} b_1 \\ b_2 \\ b_3 \\ b_4 \end{bmatrix} > \begin{bmatrix} 0 \\ 0 \\ 0 \\ 0 \end{bmatrix} \quad \text{or} \quad H^* \begin{bmatrix} b_1 \\ b_2 \\ b_3 \\ b_4 \end{bmatrix} > \begin{bmatrix} 0 \\ 0 \\ 0 \\ 0 \end{bmatrix}.$$

The matrix

$$(3.37) \qquad\qquad H^* = I^* H I^*$$

where $I^* = \begin{bmatrix} 1 & & & \\ & -1 & & \\ & & 1 & \\ & & & -1 \end{bmatrix}$ *and* H^* *is a modified Hurwitz matrix.*

The inequality of condition (3.36) involves just two extreme points (two corners) in coefficient space, namely $\underline{a}_1, \bar{a}_2, \underline{a}_3, \bar{a}_4$ and $\underline{a}_1, \bar{a}_2, \underline{a}_3, \bar{a}_4$. It can be easily shown that this idea generalises for any degree n.

REMARK 3.2. *In general it can be shown that* $P_1 = (p_{ij})$ *with*

$$(3.38) \qquad p_{ij} = \begin{vmatrix} b_{n-j} & b_{n+1-i} \\ -a_{n-j} & a_{n+1-i} \end{vmatrix} - p_{i-1,j+1}$$

where $i, j = 1, 2, \ldots, n$, $j \geq i$, $b_0 = 0, a_0 = 1, p_{0,j} = 0$, $p_{i,n+1} = 0$ *and*

$$(3.39) \qquad P_2 = P_1 + \begin{bmatrix} 0 & \cdots & 0 & a_n \\ \vdots & & \vdots & a_{n-1} \\ 0 & \cdots & 0 & \vdots \\ a_n & & a_2 & a_1 \end{bmatrix}.$$

4. Alternative analytic solution of the Lyapunov equations; second order problems.

Consider the Lyapunov equation

(4.1) $$A^T P + PA = -Q$$

with A in companion form. Arranging the diagonal and sub-diagonal entries of P and Q into vectors leads to

(4.2) $$B \text{vec} P = \text{vec} Q.$$

It can be shown by construction and using elementary row operations that B has the following form

(4.3) $$B = \left[\begin{array}{c|c} I & B^* \\ \hline O & B_{n,n} \end{array} \right]$$

where

(4.4) $$B_{n,n} = (-1)^{n+1} J^* H J^*, \quad J^* = \begin{bmatrix} & & & 0 & 1 \\ & & & -1 & \\ & & 1 & & \\ & -1 & & & \\ \cdot^{\cdot^{\cdot}} & & & & 0 \end{bmatrix}$$

and

(4.5) $$B^* = \begin{bmatrix} B_{1,2} & C_{1,1} & 0_{1,n-3} \\ B_{2,3} & C_{2,1} & 0_{1,n-4} \\ \vdots & \vdots & \vdots \\ B_{n-3,n-2} & C_{n-3,1} & 0_{n-3,1} \\ B_{n-2,n-1} & C_{n-2,1} & \\ B_{n-1,n} & & \end{bmatrix}$$

(4.6) $$B_{n-r,n-r+1} = \begin{bmatrix} -a_r & & & & -a_n \\ & \cdot^{\cdot^{\cdot}} & & 0 & \vdots \\ -a_{r-1} & & & & \vdots \\ & \cdot^{\cdot^{\cdot}} & & \cdot^{\cdot^{\cdot}} & \vdots \\ \cdot & & -a_{r-1} & -a_r & -a_{r+1} \end{bmatrix}, \quad r = 1, 2, \ldots n-1$$

(4.7) $$C_{n-r,1} = \begin{bmatrix} 0 \\ a_n \\ \vdots \\ a_{r+2} \end{bmatrix}, \quad r = 2, \ldots, n-1$$

$0_{n-r,r-2}$ is a zero matrix.

We have

(4.8) $$\det B = (-1)^{n+1} \det H.$$

But $\det H$ is the product of the two critical stability conditions. Therefore, if we choose

(4.9) $$Q = 2(\det H)I$$

then P will have elements which are polynomials in $a_1, \ldots, a_n$.

Now if $n = 2$, $\det H = a_1 a_2$. Then we have

(4.10) $$Q = 2a_1 a_2 I, \quad P = \begin{bmatrix} a_1^2 + a_2^2 + a_2 & a_1 \\ a_1 & 1 + a_2 \end{bmatrix}$$

so that P is quadratic in a_1, a_2.

From (1.6), if $Q - P > 0$ we conclude stability, i.e. for stability

(4.11) $$\begin{bmatrix} 2a_1 a_2 - 2a_1 \dot{a}_1 - 2a_2 \dot{a}_2 - \dot{a}_2 & \dot{a}_1 \\ \dot{a}_1 & 2a_1 a_2 - \dot{a}_2 \end{bmatrix} > 0.$$

As (4.11) is multi affine in $a_1, a_2, \dot{a}_1, \dot{a}_2$, we need only to check the corner values.

For $n > 2$ the procedure leads to P with some elements of higher degree than quadratic in the coefficients; therefore corner conditions are no longer sufficient.

REMARK 4.1. *One can take a positive diagonal matrix instead of the unity matrix in Q to improve on the results.*

5. Examples for second order systems.

Let $\bar{a}_1 = \bar{a}_2 = \alpha$, $\underline{a}_1 = \underline{a}_2 = \beta$, $\bar{\dot{a}}_1 = \bar{\dot{a}}_2 = \gamma$. We consider the stability problem with the different Lyapunov functions derived in the previous sections.

5.1.

(5.1) $$Q = 2 \begin{bmatrix} b_2 a_2 & 0 \\ 0 & b_1 a_1 - b_2 \end{bmatrix}, \quad P = \begin{bmatrix} b_1 a_2 + b_2 a_1 & b_2 \\ b_2 & b_1 \end{bmatrix}$$
$$\dot{P} = \begin{bmatrix} b_1 \dot{a}_2 + b_2 \dot{a}_1 & 0 \\ 0 & 0 \end{bmatrix}.$$

For $Q > 0 \quad b_1 \underline{a}_1 - b_2 > 0$, i.e. $b_1 > \frac{b_2}{\beta}$.

For $Q - \dot{P} > 0$ we require $\begin{bmatrix} 2b_2 a_2 - b_1 \dot{a}_2 - b_2 \dot{a}_1 & 0 \\ 0 & 2b_1 a_1 - 2b_2 \end{bmatrix} > 0$ which is ensured by

(5.2) $$\alpha < \frac{2 b_2 \beta}{b_1 + b_2}.$$

With $b_2 = 1$ we obtain $b_1 > \frac{1}{\beta}$ and

(5.3)
$$\alpha < \frac{2\beta}{1+\frac{1}{\beta}} = \frac{2\beta^2}{\beta+1}.$$

Hence if $\beta = 2$ then $\alpha < 2.66$, and if $\beta = 4$ then $\alpha < 6.4$ These results are independent of γ.

5.2.
$$Q = 2\begin{bmatrix} b_2 a_2 & 0 \\ 0 & b_1 a_1 - b_2 \end{bmatrix} + \begin{bmatrix} a_2^2 & a_1 a_2 \\ a_1 a_2 & a_1^2 \end{bmatrix} = Q_1 + Q_2$$

$$P = \begin{bmatrix} b_1 a_2 + b_2 a_1 & b_2 + a_2 \\ b_2 + a_2 & b_1 + a_1 \end{bmatrix}.$$

Therefore,

(5.4) $Q_1 - \dot{P} = \begin{bmatrix} 2b_2 a_2 - b_1 \dot{a}_2 - b_2 \dot{a}_1 & -\dot{a}_2 \\ -\dot{a}_2 & 2b_1 a_1 - 2b_2 a_2 - 2a_2 - \dot{a}_1 \end{bmatrix} > 0.$

To ensure $Q_1 > 0$ we have $b_1 > \frac{b_2 + \gamma}{\beta}$.

Because of the affine character of $Q - \dot{P}$ with respect to a_1 and a_2 we consider only the two vertices $(\underline{a}_1, \underline{a}_2)$ and $(\underline{a}_1, \overline{a}_2)$. From $Q_1 - \dot{P} > 0$ the positivity of the diagonal elements at the two vertices gives

(5.5)
$$\alpha < \frac{2b_2 \beta}{b_1 + b_2}, \quad \alpha < 2(b_1 - 1)\beta - 2b_2, \quad \alpha < \frac{2b_2 \gamma}{b_1 + b_2},$$
$$\alpha < 2b_1 \beta - 2\gamma - 2b_2.$$

From (5.4) and (5.5)

(5.6)
$$\alpha < \frac{2b_2 \beta^2}{b_2(1+\beta) + \gamma + \frac{\alpha}{2}} \approx \frac{2\beta^2}{\beta + 1}$$

for large b_2.

The parameters b_1 and b_2 can be chosen large enough to have (5.6) valid and $\det(Q_1 - \dot{P})$ remains positive.

Hence for $\beta = 2$ we get $\alpha < 2.66$. This is independent of γ and is a rough estimate.

5.3.
$$Q = 2a_1 a_2 I, \quad P = \begin{bmatrix} a_1^2 + a_2^2 + a_2 & a_1 \\ a_1 & 1 + a_2 \end{bmatrix},$$

$$Q - \dot{P} = \begin{bmatrix} 2a_1 a_2 - 2a_1 \dot{a}_1 - 2a_2 \dot{a}_2 - \dot{a}_2 & -\dot{a}_1 \\ -\dot{a}_1 & 2a_1 a_2 - \dot{a}_2 \end{bmatrix} > 0.$$

For positivity of $Q - \dot{P}$ at the four vertices $(\overline{a}_1, \overline{a}_2)$, $(\underline{a}_1, \overline{a}_2)$, $(\overline{a}_1, \underline{a}_2)$, $(\underline{a}_1, \underline{a}_2)$ we get $\alpha < 0.877$ if we assume $\beta = 2, \gamma = 5$.

5.4.

$$Q = 2a_1 a_2 \begin{bmatrix} c & 0 \\ 0 & d \end{bmatrix}, \quad P = \begin{bmatrix} a_1^2 c + a_2^2 d + a_2 c & a_1 c \\ a_1 c & a_2 d + c \end{bmatrix}.$$

When we demand the positivity of $(Q - \dot{P})$ at the four vertices and $d \to 0$ we get $\alpha < 1.54$, if $\beta = 2, \gamma = 5$.

5.5.

$$Q = 2a_1 a_2, \quad P = \begin{bmatrix} a_1^2 + a_2^2 + a_2 & a_1 \\ a_1 & 1 + a_2 \end{bmatrix},$$

$$\dot{P} = \begin{bmatrix} 2a_1 & 1 \\ 1 & 0 \end{bmatrix} \dot{a}_1 + \begin{bmatrix} 2a_2 + 1 & 0 \\ 0 & 1 \end{bmatrix} \dot{a}_2.$$

Now $Q - \dot{P} > 0$ will follow if $2\underline{a}_1 \underline{a}_2 - \left\| \begin{bmatrix} 2\bar{a}_1 & 1 \\ 1 & 0 \end{bmatrix} \right\|_\infty \alpha - \left\| \begin{bmatrix} 2\bar{a}_2 + 1 & 0 \\ 0 & 1 \end{bmatrix} \right\|_\infty \alpha > 0.$

This inequality will follow if $\alpha < \frac{\beta^2}{2\gamma+1} = \frac{4}{11} = 0.363$ for $\beta = 2, \gamma = 5$.

5.6. Q and P as in 5.5.

Now $Q - \dot{P} > 0$ is implied by $2a_1 a_2 - \left\| \begin{bmatrix} 2a_1 & 1 \\ 1 & 0 \end{bmatrix} \right\|_\infty \alpha - \left\| \begin{bmatrix} 2a_2 + 1 & 0 \\ 0 & 1 \end{bmatrix} \right\|_\infty \alpha > 0.$

Because of affine occurence of the parameters we get after evaluation at the four vertices $\alpha < 0.8$ for $\beta = 2$ and $\gamma = 5$.

REMARK 5.1.

The approach of 5.1 is easier to use than that of 5.2 as we deal with less parameters.

5.4 gives better results than 5.3 due to the additional free parameter.

5.5 is more conservative than 5.6

The methods of 5.1 and 5.2 can be used for $n > 2$ as they give vertex results.

6. Conclusion. Results are presented which solve at least partially the open problem in [7], where the coefficients of a linear differential equation lie in given intervals, the rates of variations of these coefficients are restricted, and exponential stability is required.

If affine or multiaffine P, Q or the Lyapunov equation are obtained, then vertex conditions are obtained which are sufficient for stability. With the Lyapunov function obtained using the critical stability conditions, vertex results are obtained only for $n = 2$. For higher order systems a complex optimization problem has to be solved. It is possible to extend the above results to discrete systems.

References

[1] V.L. Kharitonov: Asymptotic stability of an equilibrium position of a family of systems of linear differential equations. Differential' nye Uravneniya, vol. 14, pp. 1483–1485, 1979.

[2] J.-H. Su and I.-K. Fong: New robust stability bounds of linear discrete-time systems with time varying uncertainties. Int.J.Control, vol. 58, pp. 1461–1467, 1993.

[3] P.H. Bauer and K. Premaratne: Robust stability of time-variant interval matrices. Proc.29th IEEE Conf.Decision and Contro, Honolulu, HI, Dec.1990, pp. 334–335.

[4] P.H. Bauer, M. Mansour and J. Duran: Stability of polynomials with time-variant coefficients, IEEE Trans.Circuits Syst.-I, vol.40 pp. 423–426, June 1993.

[5] P.H. Bauer, K. Premanatne and J. Duran: A necessary and sufficient condition for robust stability of time-variant discrete systems. IEE Trans. Aut. Contro,, vol 38, pp. 1427–1430, Sept. 1993.

[6] K. Premaratne and M. Mansour: Robust stability of time-variant descrete-time systems with bounded parameter perturbations. IEE Trans.Circuits Syst.-I vol 42, pp. 40–45, Jan 1995.

[7] M. Mansour, S. Balemi and W. Truol, Eds.: Robustness of dynamic systems with parameter uncertainties. Birkhäuser, 1992

[8] W. Hahn: Stability of motion. Springer, 1967.

[9] B.D.O. Anderson, S. Dasgupta, P Khargonekar, F.J Kraus and M. Mansour: Robust strict positive realness: characterization and construction. IEEE Trans. Circuits, Systems, vol.37, pp. 869–876, July 1990.

[10] S. Dasgupta, G. Chockalingam, B.D.O. Anderson and M. Fu: Lyapunov functions for uncertain systems with applications to the stability of time varying systems. IEEE Trans. Circuits Syst.-I vol.41, pp. 93–106, Feb 1994.

[11] S. Dasgupta and A.S. Bhagwat: Conditions for designing strictly positive real transfer functions for adaptive output error identification. IEEE Trans. Circuits Syst, vol.34, pp. 731–737, 1987.

[12] B.D.O. Anderson, E.I. Jury and M. Mansour: On robust Hurwitz polynomials. IEEE Trans. Aut. Control, vol.32, pp. 909–913, 1987.

On the Computation of Stability Profiles

Wolfgang Sienel
DLR, German Aerospace Research Center, Institute for Robotics and System Dynamics
Oberpfaffenhofen, 82234 Wessling, Germany

Abstract. The computational effort for robust stability analysis of uncertain control systems increases with the number of uncertain parameters. However, there exists a class of so-called tree structured systems, for which stability analysis can be performed extremely fast. In this case robust stability is verified by checking zero exclusion from value sets, which are obtained by mapping the uncertainty domain via the uncertain characteristic polynomial. An extension of this approach allows to generate the stability profile. This is part of the boundary of the root set of the uncertain system containing the most important stability information. The efficiency of the method is demonstrated by a robustness analysis of the DLR plate experiment with 60 uncertain parameters.

1. Introduction. For robustness analysis of an uncertain linear plant its characteristic polynomial can be utilized.
The uncertain parameters $q = [q_1 \ q_2 \ \ldots \ q_\ell]^T$, where ℓ is the number of uncertain parameters, enter into the real coefficient functions $a_i(q)$ of the characteristic polynomial

$$(1) \quad p(s, q) = a_0(q) + a_1(q)s + \ldots + a_{n-1}(q)s^{n-1} + a_n(q)s^n, \quad q \in Q.$$

It is assumed that the coefficients of the polynomial are continuous functions of the uncertain parameters. Further, it is assumed that the uncertain parameters are real, independent of each other, and assume values within given boundaries. This results in a hyperrectangular operating domain

$$(2) \quad Q := \{q \mid q_i \in [q_i^-; q_i^+], \ i = 1, \ldots, \ell\}.$$

A control system is called robustly stable if stability of the system is fulfilled for the entire operating domain Q of the uncertain parameters.

For uncertain control systems with a small number of uncertain parameters it is still feasible to apply conventional methods for robustness analysis. One possibility is to grid the operating domain. For each grid point a stability test has to be performed, e.g. a Hurwitz test of the resulting nominal characteristic polynomials. One disadvantage of this procedure is the computational effort, which increases exponentially with the number of uncertain parameters. Another fact is that one cannot exclude the possibility that an isolated instable region in the operating domain remains undetected by a grid point [1].

In general, the analysis of control systems with a large number of uncertain parameters is an unsolved problem. However, specific system properties in the characteristic polynomial yield a drastic simplification of the robustness analysis. A result of this kind is the well-known Kharitonov Theorem [10] for interval polynomials. For affine parameter dependency the edge theorem by Bartlett, Hollot, and Huang [5] can be applied. For nonlinear parameter dependency no such simple results are known.

A novel approach to robustness analysis of uncertain systems was presented in [4]. There, it was demonstrated that a stability analysis can be performed extremely fast if the system has a so-called tree structure. The robustness analysis is performed then by construction of value sets, which are obtained by mapping the uncertainty domain Q via the characteristic polynomial. In the case of a tree structure the value sets can be constructed for arbitrary frequencies $s = \sigma + j\omega$. Therefore, the system can be checked with the same effort for both Hurwitz stability and Gamma-stability [2].

After briefly recalling the concepts of robustness analysis by value set construction and tree structures the notion of stability profile is introduced. While it is straight forward to compute the roots of a nominal characteristic polynomial for stability analysis it is still an unsolved problem to compute the root set of an uncertain polynomial $p(s, \boldsymbol{q})$, $\boldsymbol{q} \in Q$. It will be shown that the stability profile which is the right hand side of the boundary of the root set can be constructed fast for tree structured systems by iterative computation of value sets. The stability profile contains direct information about critical frequencies and the most important stability margins like maximal real part or minimal degree of damping. The efficiency of this approach will be demonstrated by a robustness analysis of the DLR plate experiment with 60 uncertain parameters.

2. Robustness Analysis by Value Set Construction. A graphically based approach to stability analysis of characteristic polynomials is the Mikhailov criterion [12], also known as Cremer-Leonhard criterion [6, 11].

Theorem 2.1 (Mikhailov, Cremer, Leonhard)
The polynomial

$$(3) \qquad p(s) = a_0 + a_1 s + a_2 s^2 + \ldots + a_{n-1} s^{n-1} + s^n$$

is stable if and only if the frequency plot of $p(j\omega)$
1. starts on the positive real axis for $\omega = 0$ and
2. encircles the origin in counterclockwise direction and passes hereby through n quadrants as the frequency increases from 0 to ∞.

□

For fixed frequency $s = j\omega$ a nominal polynomial contributes with a complex point to the frequency plot. In the case of a system with uncertain parameters $\boldsymbol{q}$ the entire operating domain Q has to be mapped via the characteristic polynomial $p(s, \boldsymbol{q})$ into the complex plane. This results in a complex value set which can be utilized for robustness analysis. If the coefficient functions $a_i(\boldsymbol{q})$ of the characteristic polynomial are continuous functions, then an extended formulation of the Mikhailov theorem, the so-called zero exclusion principle, can be applied to uncertain systems.

Theorem 2.2 (Zero exclusion principle)
The uncertain polynomial

$$p(s, \boldsymbol{q}) = a_0(\boldsymbol{q}) + a_1(\boldsymbol{q})s + \ldots + a_{n-1}(\boldsymbol{q})s^{n-1} + a_n(\boldsymbol{q})s^n, \quad a_n(\boldsymbol{q}) \neq 0 \text{ for all } \boldsymbol{q} \in Q$$

is robustly stable, if and only if
1. there exists an arbitrary stable operating point $\boldsymbol{q}_0 \in Q$ and
2. the value sets

(4) $$\mathcal{P}(\mathrm{j}\omega, Q) := \{p(\mathrm{j}\omega, \boldsymbol{q}) \mid \boldsymbol{q} \in Q\}$$

do not contain zero for all frequencies $\omega \geq 0$.

□

Item 1 in Theorem 2.2 is easy to verify. The value sets which have to be calculated for the evaluation of item 2 are simple to construct in the case of linear parameter dependency. It can be shown that for fixed frequency $s = \mathrm{j}\omega$ the value set of an affine polynomial is a convex polytope, whose vertices are formed by the image of vertices of the operating domain. An even simpler case is represented by interval polynomials. In [7] it was shown that the value sets of interval polynomials for fixed frequency $s = \mathrm{j}\omega$ are complex rectangles with edges parallel to the axes. For general nonlinear parameter dependency a construction of the value set is possible if the system has a tree structure.

3. Tree Structured Decomposition [4]. The notion of tree structures will be explained by way of the mass-spring-damper (MSD) system in Fig. 1. The differential

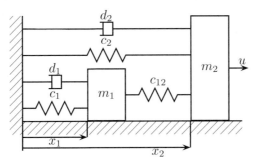

FIG. 1. *Schematic representation of a mass-spring-damper system.*

equations of the system are

(5) $$\begin{aligned} m_1 \ddot{x}_1 + d_1 \dot{x}_1 + c_1 x_1 + c_{12}(x_1 - x_2) &= 0 \\ m_2 \ddot{x}_2 + d_2 \dot{x}_2 + c_2 x_2 + c_{12}(x_2 - x_1) &= u. \end{aligned}$$

A possible way to obtain the characteristic polynomial is its direct evaluation from the Laplace transformed of the system of differential equations (5):

(6) $$\begin{bmatrix} m_1 s^2 + d_1 s + c_1 + c_{12} & -c_{12} \\ -c_{12} & m_2 s^2 + d_2 s + c_2 + c_{12} \end{bmatrix} \begin{bmatrix} x_1(s) \\ x_2(s) \end{bmatrix} = \begin{bmatrix} 0 \\ u(s) \end{bmatrix}.$$

The characteristic polynomial is the determinant of the above matrix, without further modifications it is

(7) $$p(s, q) = (m_1 s^2 + d_1 s + c_1 + c_{12}) \cdot (m_2 s^2 + d_2 s + c_2 + c_{12}) - c_{12}^2$$

with $q = [m_1\ m_2\ d_1\ d_2\ c_1\ c_2\ c_{12}]^T$. The characteristic polynomial is not displayed in its conventional notation as a polynomial ordered by orders of s. However, it seems that this representation which resulted in a natural manner from the evaluation of the determinant is much clearer than the conventional expanded representation.

A more detailed examination of the polynomial (7) together with the assumption $c_{12} = c_{12}^* = const$ reveals that each of the uncertain parameters appears just once. In Fig. 2 this property is displayed in a tree structure. The subpolynomials at the bottom line do not share common uncertain parameters. Following the operations indicated by the nodes will result in the characteristic polynomial of the MSD system.

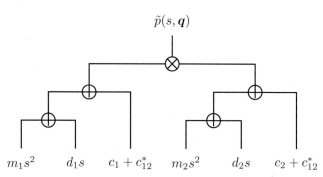

FIG. 2. Tree structure of the MSD system in Fig. 1. For completeness the number c_{12}^{*2} has to be subtracted from the polynomial $\tilde{p}(s, q)$.

The advantage of a such a tree structure is that the value sets of the terms in the bottom line can be generated independently. Then, the value set of the MSD system can be constructed very fast if the sets of the subpolynomials are added and multiplied according to the tree structure. Thus, gridding of the ℓ-dimensional operating domain is avoided. Note that in general the subpolynomials can also depend on more than one uncertain parameter. The advantages of tree structures can be fully exploited if the number of uncertain parameters in these no longer decomposable subpolynomials is low and their value sets are simple to construct.

Algorithms for decomposition of uncertain characteristic polynomials into their tree structured representation and for value set operations can be found in [2,13]. Tree structures can also be found in transfer functions and other functions suitable for stability analysis of control systems.

4. The Stability Profile. A statement about stability of the MSD system shown in Fig. 1 can be met immediately: It is a passive system and, hence, is stable for arbitrary values of masses, spring and damping constants. In most cases, however, more than Hurwitz stability is demanded from a system. The time responses of a system depend essentially on the location of the eigenvalues and can be judged by their maximal real part and minimal degree of damping. The question for both of these values is quite more difficult to answer than the one for Hurwitz stability.

In Fig. 3 a schematic representation of a root set of an uncertain characteristic polynomial is shown. For a nominal operation point $q_0 \in Q$ the roots are displayed with squares. For varying operating conditions $q \in Q$ also the roots of the uncertain characteristic polynomial $p(s, q)$ vary and generate the root set

$$\text{Roots}[p(s, q), Q] := \{s \mid p(s, q) = 0 \text{ for all } q \in Q\}.$$

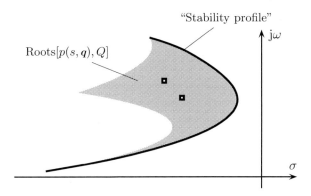

FIG. 3. *Schematic representation of a root set.*

For a robustness analysis it is not necessary to generate the entire root set, it is sufficient to know the boundary of this set. To be more specific, only that part of the boundary is required which faces the right complex half plane. In Fig. 3 this is the part of the boundary drawn with a solid line. It will be referred to as the stability profile.

The stability profile can even be computed for systems with a large number of uncertain parameters, if the system has a tree structure: The tree structure is not dependent on the frequency, i.e. the value sets can be generated for arbitrary frequencies $s = \sigma + j\omega$. If the value set is constructed for a frequency which is contained in the interior of the root set it will contain zero and vice versa. Hence, the value sets for frequencies which are located on the boundary of the root set will contain the origin exactly on its boundary. Based on this result, a definition of stability profiles in terms of value sets can be stated:

Definition 4.1 (Stability profile)
The stability profile $\mathcal{S}$ is the set

$$\mathcal{S} := \bigcup_{\omega \geq 0} (\sup\{\sigma \mid 0 \in \mathcal{P}(\sigma + j\omega, Q)\} + j\omega).$$

□

This definition can be used to develop an iterative procedure for computation of the stability profile via value sets: Starting from a point $j\omega^*$ on the imaginary axis the value sets are constructed for decreasing real part σ until the boundary of the value set $\mathcal{P}(\sigma + j\omega^*, Q)$ contains the origin. If this procedure is done for a grid along the imaginary axis it will result in the stability profile. For unstable systems the procedure may not start on the imaginary axis but on an axis shifted far enough into the right half plane. Note that it is even possible to generate the entire boundary of the root set, if for fixed ω^* all real parts σ_i are determined for which the boundary of the value set contains the origin.

5. Application: The DLR Plate Experiment.
Future space systems will be flexible light weight constructions to increase efficiency and save costs. However, this will result in rather weak damping of these structures which might even lead to their destruction while maneuvering. Therefore, the damping has to be increased by control to prevent those critical incidents. Since these light weight constructions would not bear their own weight on earth a precise identification of their system behavior is impossible. Model parameters are generally obtained by means of finite element (FE) modeling. These parameters, however, can be very inaccurate. This fact, together with the high system order obtained by FE modeling makes the controller design for these type of systems difficult.

Controllers for flexible space structures are normally designed for a nominal model not taking into account parameter variations. Crucial questions are then whether the controller robustly stabilizes the system for the entire range of uncertainties or whether it guarantees specific stability margins despite parameter variations. At DLR an experiment — the so-called DLR plate experiment — was designed for the purpose to verify and judge various controller structures.

5.1. Description of the Plate Experiment.
A sketch of the experiment assembly is given in Fig. 4.

The test structure consists of a thin rectangular sheet steel which is attached to the ceiling by ropes. It measures 2.75 m by 1.5 m and is 2 mm thin, its weight is 66 kg. For this specific experiment two collocated sensors and actuators were attached symmetrically to the test structure.

A model of the assembly of order 40 was derived by means of FE modeling. Its state space representation is

(8)
$$\begin{aligned} \dot{x} &= Ax + Bu \\ y &= Cx \end{aligned}$$

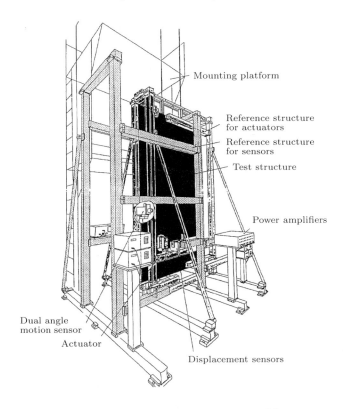

Fig. 4. *Assembly of the DLR plate experiment [8].*

with

$$A = \begin{bmatrix} 0 & I \\ -\Omega^2 & -2\xi\Omega \end{bmatrix}, \quad B = \begin{bmatrix} 0 \\ \Phi \end{bmatrix},$$

$$C = B^T, \quad \Phi = \begin{bmatrix} b_{1,1} & b_{1,2} \\ b_{2,1} & b_{2,2} \\ \vdots & \vdots \\ b_{20,1} & b_{20,2} \end{bmatrix}.$$

The diagonal matrices Ω and ξ contain the eigenfrequencies and the corresponding damping values. The two collocated sensors and actuators yield a 2×2 matrix of transfer functions

(9) $$G_0(s) = \begin{bmatrix} g_{11}(s) & g_{12}(s) \\ g_{21}(s) & g_{22}(s) \end{bmatrix}.$$

The transfer functions are formed by sums of second order terms:

$$g_{11}(s) = s \sum_{i=1}^{20} \frac{b_{i,1}^2}{s^2 + 2\xi_i \omega_i s + \omega_i^2}, \quad g_{12}(s) = s \sum_{i=1}^{20} \frac{b_{i,1}b_{i,2}}{s^2 + 2\xi_i \omega_i s + \omega_i^2}$$
$$g_{21}(s) = s \sum_{i=1}^{20} \frac{b_{i,1}b_{i,2}}{s^2 + 2\xi_i \omega_i s + \omega_i^2}, \quad g_{22}(s) = s \sum_{i=1}^{20} \frac{b_{i,2}^2}{s^2 + 2\xi_i \omega_i s + \omega_i^2}.$$
(10)

The elements $g_{12}(s)$ and $g_{21}(s)$ of the transfer function matrix are identical. Because of the symmetric arrangement of sensors and actuators

$$|b_{i,1}| \approx |b_{i,2}| =: b_i$$

is true. This causes the elements on the main diagonal of the transfer function matrix to be almost identical, i.e. $g_{11}(s) \approx g_{22}(s)$.

As already mentioned above the models obtained by FE modeling are uncertain where the uncertainty increases with the eigenfrequency. Based on this fact the uncertainty ranges for each of the FE model parameters are determined. The model parameters are the eigenfrequencies itself, the corresponding damping values, and the elements $b_{i,1}$, $b_{i,2}$ of the input matrix of the corresponding state space representation. Sorted by absolute values of the eigenfrequencies a deviation from the nominal point of 5% for small frequencies up to 20% for large frequencies is fixed:

(11)
$$\begin{array}{lll} b_i \in [b_i^-; b_i^+]; & b_i^- = (1 - \text{tol}_i)b_{i0}, & b_i^+ = (1 + \text{tol}_i)b_{i0} \\ \omega_i \in [\omega_i^-; \omega_i^+]; & \omega_i^- = (1 - \text{tol}_i)\omega_{i0}, & \omega_i^+ = (1 + \text{tol}_i)\omega_{i0} \\ \xi_i \in [\xi_i^-; \xi_i^+]; & \xi_i^- = (1 - \text{tol}_i)\xi_{i0}, & \xi_i^+ = (1 + \text{tol}_i)\xi_{i0} \end{array}$$

with

$$\begin{array}{ll} \text{tol}_i = 0.05 \text{ for } 1 \leq i \leq 5, & \text{tol}_i = 0.10 \text{ for } 5 < i \leq 10 \\ \text{tol}_i = 0.15 \text{ for } 10 < i \leq 15, & \text{tol}_i = 0.20 \text{ otherwise.} \end{array}$$

The index 0 refers to the nominal values. The number of uncertain parameters for the model of order 40 is *sixty*, altogether. A robustness analysis with conventional approaches would be impossible.

5.2. Computation of the Stability Profile for the Plate Experiment. In [9] a control design concept for flexible structures was introduced and applied to the DLR plate experiment. This approach yields a strictly positive real (SPR) controller of low order which only shifts the dominant eigenvalues of the plant. This is to prevent spill over effects, i.e. instability at high eigenfrequencies due to parameter uncertainty.

With the assumption $|b_{i,1}| \approx |b_{i,2}|$ the characteristic equation of the controlled system is

(12) $$h(s, \boldsymbol{q}) = \det\left(\boldsymbol{I} + \begin{bmatrix} g_{11}(s, \boldsymbol{q}) & g_{12}(s, \boldsymbol{q}) \\ g_{12}(s, \boldsymbol{q}) & g_{11}(s, \boldsymbol{q}) \end{bmatrix} \cdot \begin{bmatrix} k_1(s) & k_2(s) \\ k_2(s) & k_1(s) \end{bmatrix}\right).$$

The controller transfer functions from [9] are

$$k_1(s) = \frac{365.171s^3 + 2529.09s^2 + 5146.17s + 3082.61}{s^4 + 10.98s^3 + 44.82s^2 + 80.92s + 51.39}$$

$$k_2(s) = -\frac{213.437s^3 + 1595.46s^2 + 3171.96s + 2057.75}{s^4 + 10.98s^3 + 44.82s^2 + 80.92s + 51.39}.$$

With equation (12) the robust stability of the controlled system can be examined: Stability is verified if there exists at least one stable operating point and the value sets of the characteristic equation do not contain the origin for all frequencies. A stable operating point already results from the control design for nominal parameter values.

An advantage of the characteristic equation is the possibility to factorize it:

$$h(s, \boldsymbol{q}) = h_1(s, \boldsymbol{q}) \cdot h_2(s, \boldsymbol{q}) =$$

$$= [1 + (g_{11}(s, \boldsymbol{q}) + g_{12}(s, \boldsymbol{q}))(k_1(s) + k_2(s))] \cdot$$
$$[1 + (g_{11}(s, \boldsymbol{q}) - g_{12}(s, \boldsymbol{q}))(k_1(s) - k_2(s))] =$$

(13)
$$= \left[1 + (k_1(s) + k_2(s)) \sum_{i=1}^{i=20} \frac{b_{i,1}^2 + b_{i,1}b_{i,2}}{s^2 + 2\xi\omega_i s + \omega_i^2}\right] \cdot$$
$$\left[1 + (k_1(s) - k_2(s)) \sum_{i=1}^{i=20} \frac{b_{i,1}^2 - b_{i,1}b_{i,2}}{s^2 + 2\xi\omega_i s + \omega_i^2}\right].$$

Zero exclusion from the value sets of the characteristic equation is fulfilled if both value sets $\mathcal{H}_1(s, Q)$ and $\mathcal{H}_2(s, Q)$ of the two factors $h_1(s, \boldsymbol{q})$ and $h_2(s, \boldsymbol{q})$ do not contain the origin:

(14) $$0 \notin \mathcal{H}_1(s, Q) \quad \text{and} \quad 0 \notin \mathcal{H}_2(s, Q).$$

A detailed examination of these factors reveals a tree structure: Each of the uncertain parameters appears only once in one of the sum terms, the value sets of these terms are simple to construct.

The plant as well as the controller are SPR. Both together result in a stable control loop [3] which means that the plate experiment with the described controller is robustly stable. However, an interesting question is how much the eigenvalues differ from their nominal position. This question will be answered by construction of the corresponding stability profile.

The stability profile is constructed for both factors in (13) where the larger value of both contributes to the stability profile of the plate experiment. In Fig. 5 the stability profile is displayed with thick lines. The eigenvalue set of the open

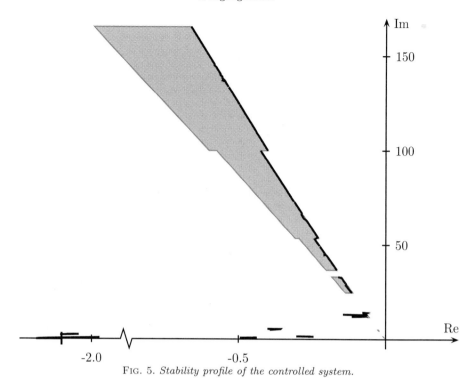

FIG. 5. *Stability profile of the controlled system.*

loop plant is shaded gray. From the graph it is obvious that the eigenvalue sets for high frequencies are nearly identical with the ones of the uncontrolled plants. This is due to the fact that the controller primarily shifts the dominant poles which are the poles with the smallest eigenfrequencies in this case.

Therefore, details for small eigenfrequencies are displayed in Fig. 6. The set of eigenvalues of the uncontrolled systems (gray) are situated very close to the imaginary axis. The eigenvalues of the controlled systems for the reference operating point are indicated by triangles. In this frequency range a significant shift of the eigenvalues occurred. The stability profile indicates that the eigenvalues differ only little from their nominal position for the considered uncertainty domain. The controller can therefore be deemed robustly stable.

6. Conclusion. Uncertain systems with nonlinear parameter dependency may be robustly unstable, though all surfaces of the operating domain are stable. Therefore, robustness analysis of uncertain systems requires in general examination of the entire operating domain. This, however, becomes more and more difficult as the number of uncertain parameters increases.

Tree structured uncertain systems can be analyzed extremely fast for robustness by construction of value sets even if the number of uncertain parameters is large. The value set approach can be utilized to compute the stability profile. This

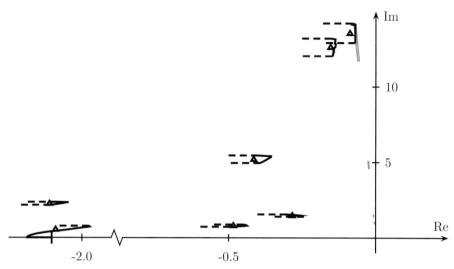

FIG. 6. *Stability profile of the controlled system (detail).*

set describes the right hand side of the boundary of the root set of the uncertain system and gives immediate information about the most important stability margins. The computational feasibility and the efficiency of the algorithms was demonstrated by a robustness analysis of the DLR plate experiment with 60 uncertain parameters.

References

[1] J. Ackermann, "Does it suffice to check a subset of multilinear parameters in robustness analysis?," *IEEE Trans. on Automatic Control*, vol. 37, pp. 487–488, 1992.
[2] J. Ackermann, A. Bartlett, D. Kaesbauer, W. Sienel, and R. Steinhauser, *Robust control: Systems with uncertain physical parameters*. London: Springer, 1993.
[3] J. Bals, *Aktive Schwingungsdämpfung flexibler Strukturen*. PhD thesis, Universität Karlsruhe (TH), 1989.
[4] B. Barmish, J. Ackermann, and H. Hu, "The tree structured decomposition: a new approach to robust stability analysis," in *Proc. Conf. on Information Sciences and Systems*, (Princeton), 1990.
[5] A. Bartlett, C. Hollot, and Huang-Lin, "Root locations of an entire polytope of polynomials: it suffices to check the edges," *Mathematics of Control, Signals and Systems*, vol. 1, pp. 61–71, 1988.
[6] L. Cremer, "Ein neues Verfahren zur Beurteilung der Stabilität linearer Regelungssysteme," *ZAMM*, p. 161, 1947.
[7] S. Dasgupta, "Kharitonov's theorem, revisited," *Systems and Control Letters*, vol. 11, pp. 381–384, 1988.
[8] DFVLR, Dornier System, "Investigation of attitude and orbit control system modelling for flexible space platforms." ESTEC Contract No. 6839/86/NL/MA(SC) performed by DFVLR, Final Report 1988.

[9] C. Goh and G. Grübel, "Robust low-order control of a generic flexible test structure," in *Proc. 1st IEEE Conference on Control Applications*, (Dayton, Ohio), pp. 206–211, 1992.

[10] V. Kharitonov, "Asymptotic stability of an equilibrium position of a family of systems of linear differential equations," *Differential Equations*, vol. 14, no. 3, pp. 26–35, 1979.

[11] A. Leonhard, "Neues Verfahren zur Stabilitätsuntersuchung," *Archiv für Elektrotechnik*, vol. 38, pp. 17–28, 1944.

[12] A. Mikhailov, "Method of harmonic analysis in control theory," *Avtomatika i Telemekhanika*, vol. 3, pp. 27–81, 1938.

[13] W. Sienel, *Analyse und Entwurf von robusten Regelungssystemen durch Konstruktion von komplexen Wertemengen*. PhD thesis, Technische Universität München, 1994.

Robust Stability of Family of Polynomials with 1-norm-bounded Parameter Uncertainties

Q.-H. Wu
Department of Automatic Control, Beijing Institute of Technology
Beijing 100081, China

M. Mansour
Automatic Control Laboratory, ETH-Zentrum, CH-8092 Zürich, Switzerland

> **Abstract.** We consider the robust stability problem of polynomial families whose coefficient vectors are affine in an uncertain parameter vector bounded by the Hölder 1-norm. We show that the value set of such a polynomial family is a convex par-polygon. Edge results are then established. Using these results we solve the robust Hurwitz stability problem for diamond of polynomials and the robust stabilization problem for control systems with diamond of plants.

1. Introduction. With respect to a given region $\mathcal{D}$ in the complex plane $\mathcal{C}$, which is symmetric with respect to the real axis, we say that a polynomial $f(s)$ is $\mathcal{D}$-stable if $f(s)$ has all roots in $\mathcal{D}$. Consider the polynomial family

$$(1) \qquad \mathcal{F}(s,\underline{\delta}) = a_n(\underline{\delta}) s^n + a_{n-1}(\underline{\delta}) s^{n-1} + \cdots + a_1(\underline{\delta}) s + a_0(\underline{\delta})$$

whose parameter vector

$$(2) \qquad \underline{a}(\underline{\delta}) := [\, a_0(\underline{\delta}) \ \ a_1(\underline{\delta}) \ \ \ldots \ \ a_n(\underline{\delta}) \,]^T$$

is affine in an uncertain vector $\underline{\delta}$, i.e.

$$(3) \qquad \underline{a}(\underline{\delta}) = \underline{a}_0 + T \cdot W \cdot \underline{\delta}$$

where $\underline{a}_0 \in \mathcal{R}^{n+1}$ is a constant vector, $T \in \mathcal{R}^{(n+1)\times M}$, $W = \mathrm{diag}\,(w_1\ w_2\ \ldots\ w_M)$ with $w_i > 0$, $\forall i$, $\|\underline{\delta}\|_p := \left(\sum_{i=1}^M |\delta_i|^p\right)^{1/p} \leq \delta$. $\mathcal{F}(s,\underline{\delta})$ is said to be robustly $\mathcal{D}$-stable if every polynomial $f(s) \in \mathcal{F}(s,\underline{\delta})$ is $\mathcal{D}$-stable. In the sequel, stability means $\mathcal{D}$-stability.

Robust stability problem has attracted tremendous research interests, see [8] and [10] for an excellent overview of the existing results. The research efforts and the results are related mainly to two problems: the stability radius problem, and the robustness checking problem.

The stability radius problem can be roughly described as follows. Assume that the nominal polynomial $f_0(s)$, obtained by setting the uncertainty $\underline{\delta} = \underline{0}$, is stable. We wish to find a maximum $\delta_{\max}$ such that for all $\|\underline{\delta}\|_p \leq \delta_{\max}$ $\mathcal{F}(s,\underline{\delta})$ is robustly stable. Soh et al. considered the case $p = 2$, $T = I_{n+1}$, $W = w \cdot I_{n+1}$ and $\mathcal{D}$ is the unit disc or the left half of the complex plane [2]. Qiu et al. solved the general problem for $p = 1, 2, \infty$ [4].

In the case of robustness checking, the norm of the uncertainty is given, say $\|\underline{\delta}\|_p \leq 1$, and we wish to find a subset $\mathcal{F}^*$ of $\mathcal{F}(s,\underline{\delta})$ such that $\mathcal{F}(s,\underline{\delta})$ is robustly stable if $\mathcal{F}^*$ is. Kharitonov showed that, in the case of $p = \infty$, $T = I_{n+1}$ and $\mathcal{D}$ is the open left-half complex plane, $\mathcal{F}^*$ is the set of only four extreme polynomials [1]. This result is then generalized to the case of polytope of polynomials. It turns out that $\mathcal{F}^*$ is the set of some edge polynomials [3]. For the same stability region, if we choose $p = 1$, $T = I_{n+1}$ and impose some restrictions on the uncertainty weightings w_i, Kharitonov et al. [9] showed that $\mathcal{F}^*$ is the set of 8 polynomials. The purpose of this paper is to solve the general robustness checking problem for the case $p = 1$, using the value set concept.

In section 2, we shall transform the value set problem to a minimum norm solution problem. In section 3 we solve the minimum norm solution problem and hence determine the boundary of the value set, and solve the robustness checking problem. Using the results obtained in section 3, we shall investigate the robust Hurwitz stability of a diamond of polynomials and solve the robust stabilization problem for control systems having diamond of plants.

2. Problem Formulation. Let us denote by $\partial \mathcal{D}$ the boundary of $\mathcal{D}$. Then for $s^* \in \partial \mathcal{D}$, the value set of $\mathcal{F}(s, \underline{\delta})$ is defined as

$$\Gamma(s^*, \underline{\delta}) := \{ f(s^*) \ : \ f(s) \in \mathcal{F}(s, \underline{\delta}) , \ \|\underline{\delta}\|_p \leq 1 \}. \tag{4}$$

Denote by $\underline{a}_i$ the ith column of the matrix $T \cdot W$, and by $\underline{s}_n$ the vector $[1 \ s \ \ldots \ s^n]^T$. Then $\mathcal{F}(s, \underline{\delta})$ can be represented as

$$\mathcal{F}(s, \underline{\delta}) = \underline{s}_n^T \underline{a}(\underline{\delta}) = f_0(s) + \sum_{i=1}^{M} f_i(s)\delta_i \tag{5}$$

where $f_i(s) = \underline{s}_n^T \underline{a}_i$. Associated with $f_i(s)$, we define the vectors:

$$\underline{f}_i := [y_i \ x_i]^T = [\,\mathrm{Im}(f_i(s^*)) \ \mathrm{Re}(f_i(s^*))\,]^T , \ i = 0, 1, 2, \ldots, M \tag{6}$$

$$\underline{x} = [x_1 \ x_2 \ \ldots \ x_M]^T , \quad \underline{y} = [y_1 \ y_2 \ \ldots \ y_M]^T$$

and the matrix $Z := [\underline{y} \ \underline{x}]$. Further, we denote by $\underline{z} = [y \ x]^T$ a vector in $\mathcal{R}^2$. It is then obvious that $\Gamma(s, \underline{\delta})$ can be represented as

$$\Gamma(s^*, \underline{\delta}) = \left\{ \underline{z} \ : \ \underline{z} = \underline{f}_0 + Z^T \underline{\delta}, \ \|\underline{\delta}\|_p \leq 1 \right\} = \underline{f}_0 + \tilde{\Gamma}(s^*, \underline{\delta}) \tag{7}$$

where

$$\tilde{\Gamma}(s^*, \underline{\delta}) := \left\{ \underline{z} \ : \ \underline{z} = Z^T \underline{\delta}, \ \|\underline{\delta}\|_p \leq 1 \right\} \tag{8}$$

It is clear that $\tilde{\Gamma}(s^*, \underline{\delta})$ is a simply connected domain in $\mathcal{R}^2$ and can be determined as long as its boundary, denoted by $\partial\tilde{\Gamma}(s^*, \underline{\delta})$, is known. We shall devote our attention to $\partial\tilde{\Gamma}(s^*, \underline{\delta})$.

Given a direction vector $\underline{t}_0 := [\sin\theta \ \cos\theta]^T$, where $\theta \in [0, \ 2\pi)$. Then, in this direction, $\partial\tilde{\Gamma}(s^*,\underline{\delta})$ is the vector $r_p(\theta)\underline{t}_0$, here $r_p(\theta)$ is the solution of the optimization problem

(9) $$r_p(\theta) := \sup\left\{\beta(\theta) \ : \ Z^T\underline{\delta} = \beta(\theta)\underline{t}_0, \ \|\delta\|_p = 1\right\}$$

Hence,

(10) $$\partial\tilde{\Gamma}(s^*,\underline{\delta}) = \{\underline{z} \ : \ \underline{z} = r_p(\theta)\cdot\underline{t}_0\}$$

where θ ranges from 0 to 2π. The optimization problem (9) can be transformed into a minimum norm solution problem:

(11) $$\rho_p(\theta) := \inf\left\{\|\underline{\tilde{\delta}}\|_p \ : \ Z^T\underline{\tilde{\delta}} = \underline{t}_0\right\}.$$

Let $\underline{\delta}_{\max}$ and $\underline{\delta}_{\min}$ be, respectively, the solution to (9) and (11). Then we have:

(12) $$r_p(\theta) = \frac{1}{\rho_p(\theta)}, \ \underline{\delta}_{\max} = \frac{\underline{\delta}_{\min}}{\rho_p(\theta)}.$$

In [6] it was shown that $\tilde{\Gamma}(s^*,\underline{\delta})$ is a parpolygon in the case of $p = \infty$. We shall show in the next section that $\tilde{\Gamma}(s^*,\underline{\delta})$ is also a parpolygon in the case of $p = 1$. Further, we shall give an algorithm for finding all the vertices of $\tilde{\Gamma}(s^*,\underline{\delta})$, hence solve the robustness checking problem for the case $p = 1$. Using this result, we shall check the robust Hurwitz stability of diamond of polynomials and the robust stabilization problem of diamond of plants.

3. The Main Results. Note that the results obtained in this section hold for all $\mathcal{D}$ and T. Let us first define the matrix

$$U_0 := \begin{bmatrix} \cos\theta & -\sin\theta \\ \sin\theta & \cos\theta \end{bmatrix},$$

and then the vectors

$$\begin{bmatrix} \underline{v}^T \\ \underline{u}^T \end{bmatrix} := U_0 \begin{bmatrix} \underline{y}^T \\ \underline{x}^T \end{bmatrix}$$

The following result is duo to Qiu *et al.* [4]:

THEOREM 3.1. *Let $p, q \geq 1$ be such that $\frac{1}{p} + \frac{1}{q} = 1$. Then*

$$\rho_p = \begin{cases} \infty, & \text{if } \operatorname{rank}\begin{bmatrix}\underline{t}_0 & Z^T\end{bmatrix} \neq \operatorname{rank}(Z^T) \\ \frac{1}{\|\underline{u}\|_q}, & \text{if } \operatorname{rank}\begin{bmatrix}\underline{t}_0 & Z^T\end{bmatrix} = \operatorname{rank}(Z^T) = 1 \\ \sup_{\alpha\in\mathcal{R}}\frac{1}{\|\underline{u}+\alpha\underline{v}\|_q}, & \text{if } \operatorname{rank}(Z^T) = 2. \end{cases}$$

In the sequel, we consider only the case $p = 1$ (this implies $q = \infty$). Denote

$$\rho(\alpha) = \frac{1}{\|\underline{u}+\alpha\underline{v}\|_\infty}, r(\alpha) = \|\underline{u}+\alpha\underline{v}\|_\infty$$

Then, $\rho_1 = \frac{1}{\inf_{\alpha \in \mathcal{R}} r(\alpha)}$. From the definition of $\|\cdot\|_\infty$ we get

$$r(\alpha) = \max\{|\alpha v_i + u_i|\} =$$

(13)
$$\begin{cases} \alpha \tilde{v}_1 + \tilde{u}_1 & \alpha \leq \alpha_{d_1} \\ \alpha \tilde{v}_i + \tilde{u}_i & \alpha_{d_{i-1}} < \alpha \leq \alpha_{d_i} \quad i = 2, 3, \ldots, l \\ \alpha \tilde{v}_{l+1} + \tilde{u}_{l+1} & \alpha_{d_l} < \alpha \end{cases}$$

where $[\tilde{v}_i \ \tilde{u}_i]^T \in \{\pm [v_i \ u_i]^T\}$. It is clear that $r(\alpha)$ is a polygonal function having the division points α_{d_i} obtained by setting

(14)
$$\alpha \tilde{v}_i + \tilde{u}_i = \alpha \tilde{v}_{i+1} + \tilde{u}_{i+1}.$$

Note that $|\alpha v_i + u_i|$ is also a polygonal function. Since $r(\alpha)$ demands selecting the one having the highest position, it is easily realized that $r(\alpha)$ is a concave polygonal function. Hence $\inf_{\alpha \in \mathcal{R}}\{r(\alpha)\} = r(\alpha_{d_j})$, herein α_{d_j} is the unique division point of $r(\alpha)$ such that $\tilde{v}_{j-1}\tilde{v}_j \leq 0$.

LEMMA 3.2. *Let α_{d_j} be the division point of the function*

$$r(\alpha) = \begin{cases} \alpha v_j + u_j & \alpha \leq \alpha_{d_j} \\ \alpha v_{j+1} + u_{j+1} & \alpha > \alpha_{d_j} \end{cases}$$

that is, α_{d_j} satisfies $\alpha_{d_j} v_j + u_j = \alpha_{d_j} v_{j+1} + u_{j+1}$. $[v_i \ u_i]^T = U_0 \underline{f}_i$, $i = j, j+1$. Further, let λ and r be such that $\underline{f}_j + \lambda\left(\underline{f}_{j+1} - \underline{f}_j\right) = r\underline{t}_0$. Then

$$\lambda = \frac{1}{1 - v_{j+1}/v_j}, \quad r = r(\alpha_{d_j}).$$

Lemma 3.2 can be proved by straightforward algebra. The details are therefore omitted. This Lemma has stated the following fact: Given the vectors $\underline{f}_j$, $\underline{f}_{j+1}$ and $\underline{t}_0$. Scale $\underline{t}_0$ by a factor r so that $r\underline{t}_0$ ends on the line passing through $\underline{f}_j$ and $\underline{f}_{j+1}$. Then the length of $r\underline{t}_0$ is equivalent to $r(\alpha_{d_j})$, i.e. the function value of $r(\alpha)$ evaluated at the division point α_{d_j}.

Now, we denote by $\text{conv}\left(\{\pm \underline{f}_i\}\right)$ the minimum convex hull of the vectors $\{\pm \underline{f}_i\}$. Here *minimum* means that $\{\pm \underline{f}_i\} \in \text{conv}\left(\{\pm \underline{f}_i\}\right)$. However, for any other convex set $\mathcal{S} \supset \{\pm \underline{f}_i\}$ we have $\mathcal{S} \supset \text{conv}\left(\{\pm \underline{f}_i\}\right)$. It is clear that $\text{conv}\left(\{\pm \underline{f}_i\}\right)$ is the convex parpolygon having the vertices

$$\left\{\pm \underline{f}_1^v, \pm \underline{f}_2^v, \ldots, \pm \underline{f}_l^v\right\} \subseteq \left\{\pm \underline{f}_1, \pm \underline{f}_2, \ldots, \pm \underline{f}_M\right\}.$$

It is readily checked that if $\underline{f}_i \in \text{conv}\left(\{\pm \underline{f}_i\}\right)$ but is not a vertex, then $[v_i \ u_i]^T = U_0 \underline{f}_i$ doesn't contribute to $r(\alpha)$.

Based on Lemma 3.2 we can prove the following result:

THEOREM 3.3. $\tilde{\Gamma}(s^*, \underline{\delta}) = conv\left(\left\{\pm \underline{f}_i\right\}\right)$.

Proof. Consider first the case $\theta \in [0, \pi)$. By replacing $\underline{f}_j$ by $-\underline{f}_j$ for $\angle \underline{f}_j - \angle \underline{t}_0 < -\frac{\pi}{2}$ or $\angle \underline{f}_j - \angle \underline{t}_0 \geq \frac{\pi}{2}$, we have

$$\angle \underline{f}_1^v < \angle \underline{f}_2^v < \cdots < \angle \underline{f}_k^v < \angle \underline{t}_0 < \angle \underline{f}_{k+1}^v < \cdots < \angle \underline{f}_l^v$$

and $-\frac{\pi}{2} \leq \angle \underline{f}_1^v - \angle \underline{t}_0$, $\angle \underline{f}_l^v - \angle \underline{t}_0 < \frac{\pi}{2}$. Denote $[\tilde{v}_i \ \tilde{u}_i]^T = U_0 \underline{f}_i^v$. From the fact that $\pm \underline{f}_i^v$ are the vertices of a convex parpolygon, we can show that at the division point at which the slope of $r(\alpha)$ changes the sign there holds

$$r(\alpha) = \begin{cases} \alpha \tilde{v}_k + \tilde{u}_k & \alpha_{d_{k-1}} < \alpha \leq \alpha_{d_k} \\ \alpha \tilde{v}_{k+1} + \tilde{u}_{k+1} & \alpha_{d_k} < \alpha \leq \alpha_{d_{k+1}} \end{cases}.$$

Hence, $\min\{r(\alpha)\} = r(\alpha_{d_k})$. From Lemma 3.2, $r(\alpha_{d_k})$ satisfies

$$\lambda_k \underline{f}_{k+1}^v + (1 - \lambda_k)\underline{f}_k^v = r(\alpha_{d_k})\underline{t}_0,$$

here $\lambda_k = \frac{1}{1 - \tilde{v}_{k+1}/\tilde{v}_k}$. The above equation shows that the vector $r(\alpha_{d_k})\underline{t}_0$ ends on the straight line passing through $\underline{f}_k^v$ and $\underline{f}_{k+1}^v$. Since $\underline{t}_0$ is arbitrary, we conclude that between $\underline{f}_k^v$ and $\underline{f}_{k+1}^v$ $\partial \tilde{\Gamma}(s^*, \underline{\delta})$ is the line segment joining $\underline{f}_k^v$ and $\underline{f}_{k+1}^v$. Note that

$$\underline{\delta} = [0 \ \ldots \ 0 \ \lambda_k \ 1 - \lambda_k \ 0 \ \ldots \ 0]^T.$$

From $\tilde{v}_{k+1} \cdot \tilde{v}_k \leq 0$ we have $0 \leq \lambda \leq 1$. Hence $\|\underline{\delta}\|_1 = 1$ and $\underline{\delta}$ is the solution to (9) in case of $p = 1$. For $\theta \in [\pi, 2\pi)$, we take $-\underline{\delta}$ and get $-\underline{z} = r(\theta)(-\underline{t}_0)$. It is obvious that $r(\theta) = r(\theta - \pi)$. Hence the result. $\square$

From Theorem 2 we know that the key step in checking the robust stability of $\mathcal{F}(s, \underline{\delta})$ is to determine the extreme points $\left\{\underline{f}_i^v\right\}$, or equivalently, the extreme polynomials $\{f_i^v(s)\}$. Having done that, we need only to check the stability of the edge polynomials $f_0(s) \pm f_i^v(s) + \lambda\left(\pm f_{i+1}^v(s) \mp f_i^v(s)\right)$. We shall now develop an algorithm for finding $\left\{\underline{f}_i^v\right\}$.

On noting that by replacing $-\underline{f}_i$ for $\underline{f}_i$ if $\angle \underline{f}_i \geq \pi$, we can renumber the vectors $\underline{f}_i$ so that $\angle \underline{f}_1 \leq \angle \underline{f}_2 \leq \cdots \angle \underline{f}_M$ and $0 \leq \angle \underline{f}_i < \pi$ $\forall i$. Furthermore, by taking $\underline{f}_i$ to be the vector $\underline{f}_{j+i}$ such that $\left\|\underline{f}_{j+i}\right\|_2 = \max_m \left\{\left\|\underline{f}_{j+m}\right\|_2\right\}$ and ignoring all the other vectors such that $\angle \underline{f}_j = \angle \underline{f}_{j+1} = \cdots = \angle \underline{f}_{j+k}$, we can assume without loss of generality that $\angle \underline{f}_1 < \angle \underline{f}_2 < \cdots < \angle \underline{f}_N$. Now, let us define the 3×3 matrices

$$S_i := \begin{bmatrix} 1 & 1 & 1 \\ y_i & y_{i+1} & y_{i+2} \\ x_i & x_{i+1} & x_{i+2} \end{bmatrix}, \quad i = 1, 2, \ldots, N$$

where $[y_{N+i} \ x_{N+i}]^T = [-y_i \ -x_i]^T$.

THEOREM 3.4. $\pm \underline{f}_1, \pm \underline{f}_2, \ldots, \pm \underline{f}_N$ are the vertices of a convex parpolygon if and only if $det(S_i) < 0$ for all $i = 1, 2, \ldots, N$.

4. Robust Hurwitz Stability and Robust Stabilization.
In this section, we shall consider the robust Hurwitz stability of diamond of polynomials and the robust stabilization problem for diamond of plants. We shall see that in these two special cases the number of the edge polynomials will be heavily reduced. In some certain cases, it suffices to check only the extreme points. Since we consider the Hurwitz stability, it is assumed throughout this section that $\mathcal{D}$ is the open left-half plane. Hence, $\partial \mathcal{D} = \{j\omega\} \cup \{0\}$.

4.1. Robust stability of diamond of polynomials.
Setting $T = I_{n+1}$ in (3), $\mathcal{F}(s, \underline{\delta})$ becomes a diamond of polynomials. Let $W = \mathrm{diag}\,[w_0 \; w_1 \; \ldots \; w_n]$. Then

$$\begin{bmatrix} \underline{f}_1 & \underline{f}_2 & \cdots & \underline{f}_{n+1} \end{bmatrix} = \begin{bmatrix} 0 & w_1\omega & 0 & w_3\omega^3 & 0 & \cdots \\ w_0 & 0 & w_2\omega^2 & 0 & w_4\omega^4 & \cdots \end{bmatrix}$$

Denote by

$$\omega_{e,1} = \min\left\{\sqrt{\frac{w_0}{w_2}}, \sqrt[4]{\frac{w_0}{w_4}}, \sqrt[6]{\frac{w_0}{w_6}}, \ldots \right\}$$

$$\omega_{o,1} = \min\left\{\sqrt{\frac{w_1}{w_3}}, \sqrt[4]{\frac{w_1}{w_5}}, \sqrt[6]{\frac{w_1}{w_7}}, \ldots \right\}$$

and

$$\omega_{e,1} = \sqrt[2I_1]{\frac{w_0}{w_{2I_1}}}, \quad \omega_{o,1} = \sqrt[2J_1]{\frac{w_1}{w_{2J_1+1}}}$$

If $\omega_{e,1}$ is achieved at more than one w_{2i}, the maximum index i is taken for I_1. The same consideration is made for J_1.

For $i = 2, 3, \ldots, m$, $j = 2, 3, \ldots, l$ we define

$$\omega_{e,i} = \min\left\{\sqrt{\frac{w_{2I_{i-1}}}{w_{2(I_{i-1}+1)}}}, \sqrt[4]{\frac{w_{2I_{i-1}}}{w_{2(I_{i-1}+2)}}}, \sqrt[6]{\frac{w_{2I_{i-1}}}{w_{2(I_{i-1}+3)}}}, \ldots \right\}$$

$$\omega_{o,j} = \min\left\{\sqrt{\frac{w_{2(J_{j-1}+1)}}{w_{2(J_{j-1}+3)}}}, \sqrt[4]{\frac{w_{2(J_{j-1}+1)}}{w_{2(J_{j-1}+5)}}}, \sqrt[6]{\frac{w_{2(J_{j-1}+1)}}{w_{2(J_{j-1}+7)}}}, \ldots \right\}$$

$$\omega_{e,i} = \sqrt[2I_i]{\frac{w_{2I_{i-1}}}{w_{2I_i}}}, \quad \omega_{o,j} = \sqrt[2J_j]{\frac{w_{2J_{j-1}+1}}{w_{2J_j+1}}}.$$

Similarly as before, we take the maximum i for I_i if $\omega_{e,i}$ is achieved at more than one w_{2i}. The same consideration is made for J_j. It is clear that

$$I_m = \begin{cases} \frac{n}{2} & n = \text{even} \\ \frac{n-1}{2} & n = \text{odd} \end{cases}, \quad J_l = \begin{cases} \frac{n}{2}-1 & n = \text{even} \\ \frac{n-1}{2} & n = \text{odd} \end{cases}.$$

Now, we define the sets of switching frequencies

$$\Omega_e = \{\omega_{e,0} \; \omega_{e,1} \; \ldots \; \omega_{e,m}\}, \Omega_o = \{\omega_{o,0} \; \omega_{o,1} \; \ldots \; \omega_{o,l}\},$$

and the sets of the indices

$$I = \{I_0, I_1, I_2, \ldots, I_m\}, \quad J = \{J_0, J_1, J_2, \ldots, J_l\}$$

herein $\omega_{e,0} = \omega_{o,0} = 0$, $I_0 = J_0 = 0$. It is apparent that corresponding an arbitrary frequency ω satisfying

$$\omega_{e,r} \leq \omega < \omega_{e,r+1}, \quad \text{and} \quad \omega_{o,s} \leq \omega < \omega_{o,s+1}$$

the value set $\tilde{\Gamma}(j\omega, \underline{\delta})$ is a diamond with the vertices $\begin{bmatrix} 0 & \pm w_{2I_r}\omega^{2I_r} \end{bmatrix}^T$ and $\begin{bmatrix} \pm w_{2J_s+1}\omega^{2J_s+1} & 0 \end{bmatrix}^T$. Correspondingly, the vertex polynomials are

$$f_1^v(s) = f_0(s) + \delta f_1(s), \quad f_2^v(s) = f_0(s) + \delta f_2(s)$$
$$f_3^v(s) = f_0(s) - \delta f_1(s), \quad f_4^v(s) = f_0(s) - \delta f_2(s)$$

here

$$\delta f_1(s) = (-1)^{I_r} w_{2I_r} s^{2I_r}, \quad \delta f_2(s) = (-1)^{J_s} w_{2J_s+1} s^{2J_s+1}.$$

When I_r ranges over I and J_s ranges over J, we obtain the set of all polynomials which determine the value set. In general, we have to check the stability of the edge polynomials of the form $\lambda_i f_i^v(s) + (1-\lambda_i) f_{i+1}^v(s)$ in order to check the robust stability of the diamond. Note that it is easy to check the Hurwitz stability of an edge polynomial, see e.g. [5].

Using Rantzer's *convex direction polynomial* concept [7], we can show the following extreme points result:

THEOREM 4.1. *The robust stability of $\mathcal{F}(s, \underline{\delta})$ is equivalent to the stability of a few extreme polynomials, if and only if for every $\omega_{e,r} \leq \omega < \omega_{e,r+1}$, $\omega_{o,s} \leq \omega < \omega_{o,s+1}$ the corresponding $I_r \in I$ and $J_s \in J$ satisfy $J_s = I_r$ or $J_s = I_r - 1$.*

The proof of this theorem is straightforward and is omitted.

4.2. Robust stabilization of control systems. Consider the control system in which the plant is described by

$$P(s, \underline{\delta}) \in \mathcal{P}(s, \underline{\delta}) := \left\{ P(s) : P(s) = \frac{\sum_{i=0}^{n} b_i(\underline{\delta}) s^i}{\sum_{i=0}^{n} a_i(\underline{\delta}) s^i} \right\}$$

where

$$\begin{bmatrix} \underline{a}(\underline{\delta}) \\ \underline{b}(\underline{\delta}) \end{bmatrix} = \begin{bmatrix} \underline{a}_0 \\ \underline{b}_0 \end{bmatrix} + \begin{bmatrix} W_a & 0 \\ 0 & W_b \end{bmatrix} \cdot \begin{bmatrix} \underline{\delta}_a \\ \underline{\delta}_b \end{bmatrix}, \quad \left\| \begin{bmatrix} \underline{\delta}_a \\ \underline{\delta}_b \end{bmatrix} \right\|_1 \leq 1$$

and

$$W_a = \text{diag}\,(w_{a,0} \quad w_{a,1} \quad \ldots \quad w_{a,n}), \quad W_b = \text{diag}\,(w_{b,0} \quad w_{b,1} \quad \ldots \quad w_{b,n}),$$

and the controller has the transfer function $C(s) = \frac{N_C(s)}{D_C(s)}$. The characteristic polynomial of the closed-loop system is $F(s,\underline{\delta}) = f_0(s) + \delta f(s)$, here

$$\delta f(s) = D_C(s) \sum_{i=0}^{n} w_{a,i} \delta_{a,i} s^i + N_C(s) \sum_{i=0}^{n} w_{b,i} \delta_{b,i} s^i.$$

The robust stabilization problem is to find a controller $C(s)$ so that the closed-loop system is stable for all $P(s,\underline{\delta}) \in \mathcal{P}(s,\underline{\delta})$, i.e. $F(s,\underline{\delta})$ is Hurwitz for all $\|\underline{\delta}\|_1 \leq 1$.

Case I: $D_C(j\omega) = 0$. In this case, $\delta f(j\omega) = N_C(j\omega) \sum_{i=0}^{n} w_{b,i} \delta_{b,i} (j\omega)^i$. The value set is a rotated and scaled version of a diamond. Consequently, the stability of four edge polynomials has to be checked.

Case II: $D_C(j\omega) \neq 0$. In this case

$$\begin{aligned}\delta f(s) &= D_C(s) \left(\sum_{i=0}^{n} w_{a,i} \delta_{a,i} s^i + C(s) \sum_{i=0}^{n} w_{b,i} \delta_{b,i} s^i \right) \\ &= D_c(s) \left(D_\delta(s) + C(s) N_\delta(s) \right).\end{aligned}$$

Note that both $D_\delta(j\omega)$ and $N_\delta(j\omega)$ have four vertices. Then, by viewing $C(j\omega)$ as a linear transformation, it follows that the vertices of $C(j\omega)N_\delta(j\omega)$ are just a rotated and scaled version of $N_\delta(j\omega)$. The value set of $\delta f(j\omega)$ can be a quadrangle, a hexagon or an octagon. As $\angle C(j\omega)$ varies, the position of the vertices of $C(j\omega)N_\delta(j\omega)$ is also varying. Hence, the edge polynomials are also changing. Checking the stability of these edge polynomials, we know whether a controller $C(s)$ robustly stabilizes $\mathcal{P}(s,\underline{\delta})$. The details are omitted. Obviously, when $C(s)$ is P-, PI- PID-, Lead-, Lag-or Lead-Lag-controller, the number of the edge polynomials will be heavily reduced. Since these controllers are widely used in the practice, the results approached in this subsection is applicable.

5. Conclusion. In this paper, we considered the robust stability problem for polynomial families whose coefficient vectors are affine in an uncertain parameter vector bounded by the Hoelder 1-norm. We show that for a fixed frequency ω, the value set of a given polynomial family is a convex parpolygon. The difference from the ∞-norm case is the following: in the ∞-norm case, all the extreme polynomials are needed for constructing the value set [6], while in the 1-norm case, the value set is determined by some subset of the extreme polynomials. The subset changes as the frequency ω varies. We show that the subset can be determined by a very simple algebraic condition. Using these results, we solved then the stability problem for a diamond of polynomials and the robust stabilization problem for control systems with diamond plants.

Acknowledgment. The first author would like to thank Professor Gao Hui-Qi for her support in the final stage of this research work.

References

[1] V.L.Kharitonov, Asymptotic stability of an equilibrium position of a family of linear differential equations, *Differential Equations*, **14**, 1483–1485, 1978.

[2] C.B.Soh, C.S.Berger and K.P.Dabke, On the stability properties of polynomials with perturbed coefficients, *IEEE Trans. Auto. Contr.*, **AC-30**, 1033–1036, 1985.

[3] A.C.Bartlett, C.V.Hollot and L.Huang, Root location of an entire polytope of polynomials: it suffices to check the edges, *Mathe. Control, Signals and Systems*, **1**, 173–179, 1988.

[4] L.Qiu and E.J.Davison, A simple procedure for the exact stability robustness computation of polynomials with affine coefficient perturbations, *Systems and Control Letters*, **13**, 413–420, 1989.

[5] E. Zeheb, Necessary and sufficient conditions for root clustering of a polytope of polynomials in simply connected domain, *IEEE Trans. Auto. Contr.*, **34**, 986–990, 1989.

[6] F.J.Kraus and W.Truoel, Robust stability of control systems with polytopical uncertainty: a Nyiquist approach, *Int. J. Control*, **53**, 967–983, 1991.

[7] A.Rantzer, Stability conditions for a polytope of polynomials, *IEEE Trans. Auto. Contr.*, **AC-37**, 79–89, 1992.

[8] B.R.Barmish and H.I.Kang, A survey of extreme point result for robustness of control systems, *Automatica*, **29**, 13–35, 1993.

[9] V.L.Kharitonov and R.Tempo, On the stability of a weighted diamond of real polynomials, *Systems and Control Letters*, **22**, 5–7, 1994.

[10] B.R.Barmish, *New Tools for Robustness of Linear Systems*, Macmillan Publishing Company, New York, 1994.

On the Characterization and Formation of Local Convex Directions for Hurwitz Stability*

Ezra Zeheb

Department of Electrical Engineering, Technion, Haifa 32000, Israel

Abstract. Let $P_o(s)$ be a real given Hurwitz polynomial. A local convex direction with respect to $P_o(s)$ is a polynomial $P_1(s)$, such that all polynomials which belong to the convex combination of $P_o(s)$ and $P_1(s)$ are Hurwitz. The main result of this paper is the characterization of local convex directions, which enables the derivation of a continuum of pertinent real polynomials $P_1(s)$, by a very simple algorithm. In arriving at this result, first a sufficient condition and then a necessary and sufficient condition are derived for a polynomial $P_1(s)$ to be a local convex direction of a given Hurwitz polynomial $P_o(s)$. These conditions are related to the property of strict positive realness of a rational function, and can be tested by tractable methods.

1. Introduction. Let $P_o(s)$ be a real stable polynomial. Then, a real polynomial $Q(s)$, which is the direction of the segment

(1) $$P_o(s) + \lambda Q(s)$$

is termed a "convex direction", if the stability of the segment (1) $\forall \lambda \in [0,1]$ is implied by the stability of the extreme polynomials $P_o(s)$ and $P_1(s) = P_o(s) + Q(s)$.

Note that the segment (1) is the convex combination of the two polynomials $P_o(s)$ and $P_1(s)$,

(2) $$(1-\lambda)P_o(s) + \lambda P_1(s), \quad 0 \leq \lambda \leq 1.$$

Numerous authors have recently treated the concept of convex directions, since it is related to problems of robust stability of systems with uncertainties. Rantzer [12] derived a condition on the argument of $Q(j\omega)$, which is necessary and sufficient for $Q(s)$ to be a convex direction for any $P_o(s)$ (real and stable) of degree larger than the degree of $Q(s)$. However, methods to test this condition in the general case are not presented. Special classes of polynomials which can be identified as convex directions (without referring to this term) were treated by Petersen [11], Hollot and Yang [10], Rantzer [13] and Fu [5]. Fu [6] also derived tractable necessary and sufficient conditions for $Q(s)$ to be a convex direction.

Recently, Hinrichsen and Kharitonov[9] pointed out that the above concept of *global* convex direction may be too restrictive and thus conservative for a variety of robust stability applications. A polynomial $Q(s)$ may ensure the stability of (1) $\forall \lambda \in [0,1]$ for a *given* polynomial $P_o(s)$, without being a convex direction. Thus, the less restrictive concept of "local convex direction" is introduced, and a graphical test for this property is derived, in [9]. A necessary and sufficient condition for a polynomial $Q(s)$ to be a "local convex direction" is readily implied by the

---
* The research was supported by the Fund for the Promotion of Research at the Technion.

results in [14]. Given two real polynomials, $P_o(s)$ (stable) and $P_1(s)$, an algebraic tractable necessary and sufficient condition, expressed on $P_o(s)$ and $P_1(s)$, ensuring the stability of (2), is derived in [14]. This can be readily translated into the following necessary and sufficient condition, expressed on the given stable $P_o(s)$ and on $Q(s)$, such that $Q(s)$ is a "local convex direction":
Let the even and odd parts of $P_o(s)$ be denoted by $m_o(s)$ and $n_o(s)$, respectively. Let the even and odd parts of $Q(s)$ be denoted by $m_q(s)$ and $n_q(s)$, respectively. Then,

THEOREM 1.1. *Let $P_o(s)$ be a given real Hurwitz polynomial. A real polynomial $Q(s)$ is a local convex direction, i.e. (1) is a family of Hurwitz polynomials $\forall \lambda \in [0,1]$, if, and only if, at every real value ω_o which is a root of*

$$(3) \qquad m_o(j\omega)n_q(j\omega) - n_o(j\omega)m_q(j\omega) = 0,$$

the value of the following expression is positive:

$$(4) \qquad m_o^2(j\omega_o) + m_o(j\omega_o) \cdot m_q(j\omega_o) > 0,$$

unless $m_o(j\omega_o) = 0$ and $m_q(j\omega_o) = 0$. In the latter case, condition (4) is replaced by

$$(5) \qquad n_o^2(j\omega_o) + n_o(j\omega_o) \cdot n_q(j\omega_o) > 0.$$

The proof of this theorem is omitted, since it is an immediate outcome of the results in [14]. Note that from the computational aspect, testing a condition such as (3),(4) or (3),(5) amounts to operating Sturm tests, which can be carried out by a final number of elementary operations [6].

Other necessary and sufficient conditions for Hurwitzness of a convex combination of two Hurwitz polynomials, are derived in [1] and in [3]. However, all these conditions, including Theorem 1.1, which are suitable for testing Hurwitzness of a convex combination of two *given* polynomials, are not suitable (at least in an easy way) for *finding* the convex directions of a single given polynomial.

In this paper, we treat the following problem: given a real Hurwitz polynomial $P_o(s)$, derive a set of real polynomials $Q(s)$, which are "local convex directions". We derive a simple characterization which allows the formation of "local convex directions".

The structure of this paper is as follows: In Section 2, we relate the concept of "positive realness" of a rational function to the property of stable convex combination of two polynomials, and derive a sufficient condition for the latter. We then "refine" the positive realness property to derive a necessary and sufficient condition for the stability of the convex combination of two polynomials. This condition is equivalent to the one given in [14], but formulated differently, proved differently, and needed for our further results on the characterization and formation of "local convex directions". In Section 3, we derive the conditions characterizing local convex directions and enabling the formation of sets of such polynomials. We conclude in Section 4.

Throughout this paper, the results are expressed in terms of the polynomials $P_o(s)$ and $P_1(s)$ as in (2). It is a trivial matter to express the results in terms of the polynomials $P_o(s)$ and $Q(s) = P_1(s) - P_o(s)$, if so desired.

2. Conditions for Stability of a Convex Combination. Let $P_o(s)$ and $P_1(s)$ be real polynomials. We will first show that strict positive realness (SPR-ness) of the rational function $P_o(s)/P_1(s)$ implies stability of the convex combination of $P_o(s)$ and $P_1(s)$, i.e.

$$(1-\lambda)P_o(s) + \lambda P_1(s)$$

is a Hurwitz polynomial for all $\lambda \in [0,1]$.

There are several formulations of the SPR condition. We will use the following:

DEFINITION 2.1. *A rational real function $P_o(s)/P_1(s)$ is SPR if $P_1(s)$ is Hurwitz stable (all the zeros of $P_1(s)$ are in the open left half complex plane), and*

$$(6) \qquad \operatorname{Re}\left[\frac{P_o}{P_1}\right]_{s=j\omega} > 0 \quad \forall \, \omega \in \mathbb{R}$$

where $\mathbb{R}$ is the set of real numbers and $\operatorname{Re}[\cdot]$ is the real part of $[\cdot]$.

THEOREM 2.2. *Let $P_o(s)$ and $P_1(s)$ be real polynomials. Then*

$$(7) \qquad \frac{P_o(s)}{P_1(s)} \quad is \quad SPR \quad \Rightarrow \quad (1-\lambda)P_o(s) + \lambda P_1(s) \in H \quad \forall \, 0 \leq \lambda \leq 1$$

where H is the set of Hurwitz polynomials.

The proof of Theorem 2.2 is based on known or easily verified facts about SPR functions, and will be omitted in this presentation.

The condition in Theorem 2.2 requires the real part of $P_o(j\omega)/P_1(j\omega)$ to be strictly positive for all $\omega \in \mathbb{R}$. We will now prove that requiring the real part of $P_o(j\omega)/P_1(j\omega)$ to be strictly positive only at some discrete values of ω, yields a necessary and sufficient condition for the Hurwitz stability of the convex combination of $P_o(s)$ and $P_1(s)$.

THEOREM 2.3. *Let $P_o(s)$ and $P_1(s)$ be real polynomials, and $P_o(s) \in H$. Then,*

$$(8) \qquad [(1-\lambda)P_o(s) + \lambda P_1(s)] \in H \quad \forall \, 0 \leq \lambda \leq 1$$

if, and only if, for every imaginary value of s, ($s = j\omega_o$, $\omega_o \in \mathbb{R}$) for which $\operatorname{Im}[P_o/P_1] = 0$, $\operatorname{Re}[P_o/P_1] > 0$.

Proof of Theorem 2.3

Necessity. Suppose, in contradiction, that there exists a real non-positive number K such that for some $s_o = j\omega_o$,

$$\left.\frac{P_o}{P_1}\right|_{s_o=j\omega_o} = K \leq 0 \tag{9}$$

Let

$$\lambda \triangleq \frac{K}{K-1} \tag{10}$$

Then,

$$0 \leq \lambda \leq 1 \tag{11}$$

where $\lambda = 0$ only if $K = 0$ and $\lambda = 1$ only if $K = \infty$.
Also,

$$\left.\frac{P_o}{P_1}\right|_{s_o=j\omega_o} = \frac{\lambda}{\lambda-1}, \tag{12}$$

so that

$$[\lambda P_1 + (1-\lambda)P_o]_{s_o=j\omega_o} = 0 . \tag{13}$$

Therefore, there exists $0 < \lambda \leq 1$ such that $\lambda P_1 + (1-\lambda)P_o \notin H$. Hence the necessity of the condition.

Sufficiency. Let

$$\left.\frac{P_o}{P_1}\right|_{s=j\omega} = a(\omega) + jb(\omega) \tag{14}$$

and let λ be a real number such that $0 < \lambda < 1$. By the conditions of the theorem, for every $\omega_o \in \mathbb{R}$ such that $b(\omega_o) = 0$, $a(\omega_o) > 0$ and hence, there is no $\omega_o \in \mathbb{R}$ such that $a(\omega_o) = b(\omega_o) = 0$. Therefore,

$$P_1(j\omega) = \frac{P_o(j\omega)}{a(\omega) + jb(\omega)}$$

and

$$[(1-\lambda)P_o + \lambda P_1]_{s=j\omega} = \left[\left(\frac{\lambda a(\omega)}{a^2(\omega)+b^2(\omega)} + 1 - \lambda\right) - j\frac{b(\omega)}{a^2(\omega)+b^2(\omega)}\right] P_o(j\omega). \tag{15}$$

Since $P_o(s) \in H$, $P_o(j\omega) \neq 0$ $\forall \omega \in \mathbb{R}$. Now, for every $\omega \in \mathbb{R}$ such that $b(\omega) \neq 0$, the imaginary part of the brackets in the right hand side of (15) does not vanish, and

(16) $$[(1-\lambda)P_o + \lambda P_1]_{s=j\omega} \neq 0 \ .$$

Moreover, for every $\omega_o \in \mathbb{R}$ such that $b(\omega_o) = 0$, since $a(\omega_o) > 0$ and $0 < \lambda < 1$, the real part of the brackets in the right hand side of (15) does not vanish, and

(17) $$[(1-\lambda)P_o + \lambda P_1]_{s=j\omega_o} \neq 0.$$

We conclude that

(18) $$[(1-\lambda)P_o + \lambda P_1]_{s=j\omega} \neq 0 \quad \forall \omega \in \mathbb{R} \ .$$

Now, since $P_o(s) \in H$ and since the zeros of a polynomial are continuous functions of its coefficients, (18) implies

(19) $$(1-\lambda)P_o(s) + \lambda P_1(s) \in H.$$

$\square$

3. The Formation of Local Convex Directions. In this section we will treat the following problem: given a real Hurwitz polynomial $P_o(s)$, derive a set of real polynomials $P_1(s)$ such that

(20) $$(1-\lambda)P_o(s) + \lambda P_1(s) \in H \quad \forall \, 0 \leq \lambda \leq 1 \, .$$

Denote the even and odd parts of $P_o(s)$ by $m_o(s)$ and $n_o(s)$, respectively. Denote the even and odd parts of $P_1(s)$ by $m_1(s)$ and $n_1(s)$, respectively.

Let $P_o(s)$ and $P_1(s)$ be such that the zeros of the polynomials $m_o m_1$ and $n_o n_1$, will be purely imaginary and will "doubly interlace" i.e., after the double zero of $n_o n_1$ at the origin, two zeros of $m_o m_1$ appear along the imaginary axis of the complex plane, then two zeros of $n_o n_1$, then again two zeros of the polynomial $m_o m_1$ and so on. (A double zero of $m_o m_1$ or of $n_o n_1$ is counted twice). A single zero of $m_o m_1$ or of $n_o n_1$ may appear at the "end" (the greatest distance form the origin) of the double interlacing.

Remark 1. Note that this double interlacing condition necessitates the degrees of $P_o(s)$ and $P_1(s)$ to differ by at most one.

Remark 2. Since $P_o(s)$ is Hurwitz, the zeros of $m_o(s)$ and $n_o(s)$ are necessarily imaginary and interlacing. Note that the "double interlacing" property then ensures that the zeros of $m_1(s)$ and $n_1(s)$ are also imaginary and interlacing, and hence, $P_1(s)$ is necessarily Hurwitz too.

Denote the set of frequencies whose values are between the values of the frequencies of two consecutive zeros of $m_o m_1$ or of $n_o n_1$, by S_R. Denote the complementary set of frequencies, i.e. those whose values are between the values of frequencies of a zero of $m_o m_1$ and a consecutive zero of $n_o n_1$ or vice versa, by S_I. (See Fig. 1). The distinct frequencies for which $m_o m_1$ is zero or $n_o n_1$ is zero belong to S_R if the corresponding zero is simple, and belong to S_I if the corresponding zero is double.

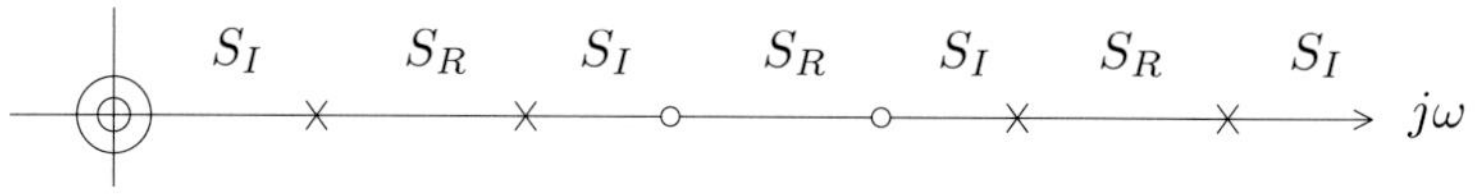

```
      S_I    S_R   S_I   S_R   S_I   S_R   S_I
   ⊕───×─────×─────o─────o─────×─────×────→ jω
```

Fig. 1. The sets of frequencies S_R and S_I.
X – a zero of the polynomial $m_o(s)m_1(s)$.
0 – a zero of the polynomial $n_o(s)n_1(s)$.

LEMMA 3.1.

(21) $$\operatorname{Im} \frac{P_o(j\omega_o)}{P_1(j\omega_o)} = 0 \quad \text{implies} \quad \omega_o \in S_I.$$

LEMMA 3.2.

(22) $$\operatorname{Re} \frac{P_o(j\omega)}{P_1(j\omega)} > 0 \quad \forall\, \omega \in S_I.$$

The proofs of Lemmas 3.1 and 3.2 will be omitted in this presentation.

THEOREM 3.3. Let $P_o(s)$ and $P_1(s)$ be Hurwitz polynomials, such that the zeros of the polynomials $m_o m_1$ and $n_o n_1$ are "doubly interlacing" in the sense explained above. Then, (20) is satisfied.

$$(1-\lambda)P_o(s) + \lambda P_1(s) \in H \quad \forall\, 0 \leq \lambda \leq 1.$$

Proof. Lemmas 3.1 and 3.2, in addition to Theorem 2.3, imply Theorem 3.3.

Remark 3. In [7] it is shown that if $m_o + n_1$ and $m_1 + n_o$ are Hurwitz polynomials, in addition to the Hurwitzness of P_o and P_1 (the notation is according to the present paper), then (20) is satisfied. A special case of the above, namely the case where $m_o = m_1$ or $n_o = n_1$, is derived in [4] and in [2]. Note that these conditions *necessitate* the "double interlacing" discussed above, and hence Theorem 3.3 also constitutes another proof to the above result in [7], including the special case in [4] and [2].

We are now in a position to indicate an algorithm to form a set of local convex directions, based on Theorem 3.3. Let a Hurwitz polynomial $P_o(s)$ be given. Denote its even part by

(23) $$m_o(s) = K_1^{(0)} \prod_i \left(s + j\omega_{ie}^{(0)}\right)\left(s - j\omega_{ie}^{(0)}\right)$$

and its odd part by

$$n_o(s) = K_2^{(0)} s \prod_i \left(s + j\omega_{io}^{(0)}\right) \left(s - j\omega_{io}^{(0)}\right). \tag{24}$$

Let

$$0 < \omega_{1e}^{(0)} < \omega_{1o}^{(0)} < \omega_{2e}^{(0)} < \omega_{2o}^{(0)} < \cdots \tag{25}$$

The possibility of the ordering in (25) is implied by the fact that $P_o(s)$ is a Hurwitz polynomial.

Denote the local convex direction (with respect to $P_o(s)$) by $P_1(s)$. Denote its even part by

$$m_1(s) = K_1^{(1)} \prod_i \left(s + j\omega_{ie}^{(1)}\right) \left(s - j\omega_{ie}^{(1)}\right) \tag{26}$$

and its odd part by

$$n_1(s) = K_2^{(1)} s \prod_i \left(s + j\omega_{io}^{(1)}\right) \left(s - j\omega_{io}^{(1)}\right) . \tag{27}$$

Then, choose the values of $\omega_{ie}^{(1)}$ and $\omega_{io}^{(1)}$ ($i = 1, 2, \ldots$) arbitrarily, but recursively, in the intervals complying with (28):

$$0 < \omega_{1e}^{(1)} < \omega_{1o}^{(0)} \tag{28.1}$$

$$\max\{\omega_{ie}^{(0)}, \omega_{ie}^{(1)}\} < \omega_{io}^{(1)} < \omega_{i+1,e}^{(0)}, \quad i = 1, 2, \ldots \tag{28.2}$$

$$\max\{\omega_{i-1,o}^{(0)}, \omega_{i-1,o}^{(1)}\} < \omega_{ie}^{(1)} < \omega_{io}^{(0)}, \quad i = 2, 3 \ldots \tag{28.3}$$

It is readily verified that (25) and (43) imply the "double interlacing" of the zeros of $m_o m_1$ and $n_o n_1$, as well as the interlacing of the zeros of $m_1(s)$ and $n_1(s)$ and the zeros of $m_o(s)$ and $n_o(s)$. This property evidently necessitates $P_o(s), P_1(s), m_o(s) + n_1(s)$ and $m_1(s) + n_o(s)$ to be Hurwitz polynomials. The conditions in Theorem 4 are satisfied, and hence, (20) is satisfied and all such polynomials $P_1(s)$ are local convex directions with respect to $P_o(s)$.

4. Conclusion. The concept of local convex direction has been related to a circuit theory concept namely, positive realness of a rational function. It has been shown that positive realness of a ratio of two real polynomials P_o/P_1 is a sufficient condition for the convex combination of $P_o(s)$ and $P_1(s)$ to be Hurwitz stable. Furthermore, a necessary condition for P_o/P_1 to be positive real, namely that the real part of $P_o(j\omega)/P_1(j\omega)$ is positive at values of ω where the imaginary value of $P_o(j\omega)/P_1(j\omega)$ is zero, is shown to be the necessary and sufficient condition for the convex combination of $P_o(s)$ and $P_1(s)$ to be Hurwitz stable. This formulation of the necessary and sufficient condition enables the derivation of yet another necessary condition for positive realness of $P_o(s)/P_1(s)$, which is also a sufficient condition for the requirement that the real part of $P_o(j\omega)/P_1(j\omega)$ is positive at values of ω where the imaginary value of $P_o(j\omega)/P_1(j\omega)$ is zero. Namely, this condition requires that not only $P_o(s)$ and $P_1(s)$ be Hurwitz polynomials, but also that the (necessarily imaginary) zeros of the polynomials $m_o(s)m_1(s)$ and $n_o(s)n_1(s)$ will be "doubly interlacing", where $m_o(s)$ ($m_1(s)$) is the even part of $P_o(s)$ ($P_1(s)$) and $n_o(s)$ ($n_1(s)$) is the odd part of $P_o(s)$ ($P_1(s)$). This latter condition makes an algorithm for deriving sets of pertinent real polynomials $P_1(s)$, where $P_o(s)$ is a given real Hurwitz polynomial, very simple. The solution of this problem of forming local convex directions in a simple way is timely and important for numerous applications in the design of robustly stable systems.

Acknowledgment. The derivation in Section 3 is partly based on some results derived in a Ph.D. thesis submitted by the author to the Senate of the Technion – Israel Institute of Technology, Haifa, Israel, 1966. The author acknowledges with thanks the guidance which he received 29 years ago from his thesis supervisor I. Navot and from his thesis adviser I. Cederbaum.

References

[1] S. Bialas, A necessary and sufficient condition for the stability of convex combinations of stable polynomials and matrices, *Bulletin Polish Acad. of Sci., Tech. Sci.*, Vol. 33, 1985, pp. 474–480.
[2] S. Bialas and J. Garloff, Convex combinations of stable polynomials, J. Franklin Inst., Vol. 319, 1985, pp. 375–377.
[3] B.S. Bollepalli and L.K. Pujara, On the stability of a segment of polynomials, *IEEE Trans. Circ. and Syst. I*, Vol. 41, 1994, pp. 898–901.
[4] N.K. Bose, A system theoretic approach to stability of sets of polynomials, Contemporary Mathematics, Vol. 47, AMS, 1985, pp. 25–34.
[5] M. Fu, A class of weak Kharitonov regions for robust stability of linear uncertain systems, *IEEE Trans. Automat. Contr.*, Vol. 36, 1991, pp. 975–978.
[6] M. Fu, Test of convex directions for robust stability, *Proc. of the 32nd Conf. on Decision and Control (CDC)*, San Antonio, Tx. 1993, pp. 502–507.
[7] B.K. Ghosh, Some new results on the simultaneous stabilization of a family of single input single output systems, *Syst. Contr. Lett.*, Vol. 6, 1985, pp. 39–45.
[8] E.A. Guillemin, *Synthesis of Passive Networks*, ch. 1, p. 34, Wiley, N.Y., 1957.
[9] D. Hinrichsen and V.L. Kharitonov, On convex directions for stable polynomials, Report 309, University of Bremen, Institute of Dynamic Systems, June 1994, pp. 1–22.
[10] C.V. Hollot and F. Yang, Robust stabilization of interval plants using lead or lag compensators, *Systems and Control Lett.*, Vol. 14, 1990, pp. 9–12.
[11] I.R. Petersen, A class of stability regions for which a Kharitonov like theorem holds, *IEEE Trans. Automat. Contr.*, Vol. 34, 1989, pp. 1111–1115.
[12] A. Rantzer, Stability conditions for polytopes of polynomials, *IEEE Trans. Automat. Contr.*, Vol. 37, 1992, pp. 79–89.
[13] A. Rantzer, Hurwitz testing sets for parallel polytopes of polynomials, *Systems and Control Lett.*, Vol. 15, 1990, pp. 99–104.
[14] E. Zeheb, Necessary and sufficient conditions for root clustering of a polytope of polynomials in a simply connected domain, *IEEE Trans. Automat. Contr.*, Vol. 34, 1989, pp. 986–990.

Application of Quantifier Elimination to Solotareff's Approximation Problem

George E. Collins*
Research Institute for Symbolic Computation, Johannes Kepler University
A-4040 Linz, Austria
e-mail: gcollins@risc.uni-linz.ac.at

Abstract. Solotareff's approximation problem is that of obtaining the best uniform approximation on the interval [-1,+1] of a real polynomial of degree n by one of degree $n-2$ or less. Without loss of generality we take the approximated polynomial to be $x^n + rx^{n-1}$, $r \geq 0$. We treat r as a parameter and seek to compute the coefficients of the best approximations, for small fixed values of n, as piecewise algebraic functions of r. We succeed only for $n \leq 4$, but also find that we can easily compute the coefficients for $n = 5$ for fixed values of r. The results serve to display the capabilities and limitations of our quantifier elimination program **qepcad**. We also prove that the coefficients of the best approximation, for any n, are continuous functions of r.

1. Introduction. In 1933 E. I. Solotareff (see [1]) posed the problem of finding the best approximation, relative to the uniform norm on the interval [-1,+1], of a real polynomial of degree n by one of degree $n-2$ or less. This is easily seen to be equivalent to obtaining the best approximation to a polynomial $x^n + rx^{n-1}$, r a non-negative real number. This may be regarded as a generalization of the problem of obtaining the best approximation to a polynomial of degree n by one of degree $n-1$ or less, a problem solved by the Chebyshev polynomials. The existence of a quantifier elimination method for the theory of real-closed fields implies that the coefficients of the best approximation for Solotareff's problem are, for any fixed $n \geq 2$, piecewise algebraic functions of r. In the following we explore the extent to which our program, **qepcad**, for quantifier elimination based on partial cylindrical algebraic decomposition, is able to compute these functions.

The Solotareff problem is just one example of the many possible applications of an efficient quantifier elimination program. Our purpose in this paper, apart from the interest inherent in our results, is to indicate by example the current capabilities and limitations of **qepcad** and to explain some of its workings.

In Section 2 we use a characterization of the best approximation by Solotareff to obtain a desirable formulation of the problem as a quantified formula. In Section 3 we present complete solutions obtained for degrees 2, 3 and 4, and show that we can easily obtain solutions for particular values of r for degree 5. In Section 4 we give an introduction to the **qepcad** program and discuss the user interaction and program behavior for this application. We also prove that the coefficients of the best Solotareff approximations are continuous functions of r. This fact was used in obtaining the solution for $n = 4$, but does not appear to be a consequence of known theorems on approximation theory.

---

* Supported by the Austrian Science Foundation (Grant P8572-PHY).

2. Formulation of the Problem.

The definition of best uniform approximation by a polynomial of degree $n-2$ or less is a condition that can be expressed within the elementary theory of real-closed fields (ETRCF) for any fixed value of n since it expresses minimality among the class of all real polynomials of degree $n-2$ or less, and this class can be parameterized by the $n-1$ coefficients of any such polynomial. But we also must express the condition that one polynomial of degree n, the difference of $x^n + rx^{n-1}$ and a competing approximation, has a uniform norm no larger than that of another. This can also be done. Let $\| A \|$ denote the uniform norm of A on $[-1, +1]$. If $A(x)$ and $B(x)$ are two polynomials then $\| A \| \leq \| B \|$ is equivalent to

(1) $\qquad (\forall x)[-1 \leq x \leq +1 \Rightarrow (\exists y)[-1 \leq y \leq +1 \wedge |A(x)| \leq |B(y)|]].$

The absolute value operator is not an accepted part of the ETRCF language but it can always be eliminated. $|x| \leq |y|$ is equivalent to $x^2 \leq y^2$ or to $[y \geq 0 \wedge -y \leq x \leq y] \vee [y \leq 0 \wedge y \leq x \leq -y]$.

To take the case $n = 3$ as an example, if we let $ax + b$ be the best approximation and $a'x + b'$ any competing approximation, we can express this by the quantified formula

(2) $\qquad (\forall a')(\forall b')(\forall x)[-1 \leq x \leq +1 \Rightarrow (\exists y)[-1 \leq y \leq +1 \wedge$
$\qquad\qquad |x^3 + rx^2 - ax - b| \leq |y^3 + ry^2 - a'y - b'|]].$

This formula has 3 free variables, r, a and b, and four bound variables, x, y, a' and b'. One can easily see that a formulation of this kind for arbitrary n will have n free variables and $n+1$ bound variables. To increase n by 1 we must add one free coefficient variable and one bound coefficient variable. However, we can reduce the number of variables, thereby enhancing the feasibility of eliminating the quantifiers, by utilizing some known theorems relating to the problem. Except where otherwise indicated these theorems can be found in [1].

The best approximation is known to be unique, and it is characterized by the existence of n points, $-1 \leq u_1 < u_2 < \ldots u_n \leq +1$ such that, if $A_n(x)$ is the best approximation and $E_n(x)$ is the difference $x^n + rx^{n-1} - A_n(x)$, then $\| E_n \| = | E_n(u_1) |$ and $E_n(u_i) = (-1)^{i+1} E_n(u_1)$ for $2 \leq i \leq n$.

We will henceforth assume that $r \geq 0$, for if $r \leq 0$ and $A_n(x)$ is the best aproximation to $x^n - rx^{n-1}$ then $A_n(-x)$ is the best appproximation to $x^n + rx^{n-1}$. Achieser shows that

(3) $\qquad E_n(x) = (-1)^n 2^{-n+1}(1 + r/n)^n T_n((-x - r/n)/(1 + r/n))$

for $0 \leq r \leq S_n = n(tan(\pi/(2n)))^2$, where $T_n(x)$ is the Chebyshev polynomial of degree n, $T_n(x) = \cos(n \arccos x)$. This is an easily computable function, from which one can read off the coefficients of $A_n(x)$, and they will be polynomial functions of r of degree n or less. So quantifier elimination is needed only for $r > S_n$.

Each of the $n-2$ points $u_2, u_3, \ldots, u_{n-1}$ must clearly be a zero of of the derivative $E'_n(x)$. Elaboration of an argument in [1] leads to the conclusion that $E_n(1) > 0$. These properties, in fact, characterize $E(x)$ and we arrive at the following quantified formula of ETRCF.

(4) $$(\exists u_2)\ldots(\exists u_{n-1})[r > S_n \wedge -1 < u_2 < \ldots < u_{n-1} < 1 \wedge E_n(1) > 0 \\ \wedge E_n(-1) = (-1)^{n+1} E_n(+1) \wedge \\ \bigwedge_{i=2}^{n-1}[E_n(u_i) = (-1)^{n-i} E_n(1) \wedge E'_n(u_i) = 0]].$$

This formula has the same n free variables but only $n-2$ bound variables, a reduction of three. It has another very important advantage, namely the occurence in it of many equations. These equations are *equational constraints*, equations occurring in, and logically implied by, the quantifier-free part of the formula, to which the quantifiers are applied, which we call the *matrix*. We will explain in the following section how equational constraints are utilized in a recently completed version of our quantifier elimination program, which is called **qepcad**. Formula 4 contains $2n-3$ equational constraints, just one less than the number of variables. We might say that it is *almost* the problem of solving a system of polynomial equations; it fails being so only because of the imposed inequalities. We can simplify the formula slightly more by using the equation $E_n(-1) = (-1)^{n+1} E_n(+1)$ to solve for one of the coefficients. Upon evaluation of $E_n(x)$ at $x = 1$ and $x = -1$ we obtain the linear equation $\sum_{i \text{ odd}} a_i = 1$, where $A_n(x) = \sum_{i=0}^{n-2} a_i x^i$. We use this equation to solve for a_0 if n is even, for a_1 if n is odd, in terms of the other a_i and substitute the result into the equations involving E_n. We are left with $2n - 4$ equations in $2n - 3$ variables.

3. Solutions. We would now like to solve the Solotareff problem for $n = 2, 3, 4, \ldots$. We take this to mean obtaining a constructive description of each coefficient of the best approximation, A_n, as a function of r in the form of a simple quantifier-free formula containing as free variables only r and the coefficient.

For $n = 2$, $S_2 = 2$ and Formula 3 produces

(5) $$a = 1/2 + r/2 - r^2/8$$

for $0 \le r \le 2$. When $n = 2$ there are no bound variables in Formula 4; it consists of only the equation $E_2(-1) = -E_2(1)$, which evaluates to $1 - r - a = -1 - r + a$. We conclude that $a = 1$ for $r > 2$. Of course we could also see this by simple reasoning.

For $n = 3$, let $A_3(x) = ax + b$. From the equation $E_3(1) = E_3(-1)$ we obtain $a = 1$. We substitute 1 for a, let $u_1 = u$, and compute $S_3 = 1$. Formula 4 becomes

(6) $$(\exists u)[r > 1 \wedge -1 < u \wedge u < 1 \wedge 3u^2 + 2ru - 1 = 0 \\ \wedge u^3 + ru^2 - u + r - 2b = 0]$$

qepcad produces the solution to this problem in 133 milliseconds. All times reported in this paper are based on the use of a Decstation 5000/240 having a 40

MHz R3400 risc processor. We find that that b is the lesser of the two real roots of the polynomial $P_3(r, b) = 27b^2 - 2r^3b - 36rb + r^4 + 11r^2 - 1$. Thus

(7) $\quad [27b^2 - 2r^3b - 36rb + r^4 + 11r^2 - 1 = 0 \land 27b - r^3 - 18r < 0]$

is a quantifier-free formula for the solution.

For $r \leq 1$ Formula 3 yields the solutions

(8) $\quad \begin{aligned} a &= 3/4 + 1/2\,r - 1/4\,r^2 \\ b &= 1/4\,r + 1/6\,r^2 - 1/108\,r^3. \end{aligned}$

We can also solve for u as a function of r for $r > 1$. All we have to do is to existentially quantify b instead of u. From the input formula we see that u must be one of the two real roots of the equation $3u^2 + 2ru - 1 = 0$ and **qepcad** tells that u is the larger of the two roots. It is not difficult to see, or to prove by quantifier elimination, that as r increases from 1 to $+\infty$, $u = u_2$ decreases monotonically from $1/3$ to 0 and u_1 decreases monotonically from -1 to $-\infty$.

Furthermore we can compute the maximum error, $e_3 = \|E_3\|$. $e_3 = E_3(1) = 1 + r - a - b = r - b$ so $b = r - e_3$ and $P_3(r, r - e_3) = 0$. By substitution $P_3(r, r - e_3) = 27e_3^2 + 2r^3 e_3 - 18re_3 - r^4 + 2r^2 - 1$. For $r > 1$ this polynomial has one positive root and one negative root; e_3 is the positive root. For $r \leq 1$, $e_3 = r - b = 3/4\,r - 1/6\,r + 1/108$.

Now consider the problem for $n=4$. Let us first compute S_4. $S_4 = 4(\tan \pi/8)^2$. Let $a = \tan \pi/8$. Substituting $\pi/4$ for x in the double angle formula $\tan 2x = 2\tan x/(1 - \tan^2 x)$, we obtain $1 = 2a/(1 - a^2)$, hence $a^2 + 2a - 1 = 0$. Let $b = S_4 = 4a^2$. Computing the resultant of $4a^2 - b$ and $a^2 + 2a - 1$ we deduce that $b^2 - 24b + 16 = 0$. Since $\tan \pi/8 < 1$, S_4 must be the lesser of the two real roots of this equation, $12 - 8\sqrt{2} = 0.6863-$. Consequently we can express $r > S_4$ by the formula $r^2 - 24r + 16 < 0 \lor [r > 1 \land r^2 - 24r + 16 \geq 0]$.

Let $A_4(x) = ax^2 + bx + c$. From $E_4(-1) = -E_4(+1)$ we obtain $c = 1 - a$. Let $u_1 = u$ and $u_2 = v$. $E_4(1) = 1 + r - a - b - (1 - a) = r - b$. Consequently the formula $E_4(u) = -E_4(1)$ simplifies to $u^4 + ru^3 - au^2 - bu - r + b + a - 1 = 0$. This factors into $(u^3 + ru^2 + u^2 - au + ru - b - a + r + 1)(u - 1) = 0$. Since $u < 1$ we can omit the factor $u - 1$. Similarly we can replace $E_4(v) = E_4(1)$ by $v^3 + rv^2 - v^2 - av - rv + v - b + a + r - 1 = 0$.

Putting this all together, we get the input formula

$$(\exists u)(\exists v)[[r^2 - 24r + 16 < 0 \lor [r > 1 \land r^2 - 24r + 16 >= 0]]$$
$$\land -1 < u \land u < v \land v < 1 \land r - b > 0$$
$$\land u^3 + ru^2 + u^2 - au + ru + u - b - a + r + 1 = 0$$
$$\land v^3 + rv^2 - v^2 - av - rv + v - b + a + r - 1 = 0$$

(9) $\quad \land 4u^3 + 3ru^2 - 2au - b = 0 \land 4v^3 + 3rv^2 - 2av - b = 0].$

We first obtain a as a function of r, $r > S_4$, by existentially quantifying Formula 9 with respect to b. **qepcad** produces the answer after 51.9 seconds of computation time. The polynomial

$$P_4(r,a) = 324a^4 + 324r^2a^3 - 2016a^3 + 108r^4a^2 - 1128r^2a^2 + 4576a^2$$
(10) $$+12r^6a - 224r^4a + 1392r^2a - 4480a - 15r^6 + 112r^4 - 608r^2 + 1600$$

has two real roots for every $r > S_4$ and a is the larger of the two.

Quantifying a instead of b and applying **qepcad** again we obtain the solution for b as a function of r for $r > S_4$. This time the answer is a bit more complicated. The polynomial

$$Q_4(r,b) = 78732b^4 + 8748r^3b^3 - 291600rb^3 + 324r^6b^2 - 18684r^4b^2$$
$$+403272r^2b^2 - 1024b^2 + 4r^9b - 616r^7b + 12692r^5b$$
(11) $$-246800r^3b + 2048rb - 3r^{10} + 280r^8 - 2708r^6 + 56332r^4 - 1024r^2$$

has two real roots for every $r > S_4$ but the two roots coincide for $r = \alpha = 6.07286+$, the unique root of $243x^8 - 8532x^6 - 15920x^4 + 2624x^2 - 1024$ between 6 and 8. b is the smaller of the two roots for $S_4 < r < \alpha$, the larger root for $r > \alpha$. **qepcad** produces the answer in 53.8 seconds.

For $r \leq S_4$ application of 3 produces the formulas

$$a = 1 + 1/2\,r - 5/16\,r^2$$
$$b = 1/2\,r + 1/4\,r^2 - 1/32\,r^3$$
(12) $$c = -1/8 - 1/8\,r + 1/64\,r^2 + 3/128\,r^3 - 1/2048\,r^4.$$

For $n = 5$ we first compute S_5. Applying the multiple-angle formula $\cos 5x = 16\cos^5 x - 20\cos^3 x + 5\cos x$ with $x = \pi/10$ we deduce that $\cos \pi/10$ is a positive real root of $16x^5 - 20x^3 + 5x = x(16x^4 - 20x^2 + 5)$. This polynomial has two positive real roots, $0.588-$ and $0.951+$. Clearly $\cos \pi/10$ is the larger of the two. By solving the quantifier elimination problem $(\exists y)[16y^4 - 20y^2 + 5 = 0 \wedge y > 3/4 \wedge x^2 + y^2 = 1 \wedge x > 0]$ we find that $\sin \pi/10$ is the positive root of $4x^2 + 2x - 1$. Then the solution to the quantifier elimination problem $(\exists y)(\exists z)[y > 0 \wedge 4y^2 + 2y - 1 = 0 \wedge z > 3/4 \wedge 16z^4 - 20z^2 + 5 = 0 \wedge xz = y]$ reveals that $\tan \pi/10$ is the root of $5x^4 - 10x^2 + 1$ between 0 and 1. Finally the quantifier elimination problem $(\exists y)[5y^4 - 10y^2 + 1 = 0 \wedge y > 0 \wedge y < 1 \wedge x = 5y^2]$ yields the solution $S_5 = 0.5278640450+$, the unique root of $x^2 - 10x + 5$ between $1/2$ and 1.

For $r < S_5$ Achieser's formula produces the results

$$a = 5/4 + 1/2\,r - 7/20\,r^2$$
$$b = 3/4\,r + 3/10\,r^2 - 1/20\,r^3$$
$$c = -5/16 - 1/4\,r + 3/40\,r^2 + 1/20\,r^3 - 1/400\,r^4$$
(13) $$d = -1/16\,r - 1/20\,r^2 - 1/200\,r^3 + 1/500\,r^4 - 1/50000\,r^5.$$

The case $n = 5$ and $r > S_5$ proves to be too difficult for the current version of our program. Perhaps after further innovations we will be able to solve it. As we went from $n = 3$ to $n = 4$ the computation time jumped from 133 milliseconds to 51.9 seconds, a factor of about 400. So it is not surprising that $n = 5$ would be very difficult. However, we have been able to solve the particular case $r = 1$ for $n = 5$. Let $A_5(x) = ax^3 + bx^2 + cx + d$. The equation $E_5(-1) = E_5(+1)$ implies that $a + c = 1$, and we substitute $1 - a$ for c. For $r = 1$, a, b and d are uniquely determined real numbers, and then it follows that $u = u_1$, $v = u_2$ and $w = u_3$ are also uniquely determined. Consequently we can solve for $a, b, d, u, v,$ and w without quantifier elimination; we input the formula

$$[1 - b - d > 0 \wedge -1 < u \wedge u < v \wedge v < w \wedge w < 1$$
$$\wedge 5u^4 + 4u^3 - 3au^2 - 2bu - (1 - a) = 0$$
$$\wedge 5v^4 + 4v^3 - 3av^2 - 2bv - (1 - a) = 0$$
$$\wedge 5w^4 + 4w^3 - 3aw^2 - 2bw - (1 - a) = 0$$
$$\wedge u^5 + u^4 - au^3 - bu^2 - (1 - a)u - d = -1 + b + d$$
$$\wedge v^5 + v^4 - av^3 - bv^2 - (1 - a)v - d = +1 - b - d$$
$$\text{(14)} \quad \wedge w^5 + w^4 - aw^3 - bw^2 - (1 - a)w - d = -1 + b + d].$$

A solution is obtained in 21.1 seconds. Following are the values of the six variables and their minimal polynomials.

$a = 1.4581775889+,$ $675x^4 - 3240x^3 + 6048x^2 - 5120x + 1600$

$b = 0.9557598789-,$ $2460375x^4 - 10060200x^3 + 15437952x^2$
$ -10535680x + 2697536$

$d = -0.1065834340+,$ $192216796875x^4 + 102160237500x^3 + 23548707018x^2$
$ +1571039500x - 117814$

$u = -0.5431806965+,$ $16875x^8 + 27000x^7 - 24300x - 49880x^5$
$ +10062x^4 + 30152x^3 + 108x^2 - 5832x - 729$

$v = 0.1803173009+,$ $135x^4 - 18x^2 + 8x - 1$

$w = 0.7730740989-,$ $16875x^8 + 27000x^7 - 24300x^6 - 49880x^5$
(15) $ +10062x^4 + 30152x^3 + 108x^2 - 5832x - 729.$

4. Solution Methods. The results we have obtained for $n = 4$ and $n = 5$ were not obtained by straightforward application of **qepcad**. Although straightforward application would, in principle, suffice, computation times would be impractically large. **qepcad** is an interactive program whose behavior can be modified by various commands, given either in advance or during the course of the computation, that may be chosen by the user, on the basis of information that the program can be caused to display, and on the basis of his knowledge of the particular problem at hand. In the following we describe some of the features of the program that were used in an important way to obtain the solutions of the previous section.

A **qepcad** computation has four sequential phases:
1. Normalization
2. Projection
3. Partial CAD construction
4. Solution formula construction

In the normalization phase, the input formula is read and then rewritten in a standard form. The input formula must be in *prenex* form, consisting of a quantifier-free *matrix* to which a string of quantifiers is applied. In the normalization phase each atomic formula is rewritten so that the right-hand side of the relation symbol is zero. The resulting polynomials on the left-hand sides are then factored into irreducibles and the formula is rewritten using just these irreducible polynomials. A linear ordering of the variables in the formula must be specified by the user that is consistent with the order of the variables in the quantifier prenex.

The projection phase begins with the set of irreducible polynomials obtained from the normalization phase. Let r be the number of variables in the input formula. The r-variate irreducible polynomials are "projected", producing a set of polynomials in $r-1$ variables, from which the rth variable has been eliminated. The resulting projection polynomials are then factored into irreducible polynomials that are added to the existing set of irreducible polynomials. Then the $r-1$-variate polynomials are projected and factored. After $r-1$ projections and factorizations the projection phase is completed, having produced, for $1 \leq i \leq r$, a set of i-variate irreducible polynomials, called *projection factors*.

The next phase proceeds with constructing a cylindrical algebraic decomposition of r-dimensional real space consisting of a finite number of connected sets, called *cells*, in each of which the matrix of the input formula is invariant in truth value. Note that this can be ensured by making each projection factor invariant in sign in each cell, the key idea in the original CAD method for quantifier elimination [2]. But in [3] the method was improved by the introduction of two methods that usually avert construction of a full sign-invariant CAD. The construction begins with a decomposition of the real line into open intervals and one-point intervals in which the univariate projection factors are sign-invariant. The construction proceeds by choosing an existing cell in i-dimensional space and building a *stack* of cells in the cylinder having the chosen cell as base, in each of which the $(i+1)$-variate projection factors are sign-invariant. However this process is restricted in three important ways. First, *trial evaluation*, introduced in [3], may reveal that the matrix is true or false in the cell due to the presence in the matrix of some necessaary condition on only some of the variables. Clearly it is unnecessary to construct a stack over any cell whose truth is already known. Second, if f is the number of free variables and $i \geq f$, the truth value of the matrix in the cell may sometimes be determined by *truth propagation*, using the quantifiers in the prenex. If, for example, it is established by trial evaluation that the matrix is true in a single cell of a stack corresponding to an existential quantifier of the prenex, then it follows that the input formula is true in the base cell of that stack. This truth value may be propagated further down by the application of other quantifiers. A third way, use of *equational constraints* was introduced more recently in [4]. If any equation is implied by the matrix, this can be used to restrict the sets of projection polynomials. This has the further effect of reducing the number of projection polynomials and thereby the size of the sign-invariant CAD determined by the projection factors. Furthermore the equational constraints can be used during the partial CAD construction to immediately assign truth values to some of the cells in a newly constructed stack. Equational constraints are particularly effective for Solotareff approximation problems.

The solution formula construction phase attempts to construct a solution formula, that is, a quantifier-free formula equivalent to the quantified input formula, one that contains as its polynomials only the projection factors, and that is as simple as possible. This is not always possible; we are currently engaged in research on suitable ways to introduce additional polynomials when necessary.

Now let's see how the program was used, and what it did, in obtaining the solutions of the previous section. For $n = 3$ the variable ordering specified for the program was (r, b, u). u must be last because free variables must precede quantified variables. We put r before b because we want to obtain b as a function of r. The two equations in the input formula are both equational constraints. The program uses the fact that the resultant of two equational constraints is also an equational constraint, and such resultants are always included in the projection set. In this example the resultant of the two initial constraint polynomials is just four times the polynomial $B(r,b) = 27b^2 - 2r^3b - 36rb + r^4 + 11r^2 - 1$ occurring in the solution formula. It turns out that the only univariate projection factors in this problem are $r + 1$, $r - 1$ and $r^2 + 3$, arising as a factor of multiplicity three of the discriminant of $B(b, r)$. Therefore the CAD of one-dimensional space consists of just five cells, namely the one-point cells $\{-1\}$ and $\{+1\}$ and the three open intervals comprising the remainder of the real line. All of the cells except the open interval $(1, +\infty)$ are immediately marked false by trial evaluation by virtue of the condition $r > 1$ in the input matrix. Then a stack is constructed over the cell $(1, +\infty)$. This stack contains five cells, two of which are sections corresponding to the two roots of $B(r, b)$ and three are sectors. The sectors are marked false because $B(r, b)$ is an equational constraint. Finally stacks are constructed over the two sections, and application of the quantifier reveals that the lower section is true, the upper section is false. In this simple example the formula construction phase of **qepcad** is unsuccessful but it is obvious that using the first derivative of $B(r, b)$ with respect to b provides the stated solution formula.

Now let's consider the solution for a when $n = 4$. The ordering of variables chosen was (r, a, b, u, v). There are four equational constraints, two in the variable v and two in u. In the first projection the resultant of the two polynomials in v is computed, which is an equational constraint polynomial in r, a and b. In the second projection the resultant of the two u polynomials produces another equational constraint polynomial in r, a and b. Each of these projections also produces some other polynomials. In the third projection the resultant of the two equational constraint polynomials in r a and b, each of which has two irreducible factors, is computed, producing an equational constraint polynomial in r and a. This polynomial has seven irreducible factors, one of which is the polynomial $P_4(r, a)$ that provides the solution. Altogether there are 10 bivariate projection factors, which lead in turn to 37 univariate projection factors. There are 89 cells in the CAD of one-space, the first 48 of which correspond to values of r less than or equal to S_4 and are consequently marked false by trial evaluation. Next the program is instructed to construct stacks over only the remaining 41 cells that are open intervals. The reason for this is that we know that a as a function of r is continuous and we intend to use this to reduce the required computations. Next the program is instructed to construct stacks over any cell other than the one-point cells of one-space. This leads to truth values for all of the cells in two-space that were constructed. Finally the program is instructed to construct stacks over the one-point cells of one-space. Display of the resulting partial CAD then reveals that the only cells marked true are those that are the second (upper) of the two sections of $P_4(r, a)$. The solution formula construction phase of **qepcad** could not be used because cells remain (some sections in the stacks over the one-point cells of one-space) having undetermined truth values. The purpose of constructing the stacks over the one-point cells of one-space was to determine whether $P_4(r, a)$ always has two distinct roots when r has these values. This being indeed the case, we can conclude by continuity that, at these values of r also, a is the larger of the two roots.

The solution for b as a function of r for $r > S_5$ was obtained in a similar manner. For $n = 5$ and $r = 1$, after constructing the CAD of one-space a few of the one-point cells, which are values of b were displayed. One of these was guessed to be the most likely value for the correct value of b, a stack was constructed over this cell, and then stacks over cells in spaces of dimension greater than 1. This resulted in a value, confirming that the guess had been correct. Without this guidance the program would have proceeded to compute truth values for all of the 23 cells in one-space.

In our solutions for $n = 4$ we made vital use of continuity. Otherwise the program would have had to establish truth values for all of the one-point cells in two-space. Some of these have coordinates which are algebraic numbers of very high degree, resulting in enormous computational times. I have not been able to find in the literature any theorem which would imply the continuity of the coefficients of Solotareff best approximation polynomials so I have constructed the proof which follows. We first prove two lemmas.

LEMMA 4.1. *Let $e_n(r)$ be the norm of the error polynomial $x^n + rx^{n-1} - A_n(r, x)$. e_n is a continuous function of r.*

Proof. Let $\delta > 0$ and assume that $\mid r_1 - r_2 \mid < \delta$. Define $P_n(r, x)$ as $x^n + rx^{n-1}$. Then $e_n(r_1) \leq \parallel P_n(r_1, x) - A_n(r_2, x) \parallel \leq \parallel P_n(r_1, x) - P_n(r_2, x) \parallel + \parallel P_n(r_2, x) - A_n(r_2, x) \parallel \leq \delta + e_n(r_2)$. Symmetrically $e_n(r_2) \leq \delta + e_n(r_1)$. So $\mid e_n(r_1) - e_n(r_2) \mid \leq \delta$. $\square$

LEMMA 4.2. *Let $a_{n,i}(r)$ be the coefficient of x^i in $A_n(r, x)$. $\mid a_{n,i}(r) \mid \leq \binom{n}{i}(n + r)$ for all i.*

Proof. $E_n(r,x)$ has $n-1$ roots in $[-1,+1]$ and another root less than -1. The sum of the roots of $E_n(r,x)$ is $-r$ so the least root is greater than $-r-(n-1) > -r-n$. $a_{n,i}(r)$ is a symmetric polynomial in the roots of $E_n(r,x)$ that is the sum of $\binom{n}{i}$ terms, each of which is the product of i of the roots. Each such product has at most one factor, the least root of $E_n(r,x)$ that is not in $[-1,+1]$, and hence the absolute value of each term is less than $n+r$. □

THEOREM 4.3. *The coefficients of $A_n(r,x)$ are continuous functions for $r > S_n$.*
Proof. Each coefficient of $A_n(r,x)$ is given by some sections in a CAD of two-dimensional space. There is one such section in each stack over a cell in the CAD of one- dimensional space. Each section over an open interval is a continuous function. Therefore each coefficient is continuous except perhaps at those isolated values of r corresponding to cells in the CAD of one- dimensional space consisting of a single point. Let $\bar{r}$ be any such point. We will show that each coefficient is also continuous at $\bar{r}$.

Let $a_{n,i}(r)$ be the coefficient of x^i in $A_n(r,x)$. $a_{n,i}(r)$ is defined and continuous in an open interval to the right of $\bar{r}$. Let $\bar{a}_i$ be the limit of $a_{n,i}(r)$ as r approaches $\bar{r}$ from the right. We must show that this limit exists and is finite. Existence follows from the fact that, as an algebraic function, $a_{n,i}(r)$ is monotone in some interval to the right of r. By Lemma 4.2 the limit must be finite.

By Lemma 1, $e(\bar{r}) = \lim_{r \leftarrow r+} e(r) = \lim_{r \leftarrow r+} \| (x^n + rx^{n-1}) - A_n(r,x) \| = \| \lim_{r \leftarrow r+} (x^n + rx^{n-1}) - \lim_{r \leftarrow r+} A_n(r,x) \| = \| (x^n + \bar{r}x^{n-1}) - \overline{A}_n(r,x) \|$. This proves that $\overline{A}_n(r,x)$ is the best approximation to $x^n + \bar{r}x^{n-1}$, from whioh the statement of the theorem follows. □

References

[1] N. I. Achieser. *Theory of Approximation.* Frederick Ungar Publishing Co., New York, 1956.
[2] G. E. Collins. Quantifier elimination for the elementary theory of real closed fields by cylindrical algebraic decomposition. In *Lecture Notes In Computer Science, Vol. 33*, pages 134–183, Springer-Verlag, New York, 1975.
[3] G. E. Collins and H. Hong. Partial cylindrical algebraic decomposition for quantifier elimination. *Journal of Symbolic Computation*, 12(3): 299–328, 1991.
[4] G. E. Collins. Quantifier elimination by cylindrical algebraic decomposition – twenty years of progress. In *Quantifier Elimination and Cylindrical Algebraic Decomposition*. Springer-Verlag, Vienna, 1995.

Stability of Time Discretization, Hurwitz Determinants and Order Stars

Rolf Jeltsch

Seminar für Angewandte Mathematik, ETH Zürich

> **Abstract.** We shall review stability requirements for time discretizations of ordinary and partial differential equations. If a constant time step is used and the method involves more than two time levels stability is always related to the location of roots of a polynomial in circular or half plane regions. In several cases the coefficients of the polynomial depend on a real or complex parameter. Hurwitz determinants allow to create a fraction free Routh array to test the stability of time discretizations. A completely different technique, called order stars, is used to relate accuracy of the schemes with their stability.

1. Introduction. When Hurwitz wrote the famous paper on the conditions which guarantee that all roots of a polynomial lie in the left half plane he was motivated by a control problem [10]. At that time numerical multi-step schemes for time discretizations such as Adams-Bashforth had already been published [2]. However since these schemes are stable by their design and they had been used for hand calculation only the problem of stiffness did not arise and the concept of stability regions was not created. To the authors knowledge most applications of the Hurwitz criterion to numerical time discretizations have been done in the second half of this century. We shall describe in Section 2 several stability problems where the Hurwitz criterion has been or could be used. However most researchers employed the Schur criterion [22] or after the transformation (2-1) the Routh algorithm. When the coefficients of the polynomial depend on parameters the Routh algorithm has the drawback that it involves divisions. Therefore scaled fraction free Routh algorithms had been introduced. It turns out that the Hurwitz determinants form a scaled fraction free Routh array with the slowest growth of the normalized degrees. This development is briefly treated in Section 3 where a Routh-type algorithm to compute the Hurwitz determinants in a fraction free way is given too.

In Section 4 we present the order star technique which relates stability to accuracy of time discretizations and thus the highly non linear Hurwitz criterion can be avoided.

2. Stability of time discretizations. Discretizations of time dependent problems involving differential equations lead always to recurrence relations. Usually the discretization is linear in the sense that it either involves the unknown function or the differential equation in a linear fashion. If the time step is constant and the differential equation is not a nonlinear partial differential equation then the growth of the solution of the recurrence relation can be analyzed by looking at the roots of polynomials. There is a large variety of examples and applications in this general frame work. Here we outline typical examples only. In the first subsection we deal with ordinary differential equations while in the second subsection we treat two examples involving partial differential equations.

2.1. Stability of discretizations of ordinary differential equations.

For simplicity we start with a scalar differential equation. The initial value problem is

$$y' = f(t,y), \quad 0 \leq t \leq T$$
$$y(0) = y_0 \text{ given}.$$

Let Δt be the time step, $t_n = n\Delta t$ and y_n is a numerical approximation to $y(t_n)$. A linear k-step scheme computes recursively y_{n+k} from the relation

$$\sum_{i=0}^{k} \alpha_i\, y_{n+i} = \Delta t \sum_{i=0}^{k} \beta_i\, f(t_{n+i}, y_{n+i}) \quad n = 0, 1, 2, \ldots .$$

α_i, β_i are fixed real numbers independent on t, Δt and y. A necessary condition for such a scheme to be **convergence**, i.e. $y_n \to y(t)$ as $\Delta t = t/n \to 0$, is that the scheme is **stable**. A linear multi-step method is **stable** if all roots of the characteristic polynomial

$$\rho(\zeta) = \sum_{i=0}^{k} \alpha_i \zeta^i$$

are inside the unit disk and the roots of modules 1 are simple. The transformation

(2-1) $$z = \frac{\zeta+1}{\zeta-1}, \quad \zeta = \frac{z+1}{z-1}$$

maps the unit disk of the ζ-plane into the left half plane. Hence a method is stable if and only if all roots of

$$r(z) = (z-1)^k\, \rho\!\left(\frac{z+1}{z-1}\right)$$

satisfy $\operatorname{Re} z \leq 0$ and if one root has $\operatorname{Re} z = 0$ then it is a simple root. Hence $r(z)$ is the first example where the Hurwitz criteria could be applied. For more details see [7], [9].

While the above stability is necessary for convergence, i.e. if $\Delta t \to 0$, for positive Δt an additional concept called **stability region** has to be introduced. This concept is best explained when applying a general linear method to the system

(2-2) $$y' = f(t,y), \quad t \geq 0, \quad y \in \mathbb{R}^m .$$

The class of general linear methods is very large. It contains among others the most popular methods such as linear multi-step and Runge-Kutta methods, see [8], p 313. For all these methods it makes sense when studying the growth of errors to consider the behavior of the scheme when applied to the leading term of the variational equation of (2-2). Let $y(t)$, $\varphi(t)$ be two exact solutions of (2-2). Then the difference $e(t) = y(t) - \varphi(t)$ satisfies

$$e' = J(t)\,e + \ldots \quad \text{where the Jacobian is} \quad J(t) = \frac{\partial f}{\partial y}(t, \varphi(t)) .$$

Hence one studies the application of a method to

$$y' = J(t)\, y \qquad y \in \mathbb{R}^m,\ J \in \mathbb{R}^m \times \mathbb{R}^m.$$

To simplify the investigation one assumes J to be constant. If J is diagonalizable then it is enough to study first the scalar equation

(2-3) $$y' = \lambda y, \quad \lambda \in \mathbb{C}.$$

Applying a general linear method leads to a recurrence relation where the growth of the solutions is governed by the roots ζ_i, $i = 1, 2, \ldots, k$ of a characteristic polynomial

$$\Phi(\zeta, \mu) = \sum_{i=0}^{k} \sum_{j=0}^{s} \alpha_{ij} \mu^j \zeta^i$$

where $\mu = \Delta t\, \lambda$ and $\alpha_{i,j}$ are method dependent real numbers. One defines the **stability region** S to be the following set

$$S := \left\{ \mu \in \overline{\mathbb{C}} \,\bigg|\, \Phi(\zeta, \mu) = 0 \implies \begin{cases} |\zeta| \leq 1 \\ |\zeta| = 1 & \text{then this is a simple root} \end{cases} \right\}.$$

With this definition the numerical solution of the scheme applied to (2-3) stays bounded for a fixed $\Delta t > 0$ as $n \to \infty$ if and only if $\mu \in S$. Again one can map the unit disk of the ζ-plane in the left half plane by the transformation (2-1) and introduces

$$\Psi(z, \mu) = (z-1)^k\, \Phi\!\left(\frac{z+1}{z-1},\, \mu\right) = \sum_{i=0}^{k} a_{k-i}(\mu)\, z^i.$$

Hence

$$S = \left\{ \mu \in \overline{\mathbb{C}} \,\bigg|\, \Psi(z, \mu) = 0 \implies \begin{cases} \operatorname{Re} z \leq 0 \\ \operatorname{Re} z = 0, & \text{then this is a simple root} \end{cases} \right\}.$$

We therefore need to test for all $\mu \in \overline{\mathbb{C}}$ whether $\mu \in S$ or not, i.e. we need to show that all roots of $\Psi(\cdot, \mu)$ are in the left half plane. Again one can apply the Hurwitz criteria. In order to determine S one can plot candidates for boundary points ∂S by solving $\Psi(iy, \mu) = 0$ for μ for all $y \in \mathbb{R}$. This is a good way if one wants to obtain S numerically but it is impractical if one should prove results on S. in Section 4 we shall present the order star technique which can be used to prove properties of S. In many cases one has information on the spectrum of J and therefore it is often enough to show that a certain subset of $\overline{\mathbb{C}}$ belongs to S. This leads to different stability definitions such as

$$\begin{aligned}
A_0\text{-stability} &\iff (-\infty, 0] \subset S,\ \text{see Section 3} \\
A(\alpha)\text{-stability} &\iff \left\{ \mu \in \mathbb{C} \,\big|\, |\arg(-\mu)| \leq \alpha \right\} \subset S \\
A\text{-stability} &\iff \left\{ \mu \in \mathbb{C} \,\big|\, \operatorname{Re} \mu \leq 0 \right\} \subset S.
\end{aligned}$$

Also one might be interested to have $D_r \subset S$ where D_r is the generalized circle

$$D_r = \begin{cases} \{\mu \in \mathbb{C} \mid |\mu + r| \leq r\} & \text{if } r > 0 \\ \{\mu \in \overline{\mathbb{C}} \mid |\mu + r| \geq -r\} & \text{if } r < 0 \end{cases}.$$

In Section 2.2 we shall discuss stability of semi-discretizations of convection-diffusion equations. There one might be interested in the following sets included in S:

$$\text{ellipses} \quad \left\{\mu = \xi + i\eta \,\Big|\, \left(\frac{\xi}{g} + 1\right)^2 + \left(\frac{\eta}{c}\right)^2 \leq 1\right\}$$

$$\text{ovals} \quad \left\{\mu = \xi + i\eta \,\Big|\, \left(\frac{\xi}{g} + 1\right)^2 + \left(\frac{\eta}{c}\right)^4 \leq 1\right\}$$

$$\text{parabolas} \quad \left\{\mu = \xi + i\eta \,\Big|\, \xi + \left(\frac{\xi}{c}\right)^2 \leq 0\right\}.$$

In general if a set $\Omega \subset \overline{\mathbb{C}}$ is simply connected one has by the maximum principle that $\Omega \subset S$ if and only if $\partial \Omega \subset S$ and $\zeta(\mu)$ has no poles in Ω. Here $\zeta(\mu)$ is the algebraic function defined by

$$\Phi(\zeta(\mu), \mu) \equiv 0.$$

Using again the transformation (2-1) we find for a simply connected Ω that $\Omega \subset S$ if and only if

i) $\partial \Omega \subset S$ and ii) $\sum_{i=0}^{k} a_i(\mu) \neq 0$ for all $\mu \in \Omega$.

Again i) can be checked by an extension of the Hurwitz criteria to polynomials with complex coefficients, see e.g. [23], p. 179.

2.2. Stability of discretizations of partial differential equations. Here we shall treat two examples. The first is concerned with a general convection-diffusion equation in several space dimensions while the second one will be needed in the discussion, Section 4.

Example 1: Semi discretizations of convection-diffusion equations.

In this example we follow the article by Wesseling [26]. The convection-diffusion equation is given by

$$u_t + Lu = 0, \quad u = u(x, t), \quad x(x_1, x_2 \ldots, x_m) \in \mathbb{R}^m, \quad t \geq 0$$

where L is the linear operator

$$Lu = \sum_{\alpha=1}^{m} \left(c_\alpha \frac{\partial}{\partial x_\alpha} - d \frac{\partial^2}{\partial x_\alpha^2}\right) u.$$

Let us discretize (2-4) first in space-direction. Δx_α is the step-size and $e_\alpha = (0, 0, \ldots, 1, 0, \ldots, 0)$ the unit vector in direction x_α. Using for example central differences one obtains a system of infinitely many ordinary differential equations

(2-4) $$\frac{du_j}{dt} = -L_{\Delta x} u_j, \quad j = (j_1, j_2, \ldots, j_m) \in \mathbb{Z}^m$$

where

$$L_{\Delta x} u_j = \frac{1}{\Delta t} \left[\sum_{\alpha=1}^{m} \frac{\nu_\alpha}{2} (u_{j+e_\alpha} - u_{j-e_\alpha}) - \delta_\alpha (u_{j+e_\alpha} - 2u_j + u_{j-e_\alpha}) \right],$$

the Courant number in direction x_α is $\nu_\alpha = c_\alpha \Delta t / \Delta x_\alpha$ and the diffusion number is $\delta_\alpha = d \Delta t / \Delta x_\alpha^2$. If δ_α is chosen differently the central difference for the advection part $c_\alpha \partial / \partial x_\alpha$ is replaced by the upwind differences. To make a Neumann stability analysis one defines the symbol $\hat{L}_{\Delta x}(\theta) = e^{-ij\theta} L_{\Delta x} e^{i\langle j, \theta \rangle}$ where $\theta = (\theta_1, \theta_2, \ldots, \theta_m)$ and $\langle, \rangle$ is the usual scalar product. Substitution of each mode $u_j = y(t) e^{i\langle j, \theta \rangle}$ in (2-4) gives

$$\frac{dy}{dt} = -\hat{L}_{\Delta x}(\theta) y.$$

One introduces the set

$$S_L = \{ -\Delta t \, \hat{L}_{\Delta x}(\theta) \in \mathbb{C} \mid \text{for all } \theta \in \mathbb{R}^m \}.$$

THEOREM 2.1. *If one applies a time discretization with a stability region S to (2-4) then*

$$S_L \subset S$$

is sufficient for Neumann stability of the overall scheme.

This result is applied in the following way. For simple set Ω e.g. ellipses, ovals, parabolas, one shows
 a) $S_L \subset \Omega$
 b) $\Omega \subset S$.

This is one reason why we claimed in Section 2.1 that one is interested to show $\Omega \subset S$. In [26] Wesseling shows a) for many different sets Ω.

Example 2: Full discretization of the advection equation.
When analyzing stability of difference schemes for solving initial boundary value problems of systems of partial differential equations

$$u_t = A(x, t) \frac{\partial u}{\partial x} + B(x, t) u + f(x, t) \quad x \in [0, 1], \, t \geq 0, \, u \in \mathbb{R}^N$$

it was shown in a series of papers [6], [20], [21] that it is enough to study the initial value problem

(2-5)
$$\frac{\partial u}{\partial t} = c\,\frac{\partial u}{\partial x}, \quad x \in \mathbb{R},\ t \geq 0,\ u \in \mathbb{R}$$
$$u(x,0) \text{ given}.$$

We discretize simultaneously space and time. Let Δx, Δt be the step-sizes and $x_m = m \cdot \Delta x$, $t_n = n \cdot \Delta t$ the grid points. u_{nm} is a numerical approximation to the exact solution $u(x_m, t_n)$. A general k-step difference scheme has the form

(2-6)
$$\sum_{\ell=-r_k}^{S_k} a_{k\ell}\, u_{n+k,m+\ell} + \sum_{j=0}^{k-1}\sum_{\ell=-r_j}^{s_j} a_{j\ell}\, u_{n+j,m+\ell} = 0$$

where $a_{j\ell}$ are method dependent coefficients which usually depend on the Courant number $\nu = c\Delta t/\Delta x$. A scheme is called explicit if $a_{k0} \neq 0$, $a_{k\ell} = 0$ if $\ell \neq 0$. Let $\tilde{u}_n$ be the Fourier transform of u_n where

$$u_n(x) = u_{nm} \quad \text{if } |x - x_m| < \Delta x/2.$$

With a similar interpretation (2-6) is an equation for all x and we can Fourier transform it. This gives

(2-7)
$$\alpha_k(e^{i\Delta x \theta})\tilde{u}_{n+k} + \sum_{j=0}^{k-1} \alpha_j(e^{i\Delta x \theta})\tilde{u}_{n+j} = 0$$

where

$$\alpha_j(\mu) = \sum_{\ell=-r_j}^{s_j} a_{j\ell}\, \mu^\ell.$$

In order that one can solve (2-7) for $\tilde{u}_{n+k}$ it is necessary and sufficient that $\alpha_k(\mu) \neq 0$ for $|\mu| = 1$. This makes it possible that we can always request the following **normalization conditions**:

a) $\alpha_k(1) = 1$

b) r_k, s_k are such that

$$r_k = \text{the number of roots of } \alpha_k(\mu) \text{ with } |\mu| < 1$$
$$s_k = \text{the number of roots of } \alpha_k(\mu) \text{ with } |\mu| > 1.$$

(2-7) is a linear recurrence relation. Hence to study the growth of its solution one has to introduce the characteristic function

$$\Phi(\zeta, \mu) = \sum_{j=0}^{k} \alpha_j(\mu)\, \zeta^j.$$

Definition 1 *The method is* **stable** *if*

$$\left.\begin{array}{rcl}\Phi(\zeta,\mu) & = & 0 \\ |\mu| & = & 1\end{array}\right\} \Longrightarrow \left\{\begin{array}{l}|\zeta| \leq 1 \\ \text{if } |\zeta| = 1, \text{ then } \zeta \text{ is a simple root}.\end{array}\right.$$

As in Section 2.1 we map the unit disk of the ζ-plane into the left half plane using the transformation (2-1). Let

$$\Psi(z) = (z-1)^k \, \Phi\left(\frac{z+1}{z-1}, \mu\right) = \sum_{i=0}^{k} a_{k-i} \, z^i \, ,$$

then the method is stable if

$$\left.\begin{array}{rcl}\Psi(z) & = & 0 \\ |\mu| & = & 1\end{array}\right\} \Longrightarrow \left\{\begin{array}{l}Re\, z \leq 0 \\ Re\, z = 0, \text{ then } z \text{ is a simple root}.\end{array}\right.$$

a_i are polynomials in $\mu \in \mathbb{C}$ with $|\mu| = 1$ and are functions of ν with $\nu \in (0, \nu_0]$.

2.3. General stability problem. In the two previous subsections we have given several examples of discretizations of time dependent problems. In all cases the stability analysis leads to a polynomial

$$\Psi(z) = \sum_{i=0}^{k} a_{k-i} \, z^i = a_0 \, z^k + a_1 \, z^{k-1} + \ldots + a_k$$

where a_i is a polynomial in $\mu \in \mathbb{C}$ and possibly also a function of a real parameter ν. One needs to show that the roots of $\Psi(z)$ are in the left half plane for all μ on a curve $C \subset \mathbb{C}$. Such curves can be straight lines e.g. $\{\mu \in \mathbb{C} \,|\, Im\,\mu = 0, \, Re\,\mu \leq 0\}$, circles, boundaries of ellipses, parabolas and ovals. In the next section we shall explain in a simple example how the Hurwitz determinants can be used to test for stability.

3. Hurwitz determinants and fraction free Routh algorithms. In this section we show how the Hurwitz criterion can be used to test stability of schemes. We take the particularly simple example of testing A_0-stability for a general linear method. This was also a part of criterias to test for stiff stability the author had devised in [12], [13], [14]. We therefore consider the special case.

(3-1) $$\Psi(z) = a_0 \, z^n + a_1 \, z^{n-1} + \ldots + a_n$$

where all a_i are real. The Routh-algorithm is then defined as follows:

Routh Algorithm Let $a_j = 0$ if $j > n$.

The first two rows of the Routh array are defined as follows

$$\begin{array}{rcl}r_{0j} & = & a_{2j-2} \quad \text{for} \quad j = 1, 2, 3, \ldots \\ r_{1j} & = & a_{2j-1} \quad \text{for} \quad j = 1, 2, 3, \ldots \, .\end{array}$$

For $i = 2, 3, \ldots$ one computes

$$(3\text{-}2) \qquad r_{ij} = -\frac{1}{r_{i-1,1}} \begin{vmatrix} r_{i-2,1} & r_{i-2,j+1} \\ r_{i-1,1} & r_{i-1,j+1} \end{vmatrix}.$$

□

Let us assume that the Routh array is regular, i.e. $r_{i1} \neq 0$ for $i = 0, 1, 2, \ldots$. The following Routh criterion is an easy way to test stability.

THEOREM 3.1. *Routh-Criterion*

$\Psi(z)$ *has all zeros in the left half plane if and only if* $r_{01}, r_{11}, r_{21}, r_{31}, \ldots$ *have the same sign.* □

The disadvantage of this algorithm is that r_{ij} with $i > 1$ become rational functions of a_i, $i = 0, 1, 2, \ldots$. Hence if a_i are elements of a polynomial ring $J[\mu]$ then r_{ij} become rational functions. If one implements the algorithm with a symbol manipulation language computations become rather involved. Hence fraction free algorithms are introduced. Let $P[a]$ be the polynomial ring in the $n+1$ variables $a_0, a_1, \ldots, a_n$ and $R[a]$ the rational functions in the same variables.

Definition 2 $\{m_{ij}\}$ *is called a* **scaled fraction free Routh array** *if there exist scaling factors* $K_i \in R[a]$ *such that*

$$m_{ij} = K_i\, r_{ij} \in P[a].$$

Example 3: Barnett in [1] suggested to multiply (3-2) through by the numerator. Hence one obtains

$$m_{ij} = - \begin{vmatrix} m_{i-2,1} & m_{1-2,j+1} \\ m_{i-1,1} & m_{i-1,j+1} \end{vmatrix} \quad \begin{matrix} i = 2, 3, \ldots \\ j = 1, 2, \ldots \end{matrix}.$$

Clearly $m_{ij} \in P[a]$. Let $\nu_i = \max_j\{\text{degree of } m_{ij} \in P[a]\}$. It is easy to see that ν_i are the Fibonacci numbers which grow exponentially.

Example 4. The **reduced PRS algorithm and the sub-resultant PRS algorithm** are fraction free versions of Euclid's algorithm [3], [4]. Hence they yield a fraction free Routh algorithm, see [15]. Contrary to Example 3 the normalized degrees ν_i grow only linearly. In fact $\nu_i = 2i - 1$, $i = 1, 2, \ldots$. See [15] for details.

The question is which scaled fraction free Routh array shows the slowest growth in ν_i. Here comes the relation to the famous paper by Hurwitz which is celebrated in this volume. Using Hurwitz's original notation the so called Hurwitz matrix is

$$H = \begin{bmatrix} a_1 & a_3 & a_5 & \cdots & a_{2n-1} \\ a_0 & a_2 & a_4 & & a_{2n-2} \\ 0 & a_1 & a_3 & & \\ 0 & a_0 & a_2 & & \\ 0 & 0 & a_1 & & \\ \vdots & \vdots & \vdots & & \end{bmatrix}.$$

Let H_{ij} be the minor formed from the first i rows, the first $i-1$ columns and the $i-1+j$th column of H. It is well known that

(3-3)
$$\begin{aligned} H_{ij} &= H_{i-1,1}\, r_{ij} & &\text{for } i = 2, 3, \ldots, \; j = 1, 2, 3, \\ H_{1j} &= r_{1j} = a_{2j-1} & & j = 1, 2, 3, \ldots \\ H_{0j} &= r_{0j} = a_{2j-2} & & j = 1, 2, 3, \ldots \;. \end{aligned}$$

Note that $H_{i-1,1}$ are called Hurwitz determinants. Due to (3-3) H_{ij} is a scaled fraction free Routh array. From the definition of H_{ij} if is obvious that $\nu_i = i$, $i = 1, 2, \ldots$. Hence one has linear growth and slower growth than in Example 4. Since Sylvester has shown that if n is odd and i even then $H_{i1}, H_{i2}, H_{i3}, \ldots$ have no common factor in $P[a]$ one cannot have a slower growth than $\nu_i = i$. Hence H_{ij} is the scaled fraction free Routh array where the growth of the normalized degrees ν_i is minimal.

Let us consider the following algorithm, see [14]

(3-4)
$$\begin{aligned} n_{0j} &= r_{0j} = a_{2j-2} & j = 1, 2, 3 \ldots \\ n_{1j} &= r_{1j} = a_{2j-1} & j = 1, 2, 3 \ldots \end{aligned}$$

(3-5)
$$n_{ij} = -\frac{1}{d_i} \begin{vmatrix} n_{i-2,1} & n_{i-2,j+1} \\ n_{i-1,1} & n_{i-1,j+1} \end{vmatrix} \quad \begin{array}{l} i = 2, 3, \ldots \\ j = 1, 2, 3 \ldots \end{array}$$

$$d_i = \begin{cases} 1 & \text{for } i = 2, 3 \\ n_{i-3,1} & \text{for } i = 4, 5, 6, \ldots \end{cases}.$$

This algorithm can be used to compute H_{ij} for one has the following

THEOREM 3.2. *[15], [5].* Assume $n_{i1} \neq 0$ and the n_{ij} are computed by (3-4), (3-5). Then $n_{ij} = H_{ij}$.

During the conference J. Garloff pointed out to the author that the algorithm (3-4), (3-5) and the theorem has been proved much earlier than in [15] by G. Fichera [5].

4. Order star technique.

The Routh-Hurwitz criterion is very good to test stability of a particular scheme if no parameter is involved in the scheme. However if the scheme depends on parameters such as μ and ν discussed in Section 2 the high level of nonlinearity involved in the Routh-Hurwitz criterion makes it extremely difficult, if not impossible to prove general results. One disadvantage is that the technique does not take in account that the numerical schemes approximate the exact solutions. For example, when solving ordinary differential equations $y' = f(t, y)$ we have seen in Section 2.1 that it makes sense to consider the equation $y' = \lambda y$. The exact solution is $y(t) = y_0 e^{\lambda t}$. Hence the exact solution satisfies the simple recurrence relation

$$y(t_n + \Delta t) = e^{\mu} y(t_n), \quad \text{where } \mu = \lambda \Delta t .$$

Thus the dominant root ζ_1 of $\Phi(\zeta, \mu)$ should approximate e^{μ} for μ close to zero. In a similar fashion in Example 2 the dominant root of $\Phi(\zeta, \mu)$ should approximate μ^{ν} for μ close to 1. In 1978 Wanner, Hairer, Nørsett [25] developed the so called order star technique which relates this approximation property to the stability requirements. The basic idea is that for a special method or the exact solution stability and accuracy is formulated in a function theoretical frame work. If it is done for a special method this involves the algebraic function defined by $\Phi(\zeta, \mu) \equiv 0$. Then one compares the **general** scheme to this special method or the exact solution. This comparison usually calls for an application of the argument principle. The technique was very successful and many open problems concerning time discretization of ordinary and partial differential equations could be solved, [11], [8]. To explain the idea in more detail we restrict ourselves to the full discretization of the advection equation which has already be treated in Example 2.

4.1. Order stars of full discretizations of the advection equation.

We have already formulated the stability of a method given by (2-6) in terms of Φ in Definition 1. The accuracy is measured by the error order p which is defined as follows. One substitutes any sufficiently differentiable function $u(x, t)$ into the left hand side of (2-6) and expand this in powers of Δx with the requirement that $\nu = c\Delta t/\Delta x$ is kept fixed, i.e.

$$(4\text{-}1) \qquad \sum_{j=0}^{k} \sum_{\ell=-r_j}^{s_j} a_{j\ell}\, u(x_{m+\ell}, t_{n+j}) = O\!\left((\Delta x)^{p+1}\right).$$

THEOREM 4.1. *[16], [24]. Let the method be stable and satisfy $\Phi(1,1) = 0$. Then the following three conditions are equivalent:*

  i) *The scheme has order p.*

  ii) $\Phi(\mu^{\nu}, \mu) = O\!\left((\mu - 1)^{p+1}\right)$ *as $\mu \to 1$.*

  iii) *The algebraic function ζ given by $\Phi(\zeta(\mu), \mu) \equiv 0$ has exactly one branch $\zeta_1(\mu)$ which is analytic in a neighborhood of $\mu = 1$ and satisfies*

$$(4\text{-}2) \qquad \mu^{\nu} - \zeta_1(\mu) = O\!\left((\mu - 1)^{p+1}\right) \quad \text{as } \mu \to 1 .$$

Hence we have expressed stability and accuracy both in terms of the algebraic function defined by $\Phi(\zeta,\mu) \equiv 0$.

To motivate the results one would like to prove we observe that the exact solution of $u_t = cu_x$ is constant along the characteristic lines $x + ct = \text{const}$. We introduce the stencil of a scheme by the set of indices:

(4-3) $\quad\quad I = \{(j,\ell) \in \mathbb{Z} \times \mathbb{Z} : 0 \leq j \leq k,\ -r_j \leq \ell \leq s_j\}.$

A difference stencil is called regular if the characteristic line through (x_m, t_{n+k}) does not pass through any other point $(x_{m+\ell}, t_{n+j})$, $(j,\ell) \in I$ [16]. The advection equation is of hyperbolic type and hence we know that information does not travel across characteristic lines. It turns out that due to the normalization conditions the characteristic line through the point (x_m, t_{n+k}) plays an important role. Let R, S be the number of stencil points to the left and right, respectively of this crucial characteristic line. With this definition the following conjecture has been made:

Conjecture: *Let a method of form (2-6) for the linear advection equation (2-5) be normalized and have a convex stencil. Then if the method is stable it implies that*

(4-4) $\quad\quad p \leq 2\min\{R, S\}.$

$\square$

A method is **convex** if the convex hull of the set of indices does not contain a point $(j,\ell) \in \mathbb{Z} \times \mathbb{Z}$ which is not a element of I. For $k = 1$ this result has been proved using the order star technique, [18]. For $k = 2$ [19], [17] have proved for certain subclasses of schemes this conjecture.

4.2. Outline of the order star technique. We have already expressed stability and accuracy of the "general" scheme in terms of the algebraic function $\zeta(\mu)$ which is the solution of $\Phi(\zeta,\mu) \equiv 0$. We do the same with the "good" scheme $\hat{\zeta}(\mu)$ or the exact solution. Now one considers

$$\varphi(\mu) = \frac{\text{"general" scheme}}{\text{"good" or exact}} = \begin{cases} \dfrac{\zeta(\mu)}{\hat{\zeta}(\mu)} & \text{if } k = 1 \\[2pt] \dfrac{\zeta(\mu)}{\mu^\nu} & \text{if } k > 1 \end{cases}$$

and define the order star Ω by

$$\Omega = \{\mu \mid |\varphi(\mu)| > 1\}.$$

If $k = 1$ then Ω is a set of $\mathbb{C}$ and for $k > 1$ it has to be defined on the Riemann surface of the algebraic function $\zeta(\mu)$. Let Δ be the "unit" disk. Then stability can be formulated as follows.

LEMMA 4.2. *The scheme is stable if and only if $\Omega \cap \partial\Delta = \emptyset$.* $\square$

The accuracy can also be expressed in terms of the order star Ω using the principle branch $\zeta_1(\mu)$ and (4-2).

LEMMA 4.3. *Assume the scheme is stable and has $\Phi(1,1) = 0$. Then it is of order p if and only if the order star Ω consists as the principle branch close to $\mu = 1$ of $p+1$ sectors with angle $\pi/(p+1)$ separated by $p+1$ sectors of the complement of Ω with the same angle.* □

As an illustration we give such an order star in the case $k=2$ with $p=8$ in Fig. 1.

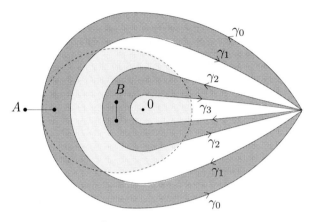

Fig. 1: Order star in the the case $k = 2$, $p = 8$. A and B denote cuts where the two sheets of the Riemann surface are connected

Let m be the number of sectors of Ω inside Δ. Hence Ω has at most $m+1$ sectors outside Δ. By Lemma 4.3 one obtains $p+1 \leq 2m+1$. Hence if we obtain a bound for m we obtain a bound for p. The argument principle is used to relate m to the location of poles of φ in Ω. To show this we do this considering the component Ω_1 given in Fig. 2.

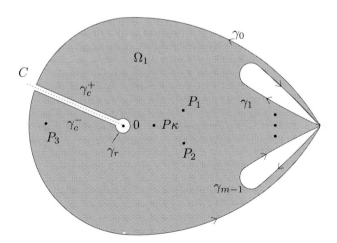

Fig. 2: Simple component Ω_1.

The important part of the order star theory is that by the Cauchy-Riemann differential equations it follows from the fact that $|\varphi| = $ constant along $\partial\Omega$ the argument of φ decreases along $\partial\Omega$. Hence

$$\frac{1}{2\pi i}\int_{\gamma_j}\frac{\varphi'}{\varphi}\,d\varphi \leq -1, \quad j=1,2,\ldots,m-1\ .$$

Since $\varphi(\mu) = \zeta(\mu)/\mu^\nu$ has an essential singularity at $\mu = 0$ we have to make the cut c indicated in Fig. 2. Let α_1 be the leading exponent of φ at $\mu = 0$, i.e.

$$\varphi(\mu) = \frac{\zeta(\mu)}{\mu^\nu} = \mu^{\alpha_1}g(\mu), \quad g(0) \neq 0\ .$$

We now apply the argument principle to the set $\Omega_1\backslash\{cut\}$.

$$-\kappa = \frac{1}{2\pi i}\int_{\partial(\Omega_1\backslash\{cut\})}\frac{\varphi'}{\varphi},$$

$$= \underbrace{\frac{1}{2\pi i}\int_{\gamma_0}}_{<0} + \underbrace{\frac{1}{2\pi i}\left(\int_{\gamma_c^+} + \int_{\gamma_c^-}\right)}_{\substack{=0\\ \text{as } r\to 0}} + \underbrace{\frac{1}{2\pi i}\int_{\gamma_r}}_{\leq \alpha_1} + \underbrace{\frac{1}{2\pi i}\int_{\gamma_1+\ldots\gamma_{m-1}}}_{\leq -(m-1)}.$$

Hence

$$m \leq \lceil\alpha_1\rceil + 1 + \kappa\ .$$

It remains to determine α_1 and κ. However due to the normalization we know that exactly r_k poles one in Δ. α_1 can be determined easily from the stencil of the scheme using the Newton-Puiseux diagram. Hence one gets a bound for m and thus for p. Unfortunately components of Ω can become rather complicated and thus one is able to prove the conjecture in certain cases only.

References

[1] Barnett S., *New reductions of Hurwitz determinants*. Int. J. Control, 18, 1973, 977–991.
[2] Bashforth F., *An attempt to test the theory of capillary action by comparing the theoretical and measured forms of drops of fluid. With an explanation of the method of integration employed in constructing the tables which give the theoretical form of such drops*. By J.C. Adams, Cambridge University Press 1883.
[3] Brown W.S., Traub J.F., *On Euclid's Algorithm and the Theory of Subresultants*. J. Assoc. Comput. Mach., **18**, 1971, 505–514.
[4] Collins G.E., *Subresultants and Reduced Polynomial Remainder Sequences*. J. Assoc. Comput. Mach., **14**, 1967, 128–142.
[5] Fichera G., *Alcune osservazioni sulle condizioni di stabilità per le equazioni algebriche a coefficienti reali*. Bolletino della Unione Matematica Italia, Ser. III, **2**, 1947, 103–109.
[6] Gustafsson B., Kreiss H.-O., Sundström A., *Stability theory of difference approximations for mixed initial boundary value problems II*. Math. Comp. **26**, 649–686, 1972.

[7] Hairer E., Nørsett S.P., Wanner G., *Solving ordinary differential equations I.* Springer, 1987.
[8] Hairer E., Wanner G., *Solving ordinary differential equations II.* Springer, 1991.
[9] Henrici P., *Discrete variable methods in ordinary differential equations.* J. Wiley & Sons, 1992.
[10] Hurwitz A., *Ueber die Bedingungen, unter welchen eine Gleichung nur Wurzeln mit negativen reellen Theilen besitzt.* Mathematische Annalen XLVI, 273–284, 1895.
[11] Iserles A., Nørsett S.P., *Order stars.* Chapman & Hall, 1991.
[12] Jeltsch R., *Stiff stability and its relation to A_0- and $A(0)$-stability.* SIAM J. Numer. Anal. **13**, 1976, 8–17.
[13] Jeltsch R., *Stiff stability of multi-step multi-derivative methods.* SIAM J. Numer. Anal, **14**, 1977, 760–772.
[14] Jeltsch R., *Corrigendum to "Stiff stability of multi-step multi-derivative methods".* SIAM J. Numer. Anal, **16**, 1979, 339.
[15] Jeltsch R., *An optimal fraction free Routh array.* Int. J. Control, **30**, 1979, 653–660.
[16] Jeltsch R., Kiani P., Raczek K., *Counterexamples to a stability barrier.* Numer. Math., **52**, 301–316, 1988.
[17] Jeltsch R., Renaut R.A., Smit J.H., *An accuracy barrier for stable three-time-level difference schemes for hyperbolic equations.* Research Report No 95-01, 1995, Seminar für Angewandte Mathematik, ETH Zürich.
[18] Jeltsch R., Smit J.H., *Accuracy barriers of difference schemes for hyperbolic equations.* SIAM J. Numer. Anal. **24**, 1–11, (1987).
[19] Jeltsch R., Smit J.H., *Accuracy barriers of three-time-level difference schemes for hyperbolic equations.* Ann. University of Stellenbosch, 1992/2, 1–34, 1992.
[20] Kreiss H.-O., *Difference approximations for the initial-boundary value problem for hyperbolic differential equations,* in: Numerical Solutions of Nonlinear Differential equations. Proc. Adv. Sympos., Madison, Wisconsin, 141–166, 1966.
[21] Kreiss H.-O., *Stability theory for difference approximations of mixed initial-boundary value problem I.* Math. Comput. **22**, 703–714, 1968.
[22] Lambert J.D., *Computational methods in ordinary differential equations,* J. Wiley & Sons, 1973.
[23] Marden M., *Geometry of polynomials.* American Mathematical Society, 1966.
[24] Strang G., Iserles A., *Barriers to stability.* SIAM J. Numer. Anal. **20**, 1251–1257, 1983.
[25] Wanner G., Hairer E., Nørsett S.P., *Order stars and stability theorems.* BIT 18, 475–489, 1978.
[26] Wesseling P., *A method to obtain Neumann stability conditions for the convection-diffusion equation.* In: Numerical Methods for Fluid Dynamics V, K W Morton and M J Baines, editors, 211–224, Clarendon Press, 1995.

Solving Stability Problems Using Quantifier Elimination*

Richard Liska
Faculty of Nuclear Sciences and Physical Engineering
Czech Technical University in Prague, Břehová 7, 115 19 Prague 1, Czech Republic,
liska@siduri.fjfi.cvut.cz

Stanly Steinberg
Department of Mathematics and Statistics, University of New Mexico
Albuquerque NM 87131-1141, USA
stanly@math.unm.edu

> **Abstract.** Stability problems for initial-value problems for partial differential equations and their discretizations can be written as quantifier elimination problems. Such problems can be solved by general quantifier elimination algorithms. However, if is far more efficient to use the Routh-Hurwitz or related criteria to eliminate one of the quantified variables before the general algorithms are applied.

1. Introduction. To solve a stability problem using quantifier elimination techniques, the stability condition being analyzed is written as a statement in the first-order theory of real closed fields, and then a quantifier-elimination algorithm is used to eliminate the quantifiers from this statement. The key to solving stability problems for initial-value problems is: first use either the Routh-Hurwitz and Liénard-Chipart criteria, that are used extensively in control theory, to remove one of the quantified variables; and then use the *quantifier elimination using partial algebraic decomposition* (QEPCAD) algorithm to remove the remaining quantifiers.

All of the steps in this procedure can be done using a computer algebra system, the most important of which is the QEPCAD program [2]. However, there are a number of other useful computer algebra programs that solve special quantifier-elimination problems, or are used to pre-process problems and post-process results, which are described in Liska and Steinberg [4]. These tools are capable of solving textbook examples in a few seconds and some simple problems in the literature in a few minutes. Serious research problems can take from a few minutes to a few hours, or they can easily exceed the available computing resource. Fortunately, many new faster algorithms are being developed [4].

Quantifier elimination techniques are also applicable to problems for ordinary differential equations and their discretization (see Liska and Steinberg [4]), to control theory problems (see Abdallah, Dorato, Liska, Steinberg, and Yang [1]), and initial-boundary value problems and their discretizations [4]. These papers also has extensive references to the literature.

2. Initial Value Problems for PDEs. A system of PDEs is called *unstable* if it possesses a solution that is unbounded in time but bounded in space, it is *stable* if

---

* This research was supported by the National Science Foundation International Programs Grant INT-9212433. In addition, R. Liska maritally by the Czech Grant Agency grant 201 / 94 / 1209 and Czech Technical University grant 48210.

all solutions that are bounded in space are also bounded in time, and *asymptotically stable* if all solutions that are bounded in space also converge to zero as as time goes to infinity. There are discussions of well-posedness in chapter 2 of Kreiss and Lorenz [3] and Chapter 9 of Strikwerda [5].

The stability testing method described in this paper is applicable to general system of partial differential equations,

$$\mathbf{S}(\partial_t, \partial_{\vec{x}}) \, \vec{f}(t, \vec{x}) = 0 \,, \tag{1}$$

where there are d spatial variables $\vec{x} = (x_1, \cdots, x_d)$, N is a positive integer, $\mathbf{S}$ is a $N \times N$ matrix of complex polynomials, and $\vec{f}$ is a N-component vector of complex-valued functions.

The stability of the initial value problem is tested by looking for exponential solutions of the form

$$\vec{f}(t, \vec{x}) = e^{\lambda t + i \vec{\xi} \cdot \vec{x}} \, \vec{v}(\lambda, \vec{\xi}) \,, \quad \vec{\xi} = (\xi_1, \cdots, \xi_d) \,,$$

where $\lambda \in \mathbb{C}$ and $\xi \in \mathbb{R}$. Only real ξ are used because it is assumed that the solutions are bounded in space. The function $\vec{f}$ can be a solution of the homogeneous system (1), if and only if

$$C(\lambda, \vec{\xi}) = \det \mathbf{S}(\lambda, i\vec{\xi}) = 0 \,,$$

where C the characteristic polynomial of the system of PDEs.

Here is a sample of the interesting stability questions for the systems of PDEs (1) that can be stated as quantifier elimination problems. First, assume that the coefficients of the system of PDEs depend on some parameters $\vec{\alpha}$, and that the characteristic polynomial for the system is written as $C(\lambda, \vec{\xi}, \vec{\alpha})$. Then the system of PDEs is *asymptotically stable* for given values of the parameters $\vec{\alpha}$ if and only if

$$\forall \vec{\xi} \in \mathbb{R}^d \, \forall \lambda \in \mathbb{C} \, \{C(\lambda, \vec{\xi}, \vec{\alpha}) = 0 \Rightarrow \Re \lambda < 0\} \,, \tag{2}$$

that is, for all $\vec{\xi}$ the polynomial is a *Hurwitz* polynomial.

Next, if the system of PDEs is stable then its characteristic polynomial must be *neutrally stable*:

$$\forall \vec{\xi} \in \mathbb{R}^d \, \forall \lambda \in \mathbb{C} \, \{C(\lambda, \vec{\xi}, \vec{\alpha}) = 0 \Rightarrow \Re \lambda \leq 0\} \,.$$

On the other hand, if the characteristic polynomial of a system of PDEs is *neutrally stable* and has only simple purely imaginary roots,

$$\forall \vec{\xi} \in \mathbb{R}^d \, \forall \lambda \in \mathbb{C}$$
$$\{C(\lambda, \vec{\xi}, \vec{\alpha}) = 0 \Rightarrow \{\Re \lambda \leq 0 \wedge \{\Re \lambda = 0 \Rightarrow \frac{\partial \, C(\lambda, \vec{\xi}, \vec{\alpha})}{\partial \, \lambda} \neq 0\}\}\} \,,$$

then the system of PDEs is *neutrally stable*.

Example (See Liska and Steinberg [4].) The easiest equations to analyze are those that are first order in time:

$$\text{(3)} \quad \frac{\partial f}{\partial t} + a \frac{\partial^4 f}{\partial x^4} + b \frac{\partial^3 f}{\partial x^3} + c \frac{\partial^2 f}{\partial x^2} + d \frac{\partial f}{\partial x} + g f = 0,$$

where $f = f(t, x), a, b, c, d, g \in \mathbb{R}$. The characteristic polynomial for equation (3) is

$$C(\lambda, \xi) = \lambda + a\xi^4 - b\imath\xi^3 - c\xi^2 + d\imath\xi + g,$$

where the dependence of $C(\lambda, \xi)$ on the parameters has been suppressed. The equation $C(\lambda, \xi) = 0$ can be solved for λ obtaining

$$\Re \lambda = -a\xi^4 + c\xi^2 - g,$$

This is a quadratic expression in ξ^2, so the analysis can be done using the special QE technique for low degree polynomials on semi-infinite interval that are described Liska and Steinberg [4]. The result is that the initial-value problem for (3) is asymptotically stable if and only if

$$[a > 0 \wedge g > 0 \wedge (c \geq 0 \vee (c > 0 \wedge c^2 - 4ga < 0))] \vee [a = 0 \wedge g > 0 \wedge c \leq 0].$$

QEPCAD produced, for this simple QE problem, an equivalent but simpler formula

$$(a \geq 0 \wedge c \leq 0 \wedge g > 0) \vee (4ag - c^2 > 0 \wedge g \geq 0)$$

in 0.5 s CPU time. By hand, one can derive that the initial-value problem for (3) is well posed if and only if

$$a > 0 \vee (a = 0 \wedge c \leq 0).$$

QEPCAD obtained this result in 1.2 s.

3. Discretized Initial Value Problems. This section considers the stability analysis of difference schemes that approximate initial-value problems (IVP) for linear homogeneous systems of PDEs like (1). There are discussion of such stability problems in Chapter 4 of Strikwerda [5]. The system PDEs is discretized on a uniform grid where

$$\text{(4)} \quad \vec{f}^n_{\vec{j}} \approx \vec{f}(n\Delta t, \vec{j} * \Delta \vec{x}),$$

for $n \geq 0$ and $\vec{j}$ and arbitrary multi-index. If the shift operators are defined by

$$\text{(5)} \quad \delta_t \vec{f}^n_{\vec{j}} = \vec{f}^{n+1}_{\vec{j}}, \quad \delta_{x_k} \vec{f}^n_{\vec{j}} = \vec{f}^n_{\vec{j}+\vec{1}_k}, \quad k = 1, \cdots, d,$$

where

$$\vec{1}_1 = (1, 0, \cdots, 0), \ \vec{1}_2 = (0, 1, \cdots, 0), \ \cdots, \vec{1}_d = (0, 0, \cdots, 1),$$

then a general linear finite difference scheme approximating (1) can be written as

$$\mathbf{S}(\delta_t, \delta_{\vec{x}}) \vec{f}_j^n = 0, \tag{6}$$

where as before S is a $N \times N$ matrix of polynomials, but now the shift operators δ_t and $\delta_{\vec{x}} = (\delta_{x_1}, \cdots, \delta_{x_d})$ are substituted for the variables.

The stability of the difference scheme is investigated by searching for solutions in the form

$$\vec{f}_j^n = s^n \vec{z}^{\vec{j}} \vec{v}, \quad |z_k| = 1, \quad k = 1, \cdots d, \tag{7}$$

where $n \geq 0$ and $s \in \mathbb{C}$, $z \in \mathbb{C}^d$. Substituting such a solution into the scheme (6) and dividing by $\vec{f}^n$ gives

$$\mathbf{S}(s, \vec{z}) \vec{v} = 0. \tag{8}$$

Thus (8) can have a nontrivial solution if and only if the characteristic polynomial C of the difference scheme is zero:

$$C(s, \vec{z}) = \det \mathbf{S}(s, \vec{z}) = 0. \tag{9}$$

The variable $\vec{z}$ is required to lie on the unit torus

$$\text{Torus}(\vec{z}) = \bigwedge_{k=1}^{d} |z_k| = 1, \tag{10}$$

because the solutions are to be bounded in the spatial index $\vec{j}$.

Stability requires that solutions of the difference scheme that are bounded in the spatial index $\vec{j}$, are also bounded with increasing time index n. time, that is,

$$\forall \vec{z} \in \mathbb{C}^d \, \forall s \in \mathbb{C} \, \{\{\text{Torus}(\vec{z}) \wedge C(s, \vec{z}, \vec{\alpha}) = 0\} \Rightarrow |s| \leq 1\}, \tag{11}$$

where $\vec{\alpha}$ are the parameters of the difference scheme. This is called the von Neumann necessary condition for stability. Typically, the parameters include grid steps sizes and the parameters from the original PDEs (1).

For the case of a scalar equation ($N = 1$), $\mathbf{S}$ has one entry which is the characteristic function C. For a one-step scheme, C is linear in s,

$$C(s, \vec{z}) = P(\vec{z}) s + Q(\vec{z}), \tag{12}$$

so there is only one solution of (9):

$$s = g(\vec{z}) = \frac{Q(\vec{z})}{P(\vec{z})},$$

where g is called the amplification factor. So for scalar one-step difference schemes, the necessary and sufficient von Neumann stability condition can be stated as the QE problem

$$\forall \vec{z} \in \mathbb{C}^d \, \{Torus(\vec{z}) \Rightarrow |g(\vec{z}, \vec{\alpha})| \leq 1\} \,. \tag{13}$$

The solution of the QE problem (11) gives useful conditions on parameters $\vec{\alpha}$ of the difference scheme so that the scheme is stable. However, before powerful QE algorithms such as QEPCAD can be used, several transformations need to be performed. The $\vec{z}$ variable in the QE problem is transformed to real variables by using trigonometric functions:

$$z_k = \cos(\xi_k \Delta x_k) + \imath \sin(\xi_k \Delta x_k)\,, \quad k = 1, \cdots, d\,. \tag{14}$$

and the trigonometric functions are then replaced by real variables with a polynomial constraint

$$s_k = \sin(\xi_k \Delta x_k)\,, \quad c_k = \cos(\xi_k \Delta x_k)\,, \quad s_k^2 + c_k^2 = 1\,. \tag{15}$$

In some problems it is useful to eliminate the two trigonometric functions by using the tangent function,

$$t_k = \tan(\xi_k \Delta x_k/2)\,, \quad k = 1, \cdots, d\,. \tag{16}$$

Note that the cases where the tangent is undefined, $\xi_k \Delta x_k/2 = \pm \pi/2$, must be analyzed separately.

For systems of equations or multi-step schemes, the linear fractional map described in Liska and Steinberg [4] is used to transform question about $|s| < 1$ to questions about $\Re \lambda < 0$. Then the Liénard-Chipart Criterion is used to eliminate the variable λ and its quantifier from the QE problem (11). This results in a new QE problem containing the quantified variables $\vec{z}$ (or s_k, c_k or t_k). We note that the choice of how to represent $\vec{z}$ is critical to the performance of QEPCAD. Reducing the number of variables, and thus using the tangent parameterization, if possible, is a good strategy.

Example (See Liska and Steinberg [4].) This simple example is the forward-time central-space difference scheme with smoothing (see Strikwerda [5, p. 47]):

$$f_j^{n+1} = \frac{1}{4}(\hat{f}_{j+1}^{n+1} + 2\hat{f}_j^{n+1} + \hat{f}_{j-1}^{n+1})\,,$$

$$\hat{f}_j^{n+1} = f_j^n - \frac{1}{2}\alpha(f_{j+1}^n - f_{j-1}^n)\,,$$

which approximates the one-way wave equation

$$\frac{\partial f}{\partial t} + a\frac{\partial f}{\partial x} = 0\,.$$

The use of the dimensionless Courant number, $\alpha = a\Delta t/\Delta x$, is important for reducing the number of variables in the problem, and is the natural variable used for describing constraints on the time step. The variable $\hat{f}$ is used to simplify the description of the scheme and may be eliminated to produce a one-step scheme. A method for calculating the amplification factor using the $\hat{f}$ is given in Strikwerda [5, p. 47].

The square of absolute value of amplification factor of the scheme is

$$|g|^2 = \frac{1}{4}\left[\alpha^2(-\cos^4(\xi\Delta x) - 2\cos^3(\xi\Delta x)\right.$$
$$\left. + 2\cos(\xi\Delta x) + 1) + \cos^2(\xi\Delta x) + 2\cos(\xi\Delta x) + 1\right].$$

After the substitution of $c = \cos(\xi\Delta x)$ into the previous, the von Neumann stability condition (13) for this scheme can be stated as the QE problem

$$\forall c \in \mathbb{R}\left\{-1 \leq c \leq 1 \Rightarrow \alpha^2(-c^4 - 2c^3 + 2c + 1) + c^2 + 2c + 1 \leq 4\right\}.$$

QEPCAD solved this QE problem in 0.45 s and returns the stability condition $\alpha^2 \leq 1/2$, which agrees with the result in Strikwerda [5, p. 48].

In the examples presented here, the quantifier corresponding to the time variable was easily eliminated. For more complex examples, such as those given in Liska and Steinberg [4], the polynomials are of higher degree in this variable, in which case the Routh-Hurwitz and Liénard-Chipart criteria play a critical role in eliminating this quantifier.

References

[1] C. Abdallah, P. Dorato, R. Liska, S. Steinberg, and W. Yang, *Applications of Quantifier Elimination to Problems in Control Theory*, Submitted to IFAC Congress, 1996.
[2] G.E. Collins and H. Hong, *Partial cylindrical algebraic decomposition for quantifier elimination*, J. Symb. Comp., **12** (1991), 299–328.
[3] H.-O. Kreiss and J. Lorenz, *Initial-Boundary Value Problems and the Navier-Stokes Equations* Academic Press, Boston, 1989.
[4] R. Liska and S. Steinberg, *Using Quantifier Elimination to Test Stability* submitted to the special issue of Journal of Symbolic Computation on Quantifier Elimination and its Applications.
[5] J.C. Strikwerda, *Finite Difference Schemes and Partial Differential Equations*, Wadsworth, Inc., Belmont, 1989.

Stability of Numerical Methods for solving Differential Equations*

Manfred R. Trummer
Department of Mathematics & Statistics and Centre for Experimental & Constructive
Mathematics, Simon Fraser University, Burnaby, B. C., V5A 1S6, CANADA
e-mail: (trummer@sfu.ca)

Abstract. We introduce some of the stability concepts for finite-difference and spectral discretizations of partial differential equations (PDEs). Many of the matrices which occur in such discretizations are non-normal, and we give a few examples for which classical eigenvalue analysis fails to give the correct stability results. This is related to the concept of Hurwitz stability radii.

1. Introduction. We would like to highlight some aspects of the stability of time discretizations of differential equations; related material can be found in this proceedings in [3]. Our model problem is the first-order wave equation (or one-way wave equation)

$$(1) \qquad u_t = u_x,$$

describing pure advection with constant wave speed -1. Propagation of energy at finite speed is characteristic of hyperbolic equations. The initial data are $u(x,0) = u_0(x)$, defined on the appropriate domain, and the exact solution is $u(x,t) = u_0(x+t)$. In case of a bounded domain, e.g. the interval $(-1,+1)$, a boundary condition is required at the right hand boundary, which we take to be zero,

$$(2) \qquad u(+1,t) = 0.$$

Let $h > 0$ and $k > 0$ be the space step and time step, respectively, and set $x_j = jh$ and $t_n = nk$. The points (x_j, t_n) define a regular **grid** (or **mesh** or **lattice**) in two dimensions, namely the subset $h\mathbb{Z} \times k\mathbb{Z} \subset \mathbb{R}^2$, in case of an infinite domain. A **grid function** $\{v_j^n\}$ is usually thought of as a discrete approximation to a continuous function u,

$$(3) \qquad v_j^n \approx u(x_j, t_n).$$

We also introduce the abbreviation $\lambda := \frac{k}{h}$, which is the the so-called *Courant number* of the scheme for wave speed 1.

In the case of time-dependent PDEs, finite difference (FD) formulas are usually applied through marching in time (i.e., with respect to t). That is, an s-step finite difference formula expresses v_j^{n+1} as a function of a finite number of grid values on s previous time levels ($n+1-s$ through n), and in the case of implicit formulas also time level $n+1$.

---

* This work was supported by the Natural Sciences and Engineering Research Council of Canada (NSERC) grant OGP0036901

TABLE 1
Finite difference formulas for $u_t = u_x$

LF: Leap Frog	$v_j^{n+1} = v_j^{n-1} + \lambda(v_{j+1}^n - v_{j-1}^n)$
BOX: Box scheme	$(1+\lambda)v_j^{n+1} + (1-\lambda)v_{j+1}^{n+1} = (1-\lambda)v_j^n + (1+\lambda)v_{j+1}^n$
LF4: 4th order Leap Frog	$v_j^{n+1} = v_j^{n-1} + \frac{4}{3}\lambda(v_{j+1}^n - v_{j-1}^n) - \frac{1}{6}\lambda(v_{j+2}^n - v_{j-2}^n)$
LXF: Lax-Friedrichs	$v_j^{n+1} = \frac{1}{2}(v_{j+1}^n + v_{j-1}^n) + \frac{1}{2}\lambda(v_{j+1}^n - v_{j-1}^n)$
UW: Upwind scheme	$v_j^{n+1} = v_j^n + \lambda(v_{j+1}^n - v_j^n)$
LW: Lax-Wendroff	$v_j^{n+1} = v_j^n + \frac{1}{2}\lambda(v_{j+1}^n - v_{j-1}^n) + \frac{1}{2}\lambda^2(v_{j+1}^n - 2v_j^n + v_{j-1}^n)$

One of the simplest formulas for our hyperbolic model (1) is the upwind formula (UW), obtained by replacing both derivatives by forward differences:

$$\text{(4)} \quad \text{UW}: \quad \frac{1}{k}(v_j^{n+1} - v_j^n) = \frac{1}{h}(v_{j+1}^n - v_j^n).$$

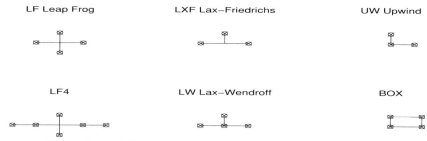

FIG. 1. *Finite difference stencils of the methods in Table 1 for $u_t = u_x$*

Table 1 lists commonly used formulas for the wave equation $u_t = u_x$, and Figure 1 shows the corresponding stencils, indicating which of the grid variables are involved in one step of the scheme.

Many numerical methods for time-dependent PDEs can be designed, implemented or analyzed as a two-stage process: First discretize in space, yielding a system of ordinary differential equations (ODEs) in time, and then apply an ODE solver.

EXAMPLE 1.1. Suppose we discretize $u_t = u_x$ in space by forward differences. Then we obtain an infinite system of ODEs

$$\text{(5)} \quad \frac{\partial w_j}{\partial t} = \frac{1}{h}(w_{j+1} - w_j), \quad j \in \mathbb{Z}.$$

Of course, on a bounded domain the system would be finite. Note that if we discretize (5) by Euler's method in time, we obtain the upwind scheme UW (4).

A system of ODEs of the kind in Example 1.1 is a semidiscrete approximation to the corresponding PDE. Solving it numerically is referred to as the "method of lines". If we use an explicit one-step finite difference formula[1], the numerical

---

[1] this framework can be extended to multistep and implicit formulas

scheme defines a marching operator $M_k = M$ advancing v^n to v^{n+1},

$$v_j^{n+1} = Mv_j^n = \sum \alpha_\mu v_{j+\mu}^n, \tag{6}$$

where the α_μ are constants specific to the FD method.

The **amplification factor** $g(\xi)$ amplifies the component of v with wave number ξ when we apply the marching operator M. Although there is a rigorous procedure based on Fourier analysis to find $g(\xi)$, it can be computed more quickly by inserting $v_j^n = g^n e^{i\xi jh}$ into the finite difference formula, and setting $g(\xi) = g^{n+1}/g^n$.

EXAMPLE 1.2. The upwind formula UW (4) is given by

$$v_j^{n+1} = Mv_j^n = v_j^n + \lambda(v_{j+1}^n - v_j^n), \qquad \lambda = k/h. \tag{7}$$

Inserting $v_j^n = g^n e^{i\xi jh}$ into (7) we obtain

$$g^{n+1} e^{i\xi jh} = g^n (e^{i\xi jh} + \lambda(e^{i\xi(j+1)h} - e^{i\xi jh})),$$

which results into

$$g(\xi) = 1 + \lambda(e^{i\xi h} - 1). \tag{8}$$

It is clear that if we march in time with the operator M, our solutions maybe amplified at each step by $||M||$ for a suitably chosen norm. Thus, M or its norm will tell us something about the stability of our numerical scheme.

For our explicit one-step scalar schemes we have

$$||M|| = ||g||_\infty, \qquad ||M^n|| = (||g||_\infty)^n = ||M||^n. \tag{9}$$

Note, that for operators we always have $||M^n|| \leq ||M||^n$, but not usually equality.

The family of marching operators $\{M_k\}_{k>0}$, and thus the corresponding numerical scheme, is said to be stable, if for every $T > 0$ there is a constant C, such that

$$||(M_k)^n|| \leq C, \tag{10}$$

for all n, k, $0 < nk \leq T$.

The famous *Lax Equivalence theorem* (e.g., [4]) states that stability of a consistent FD scheme and convergence of that scheme are equivalent. A scheme is consistent if it provides for a "good" approximation of the underlying PDE in a small time step. The technical details are in [4].

2. Stability and pseudospectra. We will now illustrate some of the stability concepts for the numerical solution of PDEs.

EXAMPLE 2.1. We apply the upwind formula (4) to the model problem $u_t = u_x$ with initial data

$$u(x,0) = \begin{cases} \cos^2(\frac{\pi}{2}x) & \text{for } |x| \leq 1, \\ 0 & \text{otherwise.} \end{cases} \tag{11}$$

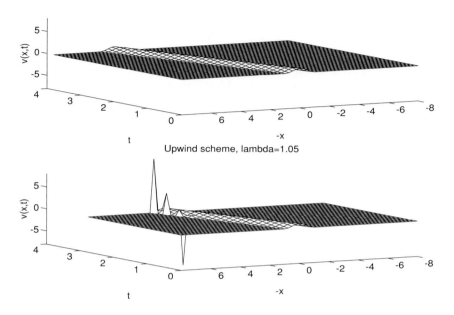

Fig. 2. *Upwind scheme applied to $u_t = u_x$ with $\lambda = 0.95$ and $\lambda = 1.05$.*

We take the space step $h = 1/128$, and set the time step $k = \lambda h$ for $\lambda = 0.95$ and $\lambda = 1.05$ (see Figure 2), although we plot on a coarser grid. It seems that everything works just fine for $\lambda = 0.95$, but for $\lambda = 1.05$ "visible" (on a plot) oscillations appear at around the time $t = 1$ which proceed to grow and very soon swamp the correct solution (see also Figure 3). This blow-up of a highly oscillatory mode is typical of a numerical instability.

This behaviour can be understood in several ways. From the experiment we expect convergence as $h, k \to 0$ for the $\lambda < 1$ case, but not for the $\lambda > 1$ case. Thus, from the Lax Equivalence theorem we conclude that UW is not stable for $\lambda > 1$ (it is consistent for all positive λ).

The CFL (Courant, Friedrichs, Lewy) condition tells us, that the numerical method must use information from the initial time level $t = 0$ which is relevant for the underlying PDE. The propagation speed for our PDE $u_t = u_x$ is 1, but in the upwind scheme, information cannot travel any faster than $1/\lambda$, i.e., $(1.05)^{-1}$ in our second calculation. Thus, we cannot expect this calculation to give us correct results.

The von Neumann condition for stability looks at the size of the amplification factor of the upwind formula. In Example 2.1 we computed

$$g(\xi) = (1 - \lambda) + \lambda e^{i\xi h},$$

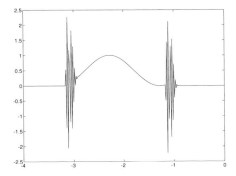
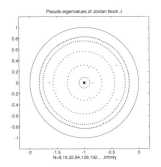

FIG. 3. Solution $v(x,t)$ at $t \approx 2.2805$ for Example 2.1 with $\lambda = 1.05$.

FIG. 4. Computed eigenvalues of QJQ^T, with Q a random orthogonal matrix, for $N = 16, 32, 64, 128, 192$

showing that $\lambda \leq 1$ is required to avoid amplification of certain Fourier components. If $\lambda > 1$ then the amplification factor becomes largest for $e^{i\xi h} = -1$, i.e., $\cos(\xi h) = -1$ or $|\xi| = \pi/h$, which explains why the instability we are observing in Figure 2 and in the more detailed Figure 3 is a mode with 2 points per wavelength.

We can also look at Example 2.1 from the "method of lines" point of view. The forward difference operator in space corresponds to a derogatory matrix which is $1/h$ times a Jordan block J with all elements on the diagonal equal to -1,

$$
(12) \qquad J = \begin{bmatrix} -1 & 1 & 0 & & & & \\ 0 & -1 & 1 & 0 & & & \\ & 0 & -1 & 1 & 0 & & \\ & & \ddots & \ddots & \ddots & \ddots & \\ & & & \ddots & \ddots & \ddots & 0 \\ & & & & 0 & -1 & 1 \\ & & & & & 0 & -1 \end{bmatrix};
$$

thus the eigenvalues are $-1/h$, and they must lie in the stability region of the ODE solver, which in this case is Euler's method. This leads to the condition $|-k/h + 1| \leq 1$, or $k/h \leq 2$. In this particular case it appears that we do not obtain a sharp bound on stability. However, the reason for this is that in dealing with non-normal matrices it is often necessary to look at the pseudospectrum (see [6]) of a matrix, to take into account finite-precision arithmetic and the sensitivity of eigenvalues to small perturbations.

The number μ is said to be ϵ-pseudoeigenvalue of an N by N matrix B if μ is an eigenvalue of a perturbed matrix $B + E$, with $||E|| < \epsilon$, or equivalently, if

$$
(13) \qquad ||(\mu I - B)^{-1}||_2 = (\sigma_N(\mu I - B))^{-1} \geq \frac{1}{\epsilon},
$$

where σ_N denotes the smallest (i.e., N-th) singular value. The set of all those μ is defined to be the ϵ-pseudospectrum of B. If B is symmetric, or, more generally, normal, this is not a very interesting set, namely just circles of radius ϵ around the eigenvalues of B. Denote by W the matrix of normalized eigenvectors of B. Then it is well known that the effect of perturbations in B on its eigenvalues may be amplified by as much as $\kappa(W)$, the condition number of W. Our matrix J is a pathological case, and can be interpreted as the limit of highly non-normal matrices (making in effect $\kappa(W) = \infty$). Of course, if we compute eigenvalues of J, we get the correct result that all of them equal -1; however, for any orthogonal matrix Q, the matrix QJQ^T has the same eigenvalues as J. If we compute those for different sizes of N (see Figure 4), we get the ϵ-pseudospectrum with $\epsilon =$ machine epsilon approaching the circle $|1 + z| \leq 1$. Thus, in the analysis above, replacing the exact spectrum of J with its pseudospectrum, we do after all obtain the correct stability criterion $k/h \leq 1$.

Of course, the previous example is somewhat extreme, as the matrix J does not even have a complete set of eigenvectors. The next example is more typical.

3. Spectral methods. For a general description and detailed analysis of spectral methods we refer to the monograph [1].

EXAMPLE 3.1 (PSEUDOSPECTRAL METHODS, SEE [7]). We now solve (1) with boundary condition (2) by a spectral method. The spatial derivative $\partial/\partial x$ of a grid function v is numerically approximated by evaluating the exact first derivative of the polynomial[2] interpolating the grid function at a prescribed set of points (the collocation points) and the boundary condition zero at $+1$. The most commonly used points are *Chebyshev* points (the zeroes or extrema of Chebyshev polynomials) or *Legendre* points, the zeroes of the Legendre polynomials on the interval $(-1, +1)$. This procedure gives rise to a so-called *spectral differentiation* matrix D, transforming the PDE $u_t = u_x$ into the ODE

$$(14) \qquad v_t = Dv.$$

Stability of the whole discretization process should be assured as long as the eigenvalues of D lie in the stability region of the ODE solver applied to (14). Not so — as in the example above, the stability region must contain the appropriate **pseudo**spectrum of D.

Figure 5 shows scaled plots of the computed eigenvalues of D_N, the N by N Legendre spectral differentiation matrix, for various values of N. In exact arithmetic, the shape of the spectra would be the same; the computed eigenvalues for $N = 56$ and $N = 70$ arise from the influence of round-off error. Figure 6 shows the pseudospectra of D_{14} for various ϵ, along with the exact eigenvalues. The matrix D_{14} is a Hurwitz matrix, i.e., all eigenvalues are in the left half plane, and it is interesting to note that although the pseudospectra become quite large, it takes

---

[2] trigonometric polynomials and equispaced collocation points are used for problems with periodic boundary conditions

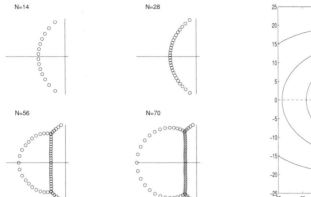
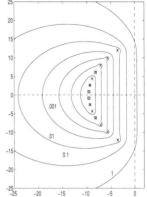

FIG. 5. *Plots of the computed eigenvalues of the Legendre spectral matrix D of dimensions $N = 14, 28, 56, 70$, scaled by $1/N$*

FIG. 6. *Pseudospectra of Legendre spectral differentiation matrix, $N = 14$, $\epsilon = 10^{-q}$, $q = 0, 1, \ldots, 8$.*

$\epsilon = O(1)$ to move part of the pseudospectrum into the right half plane. Thus, D_N has quite a large Hurwitz stability radius [2].

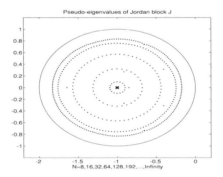

FIG. 7. *Eigenvalues of the Legendre differentiation matrix computed in double (o) and single (*) precision, and the stability region of the third-order Adams-Bashforth method for $N = 28$, $k = \frac{1}{2}N^{-1}$.*

For numerical methods employing explicit time stepping, the size of the pseudospectrum of D_N matters, even if it remains in the left half plane. For our experiment we take $N = 28$, and solve (1) with initial data (11), subject to the boundary condition (2) by Legendre spectral differentiation and time-stepping with the third-order Adams-Bashforth (AB-3) method

$$v^{n+1} = v^n + \frac{1}{12} D_N (23 v^n - 16 v^{n-1} + 5 v^{n-2})$$

with time step $k = \frac{1}{2}N^{-1} \approx .0179$. We show plots of the eigenvalues of D_N, $N = 28$, computed in double and single precision in Figure 7; superimposed is stability region for the ODE solver. In double precision all the eigenvalues are inside the stability region, and for $t = 1$, the error in our numerical solution is less than $5 * 10^{-3}$; however, some of the eigenvalues computed in single precision fall outside the stability region, and indeed, the numerical error at $t = 1$ explodes to more than 10^8.

In summary, the Legendre spectral method for $u_t = u_x$ has an $O(N^{-1})$ time step restriction in exact arithmetic, but an $O(N^{-2})$ restriction on k in finite precision. In this analysis we have dealt with stability for fixed time step k; this is not the same as stability defined in (10). It turns out, that the Lax stability restriction for this model is $O(N^{-2})$, so we have again a case where the analysis based on pseudospectra leads to more accurate conclusions (see also [5]). Moreover, it must be pointed out, that the smoothness of the initial data can have an effect on the behaviour of the scheme, and experiments indicating stability for certain initial data can be misleading. Another related point, is transient growth of solutions versus asymptotic stability. Eigenvalues (and pseudoeigenvalues) are good in predicting asymptotic stability, but may not reveal strong transient growth. This matters a lot in nonlinear stability analysis: A linear effect can lead to a large enough perturbation to invalidate stability predictions based on "small perturbations".

References

[1] C. Canuto, M. Y. Hussaini, A. Quarteroni, and T. A. Zang, *Spectral Methods in Fluid Dynamics*, Series of Comput. Physics, Springer, Heidelberg, Berlin, New York, 1988.

[2] D. Hinrichsen and A. J. Pritchard, *Real and complex stability radii: a survey*, in D. Hinrichsen and B. Mårtensson, editors, Control of Uncertain Systems, Birkhäuser, Boston, 1990.

[3] R. Jeltsch, *Stability of time discretization, Hurwitz determinants and order stars*, this proceedings.

[4] R. D. Richtmyer and K. W. Morton, *Difference Methods for Initial Value Problems*, Interscience, J. Wiley & Sons, 1967.

[5] L. N. Trefethen, *Lax-stability vs. eigenvalue stability of spectral methods*, in Numerical Methods for Fluid Dynamics III, K. W. Morton and M. J. Baines, eds., Clarendon Press, Oxford, 1988.

[6] L. N. Trefethen, *Spectra and Pseudospectra: The Behavior of Non-Normal Matrices and Operators*, in preparation.

[7] L. N. Trefethen and M. R. Trummer, *An instability phenomenon in spectral methods*, SIAM J. Numer. Anal., 24 (1987), pp. 1008–1023.

A Fast Algorithm to Compute the Real Structured Stability Radius*

J. Sreedhar[1]
Department of Electrical & Computer Engineering and the Coordinated Science Laboratory, University of Illinois at Urbana-Champaign, 1308 West Main St, Urbana, IL 61801-2307, USA.

Paul Van Dooren[2]
Université Catholique de Louvain, 1348 Louvain-la-Neuve, Belgium.
e-mail: vandooren@anma.ucl.ac.be

André L. Tits[3]
Department of Electrical Engineering and Institute for Systems Research, University of Maryland at College Park, College Park, MD 20742, USA.

Abstract. We describe a fast algorithm to compute the real structured stability radius with respect to the open left-half plane or the open unit disk. It is based on a recent formula proved by Qiu et al. (Automatica, vol. 31, pp. 879-890, 1995) and the well-known correspondence between the singular values of a transfer function matrix and the imaginary eigenvalues of a related Hamiltonian matrix. Numerical tests suggest that its local convergence is quadratic.

1. Preliminaries. The real structured stability radius measures the ability of a matrix to preserve its "stability" under a certain class of real perturbations. Consider a partitioning of the complex plane C into two disjoint sets $\mathbb{C}_g$ and $\mathbb{C}_b$ such that $\mathbb{C}_g$ is open, i.e., $\mathbb{C} = \mathbb{C}_g \dot\cup \mathbb{C}_b$. A matrix is called $\mathbb{C}_g$-stable if its spectrum $\boldsymbol{\lambda}(\cdot)$ is contained in $\mathbb{C}_g$. Let us denote the singular values of a $p \times m$ matrix, ordered nonincreasingly, by $\sigma_k(\cdot)$, $k = 1, 2, \ldots, \min\{p, m\}$. The real structured stability radius of a matrix triple $(A, B, C) \in \mathbb{R}^{n \times n} \times \mathbb{R}^{n \times m} \times \mathbb{R}^{p \times n}$, with A $\mathbb{C}_g$-stable, can be defined [1] as

$$r_{\mathbb{R}}(A, B, C) := \inf_{\Delta \in \mathbb{R}^{m \times p}} \{\sigma_1(\Delta) : \boldsymbol{\lambda}(A + B\Delta C) \not\subset \mathbb{C}_g\}.$$

When $B = I$, $C = I$, $r_{\mathbb{R}}(A, I, I)$ is usually abbreviated as $r_{\mathbb{R}}(A)$ and called the (unstructured) stability radius of A.

Let a transfer function matrix be given through

$$G(s) = C(sI - A)^{-1}B$$

---

* This paper presents research results of the Belgian Programme on Inter-university Poles of Attraction, initiated by the Belgian State, Prime Minister's Office for Science, Technology and Culture. The scientific responsibility rests with its authors.

[1] Was visiting Université Catholique de Louvain, Belgium, when this research was initiated. Research partially supported by NSF, grant CCR-9209349, and by UCL Research Board Contract FDS 729040.

[2] Research partially supported by NSF, grant CCR-9209349, and by UCL Research Board Contract FDS 729040.

[3] Was on sabbatical leave at Université Catholique de Louvain when this research was initiated. Research partially supported by NSF, grant DMI-93-13286.

and let $\partial \mathbb{C}_g$ denote the boundary of $\mathbb{C}_g$. Qiu et al. [2] proved that

$$r_{\mathbb{R}}^{-1}(A,B,C) = \sup_{s \in \partial \mathbb{C}_g} \mu_{\mathbb{R}}[G(s)], \tag{1}$$

where, for any $M \in \mathbb{C}^{p \times m}$,

$$\mu_{\mathbb{R}}(M) = \inf_{\gamma \in (0,1]} \sigma_2 \left(\begin{bmatrix} \operatorname{Re} M & -\gamma \operatorname{Im} M \\ \gamma^{-1} \operatorname{Im} M & \operatorname{Re} M \end{bmatrix} \right). \tag{2}$$

Further, they showed that the function to be minimized in (2) is unimodal; hence the computation of $\mu_{\mathbb{R}}(M)$ for a given M is reasonably straightforward. Still, efficient and accurate computation of $r_{\mathbb{R}}(A,B,C)$ is a challenging issue. For instance, in the continuous-time case, viz., $\mathbb{C}_g = \{s \in \mathbb{C} : \operatorname{Re} s < 0\}$, finding the so-called Hurwitz-stability radius

$$r_{\mathbb{R}}(A,B,C) = \left\{ \max_{\omega \in \mathbb{R}} \mu_{\mathbb{R}}[C(j\omega I - A)^{-1}B] \right\}^{-1} \tag{3}$$

requires global maximization of $\mu_{\mathbb{R}}[G(j\omega)]$ over $\omega \in \mathbb{R}$. Recently, a step was taken towards tackling this case: [3] determines a frequency range containing the global maximizer of (3).

In this paper, we propose an efficient algorithm for the computation of $r_{\mathbb{R}}(A,B,C)$ via (1) and (2) (a conceptually related algorithm can be used for the computation of an upper bound to the μ-norm of a transfer matrix [4]). For the continuous-time case, it is based on the well-known connection between the singular values of a transfer function matrix and the imaginary eigenvalues of a related Hamiltonian matrix. Typically it takes just a few iterations to converge; numerical results suggest that the rate of convergence is quadratic. We also describe a similar algorithm for the discrete-time case, viz., $\mathbb{C}_g = \{z \in \mathbb{C} : |z| < 1\}$.

2. Continuous-time case: Hurwitz-stability radius. We first assume that A is Hurwitz-stable and consider evaluating (3). We motivate our algorithm with Theorem 2.2 in Section 2.1 and give the details in Section 2.2. We illustrate its working with a numerical example in Section 2.3.

2.1. Singular values of a transfer function matrix and Hamiltonian eigenvalues. We shall use the overbar to denote complex conjugate, and superscripts T and H to denote matrix transpose and Hermitian transpose respectively; i.e., $B^{\mathrm{H}} = \overline{B}^{\mathrm{T}}$. For $\xi > 0$, define the Hamiltonian matrix

$$H(\xi,A,B,C) := \begin{bmatrix} A & BB^{\mathrm{T}}/\xi \\ -C^{\mathrm{T}}C/\xi & -A^{\mathrm{T}} \end{bmatrix}. \tag{4}$$

The following result connecting $G(j\omega)$ and $H(\xi,A,B,C)$ is well-known [5, 6, 7, 8].

PROPOSITION 2.1. *For all $\omega \in \mathbb{R}$, ξ is a singular value of $G(j\omega)$ iff $j\omega$ is an eigenvalue of $H(\xi,A,B,C)$.*

We seek to exploit this relationship to compute $r_\mathbb{R}(A,B,C)$. For $\gamma \in (0,1]$, let us use G to define a new transfer function matrix

$$(5) \qquad P(\gamma, G) := \begin{bmatrix} \operatorname{Re} G & \gamma j \operatorname{Im} G \\ \gamma^{-1} j \operatorname{Im} G & \operatorname{Re} G \end{bmatrix}.$$

From equations (1) and (2) we see that $P(\gamma, G)$ is unitarily equivalent to the matrix whose singular values are of interest in computing the real stability radius:

$$P(\gamma, G) = \begin{bmatrix} 1 & 0 \\ 0 & j \end{bmatrix} \cdot \begin{bmatrix} \operatorname{Re} G & -\gamma \operatorname{Im} G \\ \gamma^{-1} \operatorname{Im} G & \operatorname{Re} G \end{bmatrix} \cdot \begin{bmatrix} 1 & 0 \\ 0 & -j \end{bmatrix};$$

so the two have the same singular values and we can limit our attention to $P(\gamma, G)$. Our first goal is to obtain an adaptation of Proposition 2.1 involving P rather than G. To this effect, consider

$$(6) \qquad T_\gamma := \frac{1}{\sqrt{2}} \begin{bmatrix} I & \gamma I \\ \gamma^{-1} I & -I \end{bmatrix},$$

which can be factored as

$$T_\gamma = \begin{bmatrix} I & 0 \\ 0 & \gamma^{-1} I \end{bmatrix} \cdot \frac{1}{\sqrt{2}} \begin{bmatrix} I & I \\ I & -I \end{bmatrix} \cdot \begin{bmatrix} I & 0 \\ 0 & \gamma I \end{bmatrix} =: D_\gamma^{-1} U D_\gamma.$$

Note that U is symmetric and orthogonal ($U^\mathsf{T} = U = U^{-1}$) and D_γ represents diagonal scaling. Clearly, $T_\gamma^{-1} = D_\gamma^{-1} U^\mathsf{T} D_\gamma = T_\gamma$. The two matrices $P(\gamma, G)$ and T_γ are related as

$$(7) \qquad P(\gamma, G) = T_\gamma \begin{bmatrix} G & 0 \\ 0 & \overline{G} \end{bmatrix} T_\gamma,$$

where the left and right T_γ matrices are of dimensions $2p \times 2p$ and $2m \times 2m$, respectively. This is because

$$\begin{aligned}
T_\gamma \begin{bmatrix} G & 0 \\ 0 & \overline{G} \end{bmatrix} T_\gamma &= D_\gamma^{-1} U D_\gamma \cdot \begin{bmatrix} G & 0 \\ 0 & \overline{G} \end{bmatrix} \cdot D_\gamma^{-1} U D_\gamma \\
&= D_\gamma^{-1} \cdot U \begin{bmatrix} G & 0 \\ 0 & \overline{G} \end{bmatrix} U \cdot D_\gamma \\
&= D_\gamma^{-1} \begin{bmatrix} \operatorname{Re} G & j \operatorname{Im} G \\ j \operatorname{Im} G & \operatorname{Re} G \end{bmatrix} D_\gamma \\
&= \begin{bmatrix} \operatorname{Re} G & \gamma j \operatorname{Im} G \\ \gamma^{-1} j \operatorname{Im} G & \operatorname{Re} G \end{bmatrix},
\end{aligned}$$

which is $P(\gamma, G)$ by definition. This leads to the following adaptation of Proposition 2.1:

THEOREM 2.2. Let $\gamma \in (0,1]$ and $\xi > 0$ be given. Then, for all $\omega \in \mathbb{R}$, ξ is a singular value of $P(\gamma, G(j\omega))$ iff $j\omega$ is an eigenvalue of $H(\xi, \widetilde{A}, \widetilde{B}_\gamma, \widetilde{C}_\gamma)$, where

$$\widetilde{A} = \begin{bmatrix} A & 0 \\ 0 & -A \end{bmatrix}, \widetilde{B}_\gamma = \frac{1}{\sqrt{2}}\begin{bmatrix} B & \gamma B \\ -\gamma^{-1}B & B \end{bmatrix}, \widetilde{C}_\gamma = \frac{1}{\sqrt{2}}\begin{bmatrix} C & \gamma C \\ \gamma^{-1}C & -C \end{bmatrix}.$$

Proof. In view of Proposition 2.1, it is sufficient to verify that

$$(8) \qquad P(\gamma, G(j\omega)) = \widetilde{C}_\gamma (j\omega I - \widetilde{A})^{-1} \widetilde{B}_\gamma.$$

And indeed, from (7)

$$\begin{aligned}
P(\gamma, G(j\omega)) &= T_\gamma \begin{bmatrix} C(j\omega I - A)^{-1}B & 0 \\ 0 & -C(j\omega I + A)^{-1}B \end{bmatrix} T_\gamma \\
&= T_\gamma \begin{bmatrix} C & 0 \\ 0 & C \end{bmatrix} \cdot \begin{bmatrix} j\omega I - A & 0 \\ 0 & j\omega I + A \end{bmatrix}^{-1} \cdot \begin{bmatrix} B & 0 \\ 0 & -B \end{bmatrix} T_\gamma \\
&= \widetilde{C}_\gamma (j\omega I - \widetilde{A})^{-1} \widetilde{B}_\gamma,
\end{aligned}$$

since $\widetilde{C}_\gamma = T_\gamma \begin{bmatrix} C & 0 \\ 0 & C \end{bmatrix}$ and $\widetilde{B}_\gamma = \begin{bmatrix} B & 0 \\ 0 & -B \end{bmatrix} T_\gamma$. $\square$

The Hamiltonian matrix $H(\xi, \widetilde{A}, \widetilde{B}_\gamma, \widetilde{C}_\gamma)$ plays an important role in our algorithm, so we pause here for a simplified expression which will aid in its eigenvalue computation. Firstly, we note that it is a real matrix. It is mainly a function of the two scalar parameters ξ and γ, so we shall denote it sometimes by $\widetilde{H}(\xi, \gamma)$. We have, from (4),

$$\widetilde{H}(\xi, \gamma) := H(\xi, \widetilde{A}, \widetilde{B}_\gamma, \widetilde{C}_\gamma) = \begin{bmatrix} \widetilde{A} & \widetilde{B}_\gamma \widetilde{B}_\gamma^T / \xi \\ -\widetilde{C}_\gamma^T \widetilde{C}_\gamma / \xi & -\widetilde{A}^T \end{bmatrix}$$

$$= \left[\begin{array}{c|c} \begin{bmatrix} A & 0 \\ 0 & -A \end{bmatrix} & \begin{bmatrix} B & 0 \\ 0 & -B \end{bmatrix} \frac{T_\gamma T_\gamma^T}{\xi} \begin{bmatrix} B^T & 0 \\ 0 & -B^T \end{bmatrix} \\ \hline -\begin{bmatrix} C^T & 0 \\ 0 & C^T \end{bmatrix} \frac{T_\gamma^T T_\gamma}{\xi} \begin{bmatrix} C & 0 \\ 0 & C \end{bmatrix} & \begin{bmatrix} -A^T & 0 \\ 0 & A^T \end{bmatrix} \end{array}\right].$$

Expanding terms and defining

$$(9) \qquad \begin{array}{ll} \alpha := (1+\gamma^2)/2\gamma > 0, & \beta := (1-\gamma^2)/2\gamma \geq 0 \\ \Phi := \xi^{-1}BB^T \geq 0, & \Psi := \xi^{-1}C^TC \geq 0, \end{array}$$

we obtain

$$(10) \qquad \widetilde{H}(\xi, \gamma) = \left[\begin{array}{cc|cc} A & 0 & \gamma\alpha\Phi & -\beta\Phi \\ 0 & -A & -\beta\Phi & \frac{\alpha}{\gamma}\Phi \\ \hline -\frac{\alpha}{\gamma}\Psi & \beta\Psi & -A^T & 0 \\ \beta\Psi & -\gamma\alpha\Psi & 0 & A^T \end{array}\right].$$

This expression for $\widetilde{H}(\xi,\gamma)$ is poorly balanced in γ and could pose problems in computing eigenvalues accurately when γ is near zero. We can partially remedy this by scaling:

$$\begin{bmatrix} \gamma^{-1} & & & \\ & 1 & & \\ & & 1 & \\ & & & \gamma^{-1} \end{bmatrix} \widetilde{H}(\xi,\gamma) \begin{bmatrix} \gamma & & & \\ & 1 & & \\ & & 1 & \\ & & & \gamma \end{bmatrix} = \left[\begin{array}{cc|cc} A & 0 & \alpha\Phi & -\beta\Phi \\ 0 & -A & -\beta\Phi & \alpha\Phi \\ \hline -\alpha\Psi & \beta\Psi & -A^{\mathrm{T}} & 0 \\ \beta\Psi & -\alpha\Psi & 0 & A^{\mathrm{T}} \end{array}\right].$$
(11)

It is well-known that similarity transformations preserve eigenvalues. Expression (11) is better balanced in γ and should be preferred to (10). As a further simplification, which reveals additional structure when $\gamma = 1$, we can shuffle blocks around using permutation matrices and perform a second scaling similarity transformation (multiply with $\mathrm{diag}(1,1,1,-1)$ on the left, $\mathrm{diag}(1,1,1,-1)$ on the right) to obtain

$$\left[\begin{array}{cc|cc} A & \alpha\Phi & 0 & \beta\Phi \\ -\alpha\Psi & -A^{\mathrm{T}} & \beta\Psi & 0 \\ \hline 0 & -\beta\Phi & -A & -\alpha\Phi \\ -\beta\Psi & 0 & \alpha\Psi & A^{\mathrm{T}} \end{array}\right].$$

The "(1,1)-block" is now equal to the negative of the "(2,2)-block"; when $\gamma = 1$ ($\beta = 0, \alpha = 1$), each is just $H(\xi, A, B, C)$ and the big matrix is block diagonal, which means that its eigenvalue problem deflates nicely.

2.2. Algorithm outline. Theorem 2.2 suggests an iterative procedure to locate the maximum of $\mu_{\mathbb{R}}(G(\mathrm{j}\omega))$. Define

$$\Xi := \max_{\omega \in \mathbb{R}} \mu_{\mathbb{R}}[G(\mathrm{j}\omega)] = \{r_{\mathbb{R}}(A,B,C)\}^{-1} \quad \text{and} \quad \Omega := \arg\max_{\omega \in \mathbb{R}} \mu_{\mathbb{R}}[G(\mathrm{j}\omega)],$$

assuming that such a unique maximizer exists. Suppose that at each iteration, $k = 0, 1, \ldots$, we are given ω_k, the current trial frequency, and a "maximizing set", S_k, containing Ω. We maintain two quantities Ξ_k and Ω_k, our best approximations thus far of Ξ and Ω, respectively, with $\Xi_k = \mu_{\mathbb{R}}[G(\mathrm{j}\Omega_k)]$. We also form S_{k+1} by computing R_k, a union of "σ_2-intervals" of interest. The sequence of steps to be performed at iteration k is described next. Assuming that $\mu_{\mathbb{R}}(G(\mathrm{j}\omega_k))$ is achieved at $\gamma_{\omega_k}^*$, define the current "level"

(12) $\quad \xi_k := \mu_{\mathbb{R}}(G(\mathrm{j}\omega_k)) = \min_{\gamma} \sigma_2(P(\gamma, G(\mathrm{j}\omega_k))) = \sigma_2(P(\gamma_{\omega_k}^*, G(\mathrm{j}\omega_k))).$

If $\xi_k > \Xi_{k-1}$, take $\Xi_k = \xi_k$, $\Omega_k = \omega_k$, otherwise keep the old estimates, i.e., $\Xi_k = \Xi_{k-1}$, $\Omega_k = \Omega_{k-1}$. We now proceed to identify the set R_k of frequencies ω where $\sigma_2(P(\gamma_{\omega_k}^*, G(\mathrm{j}\omega)))$ exceeds Ξ_k. The significance of R_k is that, outside this set, $\mu_{\mathbb{R}}(G(\mathrm{j}\omega))$ is never larger than the current best approximation Ξ_k. This is because any frequency point $\widehat{\omega}$ not in R_k satisfies

$$\mu_{\mathbb{R}}(G(\mathrm{j}\widehat{\omega})) \leq \sigma_2(P(\gamma_{\omega_k}^*, G(\mathrm{j}\widehat{\omega}))) \leq \Xi_k.$$

This implies that, if $R_k \neq \emptyset$, the global maximizer Ω of $\mu_\mathbb{R}(G(j\omega))$ cannot lie outside R_k.

By Theorem 2.2, the imaginary eigenvalues of $\widetilde{H}(\Xi_k, \gamma^*_{\omega_k})$ are exactly the frequencies for which some singular value of $P(\gamma^*_{\omega_k}, G(j\omega))$ equals Ξ_k. The endpoints of the frequency intervals where $\sigma_2(P(\gamma^*_{\omega_k}, G(j\omega)))$ equals or exceeds Ξ_k must be among these, and can be identified using derivative information of the imaginary eigenvalues. Indeed, suppose λ_i is the ith eigenvalue, assumed simple, of $\widetilde{H}(\Xi_k, \gamma^*_{\omega_k})$ with corresponding normalized eigenvector v_i. We may differentiate λ_i with respect to ξ [9, page 81]:

$$\frac{\partial \lambda_i}{\partial \xi}(\Xi_k) = \widehat{v}_i^{\mathrm{H}} \frac{\partial \widetilde{H}}{\partial \xi}(\Xi_k, \gamma^*_{\omega_k}) v_i, \tag{13}$$

where we have defined $V = [v_1, v_2, \ldots, v_{4n}]$ and $[\widehat{v}_1, \widehat{v}_2, \ldots, \widehat{v}_{4n}] = V^{-\mathrm{H}}$. Note that V is known from the eigendecomposition of $\widetilde{H}(\Xi_k, \gamma^*_{\omega_k})$; hence the derivatives in (13) can be calculated at low cost. Now, for every i such that λ_i is pure imaginary, say, $\lambda_i = jw_i$, let $s_i : \mathbb{R} \to \mathbb{R}$ be the unique function that is equal to Ξ_k at w_i, is analytic in a neighborhood of w_i and has the property that, for all ω in a neighborhood of w_i, $s_i(\omega)$ is equal to some singular value of $P(\gamma^*_{\omega_k}, G(j\omega))$ (recall that λ_i is assumed to be a simple eigenvalue of $\widetilde{H}(\Xi_k, \gamma^*_{\omega_k})$). For such i, both sides of (13) are pure imaginary and the slope of s_i at w_i is given by

$$s'_i(w_i) = \left[\mathrm{Im}\{ \frac{\partial \lambda_i}{\partial \xi}(\Xi_k) \} \right]^{-1}. \tag{14}$$

In Fig. 1, we illustrate a possible scenario to identify R_k after (12) and (14) have been calculated: all w_i's are shown along with the sign of $s'_i(w_i)$. In practice, due to the symmetry, it is sufficient to compute these values on one side of the origin only. Since $P(\gamma^*_{\omega_k}, G(j\omega)) \to 0$ as $\omega \to -\infty$, the leftmost point among the w_i's, say, w_1, must correspond to σ_1, i.e., $\sigma_1(P(\gamma^*_{\omega_k}, G(jw_1))) = \Xi_k$ (and this is always an "upward" crossing, i.e., $s'_1(w_1)$ is always positive). Moving to the right, $s'_2(w_2)$ is found to be positive, indicating a (upward) σ_2-crossing, i.e., $\sigma_2(P(\gamma^*_{\omega_k}, G(jw_2))) = \Xi_k$, since a σ_1-crossing would yield a negative value of $s'_2(w_2)$ (downward crossing). Thus w_2 is the left endpoint of one of the σ_2-intervals we are seeking, i.e., of one of the intervals of ω over which $\sigma_2(P(\gamma^*_{\omega_k}, G(j\omega))) > \Xi_k$; we denote it $-\delta_1$. Similarly, since $s'_3(w_3) > 0$, w_3 must correspond to an upward σ_3-crossing and, since $s'_4(w_4) < 0$, w_4 must correspond to a downward σ_3-crossing, and so on. The σ_2-intervals of interest, viz., those for which $\sigma_2(P(\gamma^*_{\omega_k}, G(j\omega))) > \Xi_k$, are $(-\delta_1, -\delta_2)$, $(-\delta_3, +\delta_3)$ and (δ_2, δ_1); and R_k is their union.

Since S_k and R_k both contain Ω, it is a simple matter to bracket Ω at all times: for the next iteration, set $S_{k+1} = R_k \cap S_k$. As for the next trial frequency ω_{k+1}, if S_{k+1} contains more than one interval we set ω_{k+1} equal to the midpoint of the largest such interval; otherwise we follow a more complicated rule which sometimes involves cubic-fit (two function values and two derivatives) of $\mu_\mathbb{R}$ – please see pseudo-code below. In our numerical tests we have observed that, eventually,

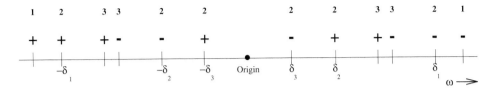

FIG. 1. *Using derivative information $s'_i(w_i)$ to pick off the σ_2-intervals $(-\delta_1, -\delta_2)$, $(-\delta_3, +\delta_3)$ and (δ_2, δ_1).*

S_k always reduces to a single interval and that the cubic-fit rule appears to lead to quadratic convergence (of Ξ_k to Ξ). We can summarize the salient features of our algorithm as follows:

Algorithm 1 (continuous-time)
Input A, B, C and tolerance $\tau > 0$.
Output $r_{\mathbb{R}}(A, B, C)$, $\Omega = \arg\max \mu_{\mathbb{R}}[G(\mathrm{j}\omega)]$.
Initialization $\omega_0 = 0, S_0 = (-\infty, \infty)$. $\Omega_0 = 0$, best ω so far,
$\Xi_0 = \mu_{\mathbb{R}}[G(\mathrm{j}\Omega_0)]$, highest "level" so far (assumed positive).

At iteration k $(= 0, 1, \ldots)$, given ω_k and S_k,
- Minimize $\sigma_2(P(\gamma, G(\mathrm{j}\omega_k)))$ w.r.t. γ, i.e., find
$$\xi_k := \mu_{\mathbb{R}}(G(\mathrm{j}\omega_k)) = \inf \sigma_2(P(\gamma, G(\mathrm{j}\omega_k)))$$
$$= \sigma_2(P(\gamma^*_{\omega_k}, G(\mathrm{j}\omega_k))).$$
- If $\xi_k > \Xi_{k-1}$, take $\Xi_k \leftarrow \xi_k$, $\Omega_k \leftarrow \omega_k$,
  else $\Xi_k \leftarrow \Xi_{k-1}$, $\Omega_k \leftarrow \Omega_{k-1}$.
- Using the imaginary eigenvalues of $\widetilde{H}(\Xi_k, \gamma^*_{\omega_k})$, find
$$R_k = \{\omega \in \mathbb{R} : \sigma_2(P(\gamma^*_{\omega_k}, G(\mathrm{j}\omega))) > \Xi_k\}.$$
- Update the maximizing set, i.e., $S_{k+1} \leftarrow R_k \cap S_k$.
  Quit if length of largest interval in S_{k+1} is less than τ.
- Define $m_{k+1} :=$ midpoint of largest interval in S_{k+1}.
- If S_{k+1} contains just one interval and there exist at least two (different) differentiable[1] members in the sequence $(\Omega_\ell, \Xi_\ell), \ell = 0, 1, \ldots$,
    Define $c_{k+1} :=$ maximizer within S_{k+1} of cubic fit through the latest two (different) differentiable points (Ω_i, Ξ_i). Set $\omega_{k+1} \leftarrow c_{k+1}$ or m_{k+1}, whichever has a larger value of $\mu_{\mathbb{R}}(\cdot)$.
  otherwise,
    Set $\omega_{k+1} \leftarrow m_{k+1}$.
  Replace $k \leftarrow k + 1$ and continue.
$r_{\mathbb{R}}(A, B, C) = \Xi_k^{-1}$, $\Omega = \Omega_k$.

---

[1] The derivative of $\mu_{\mathbb{R}}(G(\mathrm{j}\omega))$ with respect to ω at $\omega = \Omega_\ell$, whenever it exists, is equal to that of $\sigma_2(P(\gamma^*_{\Omega_\ell}, G(\mathrm{j}\omega)))$ at the same point.

2.3. Example 1. We illustrate our algorithm using a simple example. Take

$$n = 4, m = 2, p = 2,$$

$$A = \begin{bmatrix} -0.9925 & -1.1952 & -0.7372 & -0.9711 \\ -1.2178 & -0.9424 & -1.2466 & -1.5090 \\ -1.6607 & -1.6490 & -1.3853 & -1.5430 \\ -1.2727 & -1.2716 & -1.2684 & -1.3419 \end{bmatrix},$$

(15)

$$B = \begin{bmatrix} 0.9397 & 0.1238 \\ 0.9649 & 0.5263 \\ 0.2550 & 0.1601 \\ 0.0703 & 0.5177 \end{bmatrix},$$

$$C = \begin{bmatrix} 0.1052 & 0.3686 & 0.3272 & 0.5768 \\ 0.8416 & 0.4239 & 0.7135 & 0.8719 \end{bmatrix}.$$

The first graph in Fig. 2 contains three sample σ_2-curves (for $\gamma = 1, 0.0654, 0.2187$) and the $\mu_\mathbb{R}$ curve, showing that the latter is indeed the lower envelope of σ_2-curves. The remaining graphs illustrate algorithm iterations: $\mu_\mathbb{R}(G(j\omega))$ is plotted as a solid curve throughout, and the relevant σ_2-curve dashed. The level Ξ_k is marked as a horizontal line. The last three graphs also show the cubic fit curve: using 'dots' in iterations 3 and 4, and 'crosses' in iteration 5 (where it almost coincides with $\mu_\mathbb{R}$).

For this example, it turns out that R_k and S_k are intervals at every iteration. We show various quantities of interest in Table 1. As can be seen there, the cubic fit candidate (for ω_{k+1}) wins over the midpoint candidate near the optimum, and the algorithm appears to be quadratically convergent. Moreover, it takes only six iterations to shrink the size of S_{k+1} to 2×10^{-8}. The maximum (5.2389) of $\mu_\mathbb{R}(G(j\omega))$ occurs at $\Omega \approx 0.1099$ and we get $r_\mathbb{R}(A, B, C) \approx 0.1909$.

3. Discrete-time case: Schur-stability radius. We briefly consider the discrete-time case next, viz., $\mathbb{C}_g = \{z \in \mathbb{C} : |z| < 1\} := \mathbb{C}_1$ and give results analogous to those in Section 2. To proceed, we generalize Proposition 2.1 to discrete-time generalized state space systems. See [8, 10] for the first results in this area.

PROPOSITION 3.1. *Consider* $(E, A, B, C) \in \mathbb{C}^{n \times n} \times \mathbb{C}^{n \times n} \times \mathbb{C}^{n \times m} \times \mathbb{C}^{p \times n}$ *and assume that* $zE - A$ *has no generalized eigenvalues on the unit circle,* $z = e^{j\theta}$, $\theta \in [0, 2\pi]$. *Let* $M(z) = C(zE - A)^{-1}B$. *For any* $\theta_0 \in [0, 2\pi]$ *and any* $\xi > \|M(e^{j\theta_0})\|$,

$$M(e^{j\theta}) \text{ has singular value } \xi$$
$$\Updownarrow$$
$$F(\xi, E, A, B, C) - zG(\xi, E, A, B, C) \text{ has generalized eigenvalue } e^{j\theta},$$

where

$$F(\xi, E, A, B, C) = \begin{bmatrix} A & 0 \\ C^{\mathrm{H}}C/\xi & -E^{\mathrm{H}} \end{bmatrix}, \quad G(\xi, E, A, B, C) = \begin{bmatrix} E & -BB^{\mathrm{H}}/\xi \\ 0 & -A^{\mathrm{H}} \end{bmatrix}.$$

Index k	0	1	2	3	4	5
$\omega_k =$ Current "trial ω"	0	0.15024	0.10582	0.11029	0.10994	.10992408
$\Omega_k =$ "Best ω" so far	0	0.15024	0.10582	0.11029	0.10994	.10992408
$\gamma^*_{\Omega_k}$	1	0.08049	0.22862	0.16920	0.17714	0.17742
$\Xi_k = \mu_{\mathbb{R}}[G(j\Omega_k)]$	1.76957	2.01955	5.06924	5.23744	5.238936	5.238938
Slope at Ω_k	—	-28.44	77.286	-8.158	-0.282	-0.0002
$S_{k+1} =$ Updated "maximizing set"	(0, 0.30048)	(0.06141, 0.15024)	(0.10582, 0.11476)	(0.1095, 0.11029)	(0.10991, 0.10994)	(.10992406, .10992408)
Length of largest interval in S_{k+1}	0.300479	0.088831	0.008936	.000793	.000028	2×10^{-8}
$\omega_{k+1} =$ Next "trial ω" (how obtained)	0.15024 (midpt)	0.10582 (midpt)	0.11029 (midpt)	0.10994 (cubic)	.10992408 (cubic)	.10992407 (cubic)
Significant digits in Ξ_k	0	0	1	3	6	13

TABLE 1
Results showing convergence of the "maximizing set" S_k (which contains the global maximizer of $\mu_{\mathbb{R}}[G(j\omega)]$) for Example 1. Due to the symmetry, only quantities to the right of the origin are shown.

Proof. Omitted for the sake of brevity. $\square$

From equations (1), (2) and (5), we see that we must compute

$$r_{\mathbb{R}}^{-1}(A, B, C) = \sup_{\theta \in [0, 2\pi]} \inf_{\gamma \in (0,1]} \sigma_2\left(P(\gamma, G(e^{j\theta}))\right).$$

From equations (6) and (7),

$$\begin{aligned}
P(\gamma, G(e^{j\theta})) &= T_\gamma \begin{bmatrix} G(e^{j\theta}) & 0 \\ 0 & G(e^{-j\theta}) \end{bmatrix} T_\gamma \\
&= T_\gamma \begin{bmatrix} C(e^{j\theta}I - A)^{-1}B & 0 \\ 0 & Ce^{j\theta}(I - e^{j\theta}A)^{-1}B \end{bmatrix} T_\gamma \\
&= T_\gamma \begin{bmatrix} C & 0 \\ 0 & Ce^{j\theta} \end{bmatrix} \cdot \begin{bmatrix} (e^{j\theta}I - A)^{-1} & 0 \\ 0 & (I - e^{j\theta}A)^{-1} \end{bmatrix} \cdot \begin{bmatrix} B & 0 \\ 0 & B \end{bmatrix} T_\gamma \\
&= T_\gamma \begin{bmatrix} C & 0 \\ 0 & Ce^{j\theta} \end{bmatrix} \cdot \left\{ -\begin{bmatrix} -A & 0 \\ 0 & I \end{bmatrix} + e^{j\theta} \begin{bmatrix} -I & 0 \\ 0 & A \end{bmatrix} \right\}^{-1} \cdot \begin{bmatrix} -B & 0 \\ 0 & -B \end{bmatrix} T_\gamma,
\end{aligned}$$

Applying Proposition 3.1 to $P(\gamma, G(z)) = \widehat{C}_\gamma(z\widehat{E} - \widehat{A})^{-1}\widehat{B}_\gamma$, where

$$\widehat{E} = \begin{bmatrix} -I & 0 \\ 0 & A \end{bmatrix}, \quad \widehat{A} = \begin{bmatrix} -A & 0 \\ 0 & I \end{bmatrix},$$

$$\widehat{B}_\gamma = \begin{bmatrix} -B & 0 \\ 0 & -B \end{bmatrix} T_\gamma, \quad \widehat{C}_\gamma = T_\gamma \begin{bmatrix} C & 0 \\ 0 & Cz \end{bmatrix},$$

it is clear that the pencil of interest is

$$F(\xi, \widehat{E}, \widehat{A}, \widehat{B}_\gamma, \widehat{C}_\gamma) - zG(\xi, \widehat{E}, \widehat{A}, \widehat{B}_\gamma, \widehat{C}_\gamma)$$

$$= \begin{bmatrix} \widehat{A} & 0 \\ \widehat{C}_\gamma^{\mathrm{H}}\widehat{C}_\gamma/\xi & -\widehat{E}^{\mathrm{T}} \end{bmatrix} - z \begin{bmatrix} \widehat{E} & -\widehat{B}_\gamma \widehat{B}_\gamma^{\mathrm{T}}/\xi \\ 0 & -\widehat{A}^{\mathrm{T}} \end{bmatrix} = \begin{bmatrix} \widehat{A} - z\widehat{E} & z\widehat{B}_\gamma\widehat{B}_\gamma^{\mathrm{T}}/\xi \\ \widehat{C}_\gamma^{\mathrm{H}}\widehat{C}_\gamma/\xi & z\widehat{A}^{\mathrm{T}} - \widehat{E}^{\mathrm{T}} \end{bmatrix}$$

$$= \left[\begin{array}{c|c} \begin{bmatrix} -A & 0 \\ 0 & I \end{bmatrix} - z \begin{bmatrix} -I & 0 \\ 0 & A \end{bmatrix} & \begin{bmatrix} B & 0 \\ 0 & B \end{bmatrix} \frac{T_\gamma T_\gamma^{\mathrm{T}}}{\xi} \begin{bmatrix} zB^{\mathrm{T}} & 0 \\ 0 & zB^{\mathrm{T}} \end{bmatrix} \\ \hline \begin{bmatrix} C^{\mathrm{T}} & 0 \\ 0 & z^{-1}C^{\mathrm{T}} \end{bmatrix} \frac{T_\gamma^{\mathrm{T}} T_\gamma}{\xi} \begin{bmatrix} C & 0 \\ 0 & Cz \end{bmatrix} & z \begin{bmatrix} -A^{\mathrm{T}} & 0 \\ 0 & I \end{bmatrix} - \begin{bmatrix} -I & 0 \\ 0 & A^{\mathrm{T}} \end{bmatrix} \end{array}\right]$$

$$= \begin{bmatrix} zI - A & 0 & z\alpha\gamma\Phi & z\beta\Phi \\ 0 & -zA + I & z\beta\Phi & z\frac{\alpha}{\gamma}\Phi \\ \frac{\alpha}{\gamma}\Psi & -z\beta\Psi & I - zA^{\mathrm{T}} & 0 \\ -\beta z^{-1}\Psi & \alpha\gamma\Psi & 0 & -A^{\mathrm{T}} + zI \end{bmatrix}$$

where $\alpha = (1 + \gamma^2)/2\gamma > 0$, $\beta = (1 - \gamma^2)/2\gamma \geq 0$, $\Phi = \xi^{-1}BB^{\mathrm{T}} \geq 0$ and $\Psi = \xi^{-1}C^{\mathrm{T}}C \geq 0$, as defined in (9). Borrowing the γ-scaling idea from the continuous-time case, we multiply with $\mathrm{diag}(1/\gamma, 1, 1, z/\gamma)$ on the left and with $\mathrm{diag}(\gamma, 1, 1, \gamma/z)$ on the right and get a "similar" pencil

$$\begin{bmatrix} zI - A & 0 & z\alpha\Phi & \beta\Phi \\ 0 & -zA + I & z\beta\Phi & \alpha\Phi \\ \alpha\Psi & -z\beta\Psi & I - zA^{\mathrm{T}} & 0 \\ -\beta\Psi & z\alpha\Psi & 0 & -A^{\mathrm{T}} + zI \end{bmatrix}.$$

Once again, just as before, we can shuffle blocks around and obtain

$$\begin{bmatrix} zI - A & z\alpha\Phi & 0 & \beta\Phi \\ \alpha\Psi & I - zA^{\mathrm{T}} & -z\beta\Psi & 0 \\ 0 & z\beta\Phi & -zA + I & \alpha\Phi \\ -\beta\Psi & 0 & z\alpha\Psi & -A^{\mathrm{T}} + zI \end{bmatrix}$$

$$= \begin{bmatrix} -A & 0 & 0 & \beta\Phi \\ \alpha\Psi & I & 0 & 0 \\ 0 & 0 & I & \alpha\Phi \\ -\beta\Psi & 0 & 0 & -A^{\mathrm{T}} \end{bmatrix} - z \begin{bmatrix} -I & -\alpha\Phi & 0 & 0 \\ 0 & A^{\mathrm{T}} & \beta\Psi & 0 \\ 0 & -\beta\Phi & A & 0 \\ 0 & 0 & -\alpha\Psi & -I \end{bmatrix}.$$

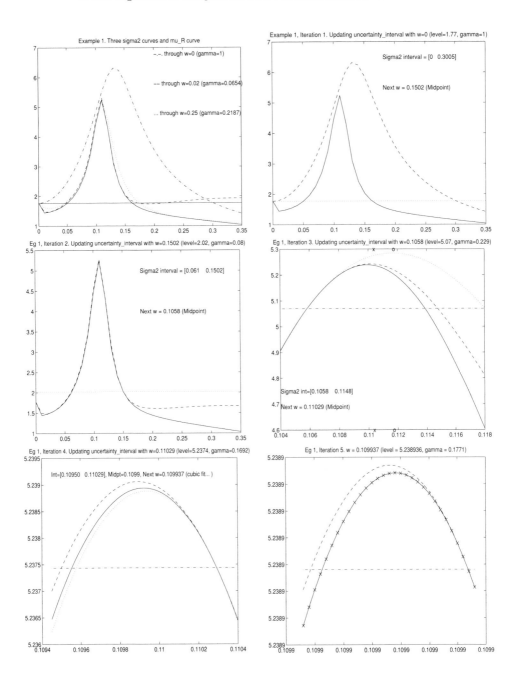

FIG. 2. *Plots showing S_k at each iteration for Example 1.*

4. Concluding remarks.

In this paper, we described an efficient algorithm to compute the real structured stability radius $r_\mathbb{R}(A, B, C)$. Numerical tests suggest that the rate of convergence is quadratic; conditions under which this can be proved are under investigation. Currently we have a convergence proof for a "pure midpoint rule" variant which begins by evaluating $\mu_\mathbb{R}$ at all its discontinuity points: the proof is essentially identical to that of the related algorithm discussed in [4]. We also have a modified updating rule for ω_k when $\mu_\mathbb{R}$ has discontinuous derivative at the optimum. Finally, a more sophisticated stopping criterion can be used which guarantees that Ξ has been computed with prescribed accuracy. All of this will be discussed elsewhere.

Acknowledgement. The authors would like to thank Prof. Barmish, Prof. Bernhardsson and Prof. Sontag for some useful remarks regarding a previous version of this manuscript.

References

[1] D. Hinrichsen and A. J. Pritchard, "Stability radius for structured perturbations and the algebraic Riccati equation," *Systems & Control Letters*, vol. 8, pp. 105–113, 1985.
[2] L. Qiu, B. Bernhardsson, A. Rantzer, E. Davison, P. Young, and J. Doyle, "A formula for computation of the real stability radius," *Automatica*, vol. 31, pp. 879–890, 1995.
[3] L. Qiu, A. L. Tits, and Y. Yang, "On the computation of the real Hurwitz-stability radius," *IEEE Transactions on Automatic Control*, vol. 40, No. 8, pp. 1475–1476, 1995.
[4] C. T. Lawrence, A. L. Tits, and P. Van Dooren, "A fast algorithm for the computation of an upper bound on the μ-norm", Proceedings of the 13th IFAC World Congress, San Francisco, California, 30 June–5 July 1996, Pergamon Press, 1996.
[5] D. Hinrichsen, B. Kelb, and A. Linnemann, "An algorithm for the computation of the complex stability radius," *Automatica*, vol. 25, pp. 771–775, 1989.
[6] S. Boyd, V. Balakrishnan, and P. Kabamba, "A bisection method for computing the H_∞ norm of a transfer matrix and related problems," *Mathematics of Control, Signals, and Systems*, vol. 2, pp. 207–219, 1989.
[7] N. A. Bruinsma and M. Steinbuch, "A fast algorithm to compute the H_∞ norm of a transfer function matrix," *Systems & Control Letters*, vol. 14, pp. 287–293, 1990.
[8] R. Byers, "A bisection method for measuring the distance of a stable matrix to the unstable matrices," *SIAM J. Sci. Stat. Comput.*, vol. 9, pp. 875–881, Sept 1988.
[9] T. Kato, *Perturbation Theory of Linear Operators*. New York, NY: Springer-Verlag, 1966.
[10] D. Hinrichsen and N. K. Son, "The complex stability radius of discrete-time systems and symplectic pencils," in *Proc. IEEE 28th Conference on Decision and Control*, (Tampa, FL), pp. 2265–2270, 1989.

Open Problems

In the discussion session organized by B.D.O. Anderson several participants have presented some open problems. We thank the authors for their contributions and do hope that they will enhance further research in the area of stability theory.

R. Jeltsch and M. Mansour

1. Stability of a Convex Set of Matrix Polynomials
   (B.D.O. Anderson)

2. Robust Stability and Passivity Connection
   (B.D.O. Anderson)

3. Continuity Properties of the Real Perturbation Values
   (B. Bernhardson)

4. A Real Perturbation Value Decomposition
   (B. Bernhardson)

5. Numerical Spectral Properties of Hermitian Toeplitz Matrices
   (Y.V. Genin)

6. A Real Counterpart to a Singular H^∞-Optimal Control Problem
   (D. Hinrichsen)

7. Time Varyingly Stable Interval Systems
   (D. Hinrichsen)

8. Order Barrier Due to Stability for Full Discretizations of the Advection Equation
   (R. Jeltsch)

9. On the Simplest Possible Property of Hurwitz Criterion
   (E.I. Jury)

10. Conversion of a Formal Power Series to a Continued Fraction
    (J. Waldvogel)

11. Degree of Robust Stability for Fuzzy Uncertainties
    (A. Weinmann)

1. Stability of a convex set of matrix polynomials. (B.D.O. Anderson)

This problem is concerned with generalizing a known result for scalar polynomials to the matrix case. A partial generalization has already been achieved.

Specifically let $p_1(s), p_2(s), \ldots, p_n(s)$ be real monic scalar polynomials of the same degree. Then the convex hull is stable if and only if there exists a real rational stable minimum phase transfer function $G(s)$ such that $G(s)p_i(s)$ is strictly positive real for $r = 1, 2, \ldots, n$, see [1].

Now suppose $P_1(s), P_2(s), \ldots, P_n(s)$ are real monic matrix polynomials of the same degree and dimension. If $n = 2$, it is known that the convex hull, viz $\lambda P_1 + (1-\lambda)P_2$ for $0 \leq \lambda \leq 1$, has stable determinant if and only if there exist real rational stable square $G(s)$ and $H(s)$ such that $G(s)P_i(s)H(s)$ is strictly positive real for $i = 1, 2$, see [2].

What can be said about $n > 2$?

References

[1] B.D.O. Anderson and S. Dasgupta, P.P. Khargonekar, F.J. Kraus and M. Mansour: Robust strict positive realness: characterization and construction, IEEE Trans Circuits and Systems Vol 37, pp 869–876, 1990.

[2] B.D.O. Anderson and S. Dasgupta: Multiplier theory and operator square roots: application to robust and time-varying stability, this volume.

2. Robust stability and passivity connection. (B.D.O. Anderson)

Let $H(s)$ be a stable scalar transfer function, and suppose that $(1+kH)^{-1}$ is also stable for $0 \leq k \leq 1$. Then the Nyquist diagram of H avoids the negative real axis between $-\infty$ and -1, and the Nyquist diagram of $1 + H$ then avoids $(-\infty, 0]$. It follows that $(1+H)^{\frac{1}{2}}$ is strictly positive real. This fact and various consequences of it are discussed in [1], which considers extension of the idea to the case when H is replaced by a time-varying impulse response or a matrix transfer function.

Suppose H is replaced by a causal nonlinear operator (with suitable smoothness properties). Can $(1+H)^{\frac{1}{2}}$ be defined? Is it passive? And if so, what can then be concluded?

References

[1] B.D.O. Anderson and S. Dasgupta: Multiplier theory and operator square roots: application to robust and time-varying stability, this volume.

3. Continuity properties of the real perturbation values. (B. Bernhardsson)

For a complex matrix M define

$$\tau_k(M) := \inf_{\gamma \in (0,1]} \sigma_{2k} \begin{bmatrix} \operatorname{Re} M & -\gamma \operatorname{Im} M \\ \gamma^{-1} \operatorname{Im} M & \operatorname{Re} M \end{bmatrix}.$$

For background see the contribution by Bernhardsson et. al. in these proceedings or [1]. Find the left- and right-hand sided Taylor expansions (or at least the first terms), of $\tau_k(M(\epsilon))$ at $\epsilon = 0$. Here $M(\cdot)$ is a smooth function.

References

[1] L. Qiu, B. Bernhardsson, A. Rantzer, E. J. Davison, P. M. Young, and J. C. Doyle: A formula for computation of the real stability radius. Automatica, to appear July 1995.

4. A real perturbation value decomposition. (B. Bernhardson)

Many properties of the singular values have parallels for the real perturbation values τ_k defined above. We have however not been able to find what would correspond to the Singular Value Decomposition. Is there a parallel to the SVD for the real perturbation values?

5. Numerical spectral properties of hermitian toeplitz matrices. (Y.V. Genin)

A matrix C_n with the algebraic structure $C_n = [c_{i,j} = c_{i-j}; i,j,= 0,1,\ldots,n;$ $c_{i,j} = \bar{c}_{j,i}]$ is said to be Hermitian *Toeplitz*. In other words, it is a Hermitian matrix whose entries on its successive diagonals are equal, i.e.

$$C_n = \begin{bmatrix} c_0 & c_1 & c_2 & c_3 & \cdot & \cdot & c_n \\ \bar{c}_1 & c_0 & c_1 & c_2 & \cdot & \cdot & c_{n-1} \\ \bar{c}_2 & \bar{c}_1 & c_0 & c_1 & \cdot & \cdot & c_{n-2} \\ \bar{c}_3 & \bar{c}_2 & \bar{c}_1 & c_0 & \cdot & \cdot & c_{n-3} \\ \cdot & \cdot & \cdot & \cdot & \cdot & \cdot & \cdot \\ \cdot & \cdot & \cdot & \cdot & \cdot & \cdot & \cdot \\ \bar{c}_n & \bar{c}_{n-1} & \bar{c}_{n-2} & \bar{c}_{n-3} & \cdot & \cdot & c_0 \end{bmatrix}.$$

Hermitian Toeplitz matrices are omnipresent in engineering applications. There exists a variety of numerical algorithms for their fast numerical processing. In spite of this, it is a noticeable fact that none of them takes benefit of the specific Toeplitz spectral properties. It is even not known how to compute the eigenvalues and the eigenvectors of a Hermitian Toeplitz matrix in the most efficient manner. Similarly, although it has been proved that the Toeplitz structure does not induce any particular constraint on the matrix spectrum in the symmetric real as well in the Hermitian situation, there exists today no constructive algorithm to build the set of Toeplitz matrices associated with an arbitrarily given spectrum.

6. A real counterpart to a singular H^∞-optimal control problem. (D. Hinrichsen)

Consider a system

$$\dot{x} = Ax + D\Delta Ex + Bu \qquad (1)$$

where $A \in \mathbb{R}^{n\times n}$, $D \in \mathbb{R}^{n\times \ell}$, $E \in \mathbb{R}^{q\times n}$, $B \in \mathbb{R}^{n\times m}$ are given and $\Delta \in \mathbb{R}^{\ell \times q}$ is an unknown disturbance matrix. We assume that A is Hurwitz stable, i.e. the spectrum $\sigma(A)$ is contained in the open left half-plane $\mathbb{C}_-$. The real stability radius of A under perturbations of the form $A \rightsquigarrow A + D\Delta E$ (see [1], [2]) is defined by

$$r_{\mathbb{R}}(A;D,E) = \inf\left\{\|\Delta\|; \Delta \in \mathbb{R}^{\ell\times q}, \sigma(A+D\Delta E) \not\subset \mathbb{C}_-\right\}.$$

The problem is:

Maximize $r_{\mathbb{R}}(A+BF;D,E)$ subject to $F \in \mathbb{R}^{m\times n}$ and $\sigma(A+BF) \subset \mathbb{C}_-$.

That is: determine the optimal value

$$\bar{r}_{\mathbb{R}}(A;D,E;B) = \sup_{F\in\mathbb{R}^{m\times n}, \sigma(A+BF)\subset \mathbb{C}_-} r_{\mathbb{R}}(A+BF;D,E)$$

and find for every $\varepsilon > 0$ a suboptimal feedback matrix $F_\varepsilon \in \mathbb{R}^{m \times n}$ such that

$$\sigma(A + BF_\varepsilon) \subset \mathbb{C}_- \quad \text{and} \quad r_\mathbb{R}(A + BF_\varepsilon; D, E) > \bar{r}_\mathbb{R}(A; D, E; B) - \varepsilon.$$

A similar problem can be stated for static or dynamic output feedback. Complex versions are solvable as singular H^∞-optimal control problems, see [3].

References

[1] D. Hinrichsen and A.J. Pritchard: Real and complex stability radii: a survey. In *Control of Uncertain Systems*, volume 6 of *Progress in System and Control Theory*, 119–162, Basel. Birkhäuser, 1990.

[2] L. Qiu, B. Bernhardsson, A. Rantzer, E.J. Davison, P.M. Young and J.C. Doyle: A formula for computation of the real stability radius. Automatica, to appear July 1995.

[3] D. Hinrichsen, A. J. Pritchard and S.B. Townley: A Riccati equation approach to maximizing the complex stability radius by state feedback. *International Journal of Control*, 52,769–794, 1990.

7. Time varyingly stable interval systems. (D. Hinrichsen)

Consider a matrix interval $[\underline{A}, \overline{A}] = \{X \in \mathbb{R}^{n \times n}; \underline{A} \leq X \leq \overline{A}\}$ where $\underline{A} = (\underline{a}_{ij})$, $\overline{A} = (\overline{a}_{ij}) \in \mathbb{R}^{n \times n}$ are given such that $\underline{A} \leq \overline{A}$, i.e. $\underline{a}_{ij} \leq \overline{a}_{ij}$ for $i, j \in \{1, 2, \ldots, n\}$. Find computable necessary and sufficient conditions for the interval $[\underline{A}, \overline{A}]$ to be time-varyingly stable in the sense that, for every piecewise continuous system matrix $A(\cdot) : [0, \infty) \to [\underline{A}, \overline{A}]$, the system $\dot{x}(t) = A(t)x(t)$ is asymptotically stable. For the case $n = 2$ the problem has been solved by Gonzalez (to my knowledge his result is not yet published, but see [1] for a description).
An interesting subproblem is obtained by choosing $\underline{A}, \overline{A}$ in companion form: Under which conditions are all linear differential equations of a given order with coefficients $a_i(t)$ varying in prescribed real intervals $[\underline{a}_i, \overline{a}_i]$ asymptotically stable?

References

[1] D. Hinrichsen and A.J. Pritchard: Destabilization by output feedback. *Differential and Integral Equations*, 5, 357–386, 1992.

8. Order barrier due to stability for full discretizations of the advection equation. (R. Jeltsch)

In Section 2.2 and 4 of the article [1] linear difference schemes of the form

$$(1) \qquad \sum_{j=0}^{k} \sum_{\ell=-r_j}^{s_j} a_{j\ell} u_{m+j, m+\ell} = 0$$

for the linear advection equation $u_t = cu_x$ have been treated. In order to formulate a slightly more general conjecture than the one formulated in [1] we introduce the following definition. A method is called **convex on the downwind side** if the convex hull of the set of indices does not contain a point $(j, \ell) \in \mathbb{Z} \times \mathbb{Z}$ which is not an element of I and $(x_{m+\ell}, t_{n+j})$ lies to the **left** of the characteristic line through the point (x_m, t_{n+k}).

Conjecture A *Let a method of form (1) for $u_t = cu_x$ be normalized and convex on the downwind side. Then stability of the method implies*

$$p \leq 2R.$$ □

Clearly Conjecture A implies the conjecture in [1]. It has been proved for $k = 1$ in [3] and for special cases for $k = 2$ in [2,4]. If one restricts the class of methods further it may be easier to prove this conjecture. Such restrictions are explicit schemes (i.e. schemes with $r_k = s_k = 0$), schemes with an increasing stencil on the downwind side; i.e. $r_j \leq r_{j+1}$, for all j, or schemes where the algebraic function defined in [1] has no branch point at $\mu = 0$. An even more challenging problem would be to get restriction on the accuracy due to stability for multidimensional advection equations, e.g. $u_t = au_x + bu_y$, $u = u(x, y, t)$.

References

[1] R. Jeltsch: Stability of time discretization, Hurwitz determinants and order stars, in this volume.
[2] Jeltsch R., Renaut R.A., Smit J.H., *An accuracy barrier for stable three-time-level difference schemes for hyperbolic equations*. Research Report No 95-01, 1995, Seminar für Angewandte Mathematik, ETH Zürich.
[3] Jeltsch R., Smit J.H., *Accuracy barriers of difference schemes for hyperbolic equations*. SIAM J. Numer. Anal. **24**, 1–11, (1987).
[4] Jeltsch R., Smit J.H., *Accuracy barriers of three-time-level difference schemes for hyperbolic equations*. Ann. University of Stellenbosch, 1992/2, 1–34, 1992.

9. On the simplest possible property of hurwitz criterion. (E.I. Jury)

Conjecture:
It is not possible to reduce the *number* and *sum* of the *degress* of the inequalities in the Liénard-Chipart criterion (which is a simpler form of Hurwitz stability criterion) if an alternate set of polynomial inequalities were used.
The above may be ascertained from the references [1], [2].

References

[1] B.D. Anderson and E.I. Jury: A simplest possible property of the generalized Routh-Hurwitz conditions. SIAM J. Cont. Opt., 15, 1977.
[2] B.D. Anderson, E.I. Jury and L.F. Chaparro: Relations between real and complex polynomials for stability and aperiodicity conditions. IEEE Trans. Auto. Cont., AC-20, 244–246, 1975.

10. Conversion of a formal power series to a continued fraction. (J. Waldvogel)

Polynomials or, more generally, rational functions play an essential role as transfer functions in automatic control and in stability theory. It turns out that their representation as continued fractions is often more useful than the usual representation as quotients of polynomials given by their coefficients.

As a model, we consider the conversion of the formal power series $f(z) := a_0 - a_1 z + a_2 z^2 - a_3 z^3 + \ldots, a_k \neq 0$ into the standard continued fraction

$$(1) \qquad f(z) = \cfrac{c_0}{1 + \cfrac{c_1 z}{1 + \cfrac{c_2 z}{1 + \ldots}}} = \frac{c_0}{1+} \; \frac{c_1 z}{1+} \; \frac{c_2 z}{1+} \ldots, \quad c_0 = a_0,$$

which may basically be carried out by means of the Euclidian algorithm. A handy version of it is the quotient-difference (QD) algorithm by H. Rutishauser [2], which is based on the QD table

$$(2) \qquad \begin{array}{cccccc}
d_0^0 & & & & & \\
 & q_0^0 & & & & \\
d_1^0 & & d_0^1 & & & \\
 & q_1^0 & & q_0^1 & & \\
d_2^0 & & d_1^1 & & d_0^2 & \\
 & q_2^0 & & q_1^1 & & \vdots \\
d_3^0 & & d_2^1 & & \vdots & \ddots \\
 & q_3^0 & & \vdots & & \\
d_4^0 & & \vdots & & & \\
\vdots & & & & &
\end{array}$$

with $d_k^0 = 0$, $q_k^0 = a_{k+1}/a_k$ $(k = 0, 1, 2, \ldots)$ as initial (leftmost) columns. The table is propagated to the right by alternatingly using the recurrence rules

$$(3) \qquad \begin{aligned} d_k^{\ell+1} &= d_{k+1}^\ell + q_{k+1}^\ell - q_k^\ell, & k &= 0, 1, \ldots \\ q_k^{\ell+1} &= q_{k+1}^\ell \, d_{k+1}^{\ell+1}/d_k^{\ell+1}, & \ell &= 0, 1, \ldots. \end{aligned}$$

If the QD table exists the continued fraction coefficients c_ℓ may be read from the first diagonal:

$$(4) \qquad c_{2\ell-1} := q_0^{\ell-1}, \quad c_{2\ell} = d_0^\ell, \quad \ell = 1, 2, \ldots.$$

Alternatively, the QD table may be expressed by means of Hankel determinants, see e.g. P. Henrici [1]. Other types of continued fractions may easily be obtained from (1), see [2].

Any direct numerical algorithm to compute the c_ℓ from the a_k is inherently unstable (A.N. Stokes [3]). Exact rational or high-precision computations yield only a finite section of the sequence $\{c_\ell\}$, unless the entire QD table can be found explicitly. An example of this kind is $a_k = k!$, $(k \geq 0)$, corresponding to the exponential integral $f(z) = z^{-1} \exp(z^{-1}) E_1(z^{-1})$, which implies $c_\ell = [\frac{\ell+1}{2}]$, $(\ell \geq 1)$. However, the slightly modified coefficients $\tilde{a}_k = k!/(k+1)$ yield a complicated QD table. It seems to exist and to depend smoothly on the subscripts k, ℓ.

Computations in a precision of 60 decimal digits yield

$$
\begin{aligned}
&c_1 = \frac{1}{2} \quad c_3 = \frac{22}{15} \quad c_9 = 4.44048 \quad c_{29} = 14.42739 \quad c_{99} = 49.42549 \\
&c_2 = \frac{5}{6} \quad c_4 = \frac{98}{55} \quad c_{10} = 4.72223 \quad c_{30} = 14.66818 \quad c_{100} = 49.62723 \;.
\end{aligned}
\tag{5}
$$

Problems: Find (i) conditions on a smooth sequence $a_k = a(k)$ for the existence and smoothness of the sequence $\{c_\ell\}$, (ii) the asymptotics of c_ℓ as $\ell \to \infty$.

Remark: The recurrence relations (3) correspond in their linear part to a discrete wave equation. This is in agreement with the necessity of two initial data sets d_k^0, q_k^0 and with the fact that c_ℓ is influenced by all q_k^0 with $k < \ell$.

References

[1] P. Henrici: Applied and Computational Complex Analysis, Vol. 1. John Wiley, New York, 682 pp, 1974.
[2] H. Rutishauser: Der Quotienten-Differenzen-Algorithmus. Mitt. Inst. Angew. Math. ETH **7**. Birkhäuser, Basel, 74 pp, 1957.
[3] A.N. Stokes: Efficient Stable Ways to Calculate Continued Fraction Coefficients From Some Series. Num. Math. **42**, 237–245, 1983.

11. Degree of robust stability for fuzzy uncertainties. (A. Weinmann)

The open problem is related to fuzzy logic in automatic control; not in the usual sense that the controller is a fuzzy one using fuzzification, inference and defuzzification. The uncertainty of the plant is assumed fuzzy and associated with a given membership function.

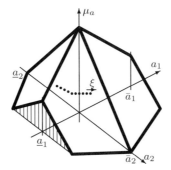

FIG. 1. *Membership function for two uncertain parameters a_1 and a_2*

Consider the set of perturbed time-invariant plants with the matrix-valued transfer function $\mathbf{G}(s,\mathbf{a})$ where $\mathbf{a}$ is the uncertain parameter. The membership function associated with $\mathbf{a}$ is μ_a. Assume, e.g., $0 < \mu_a \le 1$. Moreover, define some stability degree ρ which characterizes the dynamic behaviour of $\mathbf{G}(s,\mathbf{a})$ being connected back to back with a controller $\mathbf{K}(s)$. Presuppose that the controlled system for any $\mathbf{K}(s)$ of a given class is stable, i.e., e.g. $\rho > 0$. Let the path on the surface of the membership function be denoted ξ for $\mu_a = $ constant, see Fig. 1.

Then the desired optimum controller $\mathbf{K}^\star(s)$ has to maximize the over-all stability degree weighted by a weight $w(\mu_a)$ which penalizes the stability degree with respect to the membership function, i.e.,

$$\mathbf{K}^\star(s) = \arg\max_{\mathbf{K}(s)} \int_0^1 w(\mu_a)\bar{\rho}(\mu_a)d\mu_a \qquad \text{where} \qquad \bar{\rho}(\mu_a) = \int_{\xi_0}^{\xi_1} \rho(\xi)\,d\xi$$

is a mean value of ρ with respect to the path ξ.

Additionally, the membership function for μ_a for various (fixed, nonoptimal) controllers $\mathbf{K}(s)$ could figure out how the stability degree depends on the uncertainty. The problem can also be extended by replacing transfer matrices by time-variant or nonlinear operators.

Über die Bedingungen, unter welchen eine Gleichung nur Wurzeln mit negativen reellen Theilen besitzt

Von A. Hurwitz in Zürich

1. Auf Veranlassung meines verehrten Collegen, Herrn A. Stodola, beschäftigte ich mich vor einiger Zeit mit der Frage, wann eine Gleichung n-ten Grades mit reellen Coefficienten

$$a_0 x^n + a_1 x^{n-1} + \cdots + a_n = 0$$

nur solche Wurzeln besitzt, deren reelle Bestandtheile negativ sind. Wenn auch die Erledigung dieser Frage nach den Methoden von Sturm, Liouville, Cauchy und Hermite keine principielle Schwierigkeit bietet, so erlaube ich mir doch das Resultat, zu welchem ich gelangt bin, hier mitzutheilen, weil dasselbe wegen seiner einfachen, für die Anwendungen brauchbaren Gestalt vielleicht einiges Interesse verdient[1].

Die Herleitung des Resultates giebt mir zugleich Gelegenheit, die Methode von Hermite-Jacobi in einer Form darzustellen, in welcher sie eine Verallgemeinerung nach verschiedenen Richtungen zulässt.

Man darf sich, was hier geschehen soll, offenbar auf den Fall beschränken, wo der Coefficient a_0 positiv ist. Denn andernfalls kann man die linke Seite der Gleichung mit dem Factor -1 multipliciren. Man bilde nun die Determinante

$$(1) \quad \Delta_\lambda = \begin{vmatrix} a_1, & a_3, & a_5, & \ldots & a_{2\lambda-1} \\ a_0, & a_2, & a_4, & \ldots & a_{2\lambda-2} \\ 0, & a_1, & a_3, & \ldots & a_{2\lambda-3} \\ \cdot & \cdot & \cdot & \cdot & \\ \cdot & \cdot & \cdot & \cdot & a_\lambda \end{vmatrix}$$

nach der Maassgabe, dass die Indices in der ersten Horizontalreihe immer um zwei Einheiten wachsen, in jeder Verticalreihe immer um eine Einheit abnehmen. Dabei ist allgemein $a_\kappa = 0$ zu setzen, wenn der Index κ negativ oder grösser als n ist.

Dies vorausgeschickt, gilt der Satz:

Die nothwendige und hinreichende Bedingung dafür, dass die Gleichung

$$(2) \quad a_0 x^n + a_1 x^{n-1} + \cdots + a_n = 0,$$

---

[1] Herr Stodola benutzt mein Resultat in seiner Abhandlung über "die Regulirung von Turbinen" (Schweiz. Bauzeitung, Bd. 23, Nr. 17, 18), deren Ergebnisse bei der Turbinenanlage des Badeortes Davos mit glänzendem Erfolge Anwendung gefunden haben. — Die obige Frage wird auch, worauf mich Herr Stodola aufmerksam machte, in Thomson und Tait's Natural Philosophy (1886. Theil I, pag.390) aufgeworfen und ihre Erledigung als wünschenswerth bezeichnet. Mathematische Annalen. XLVI.

in welcher der Coefficient a_0 positiv vorausgesetzt wird, nur Wurzeln mit negativen reellen Bestandtheilen besitzt, ist die, dass die Werthe der Determinanten

(3) $$\Delta_1 = a_1, \Delta_2, \Delta_3, \ldots, \Delta_n$$

sämmtlich positiv sind.

Zu diesem Satze ist noch folgendes zu bemerken. Die Determinante Δ_n ist, wie man leicht erkennt, indem man sie nach den Elementen der letzten Verticalreihe entwickelt, gleich $a_n \cdot \Delta_{n-1}$.

Daher ist die Forderung, dass Δ_{n-1} und Δ_n positiv sein sollen, gleichbedeutend mit der anderen, dass Δ_{n-1} und a_n positiv sein sollen. Der obige Satz bleibt also richtig, wenn a_n an Stelle von Δ_n gesetzt wird. Eine andere Bemerkung ist diese:

Betrachtet man die Reihe der Determinanten

(4) $$\Delta_1, \Delta_2, \Delta_3, \ldots,$$

so verschwinden die Glieder dieser Reihe vom $(n+1)$sten ab identisch, d.h. für unbestimmt gedachte Werthe von $a_0, a_1, \ldots, a_n$. Denn die Elemente der letzten Verticalreihe von Δ_λ sind für $\lambda > n$ sämmtlich Null. Die Bedingung des Satzes kann also auch dahin ausgesprochen werden, dass die nicht identisch verschwindenden Glieder der Reihe (4) sämmtlich positiv sein müssen. Die Glieder dieser Reihe sind ausführlich geschrieben diese:

$$a_1, \quad \begin{vmatrix} a_1 & a_3 \\ a_0 & a_2 \end{vmatrix}, \quad \begin{vmatrix} a_1 & a_3 & a_5 \\ a_0 & a_2 & a_4 \\ 0 & a_1 & a_3 \end{vmatrix}, \quad \begin{vmatrix} a_1 & a_3 & a_5 & a_7 \\ a_0 & a_2 & a_4 & a_6 \\ 0 & a_1 & a_3 & a_5 \\ 0 & a_0 & a_2 & a_4 \end{vmatrix}$$

und man bildet hiernach ohne Weiteres die Bedingungen für jeden speciellen Werth von n.

Beispielsweise lauten die Bedingungen für die Gleichung 4ten Grades ($n = 4$):

$$a_1 > 0, \quad \begin{vmatrix} a_1 & a_3 \\ a_0 & a_2 \end{vmatrix} > 0, \quad \begin{vmatrix} a_1 & a_3 & 0 \\ a_0 & a_2 & a_4 \\ 0 & a_1 & a_3 \end{vmatrix} > 0, \quad a_4 > 0.$$

Herr Stodola hat bemerkt, dass eine *nothwendige* Bedingung dafür dass die Gleichung (2) nur Wurzeln mit negativen reellen Bestandtheilen besitzt, die ist, dass sämmtliche Coefficienten $a_0, a_1, \ldots a_n$ positiv sind. In der That: wenn die reellen Bestandtheile aller Wurzeln der Gleichung (2) negativ sind, so hat jeder reelle Linearfactor der linken Seite der Gleichung die Form $x + p$ und jeder reelle quadratische Factor die Form $(x+p_1)^2 + p_2^2 = x^2 + p'x + p''$, wo p, p_1, p_2, p', p'' positive Grössen bezeichnen. Da aber das Product von ganzen Functionen mit positiven Coefficienten ebenfalls positive Coefficienten besitzt, so wird auch die linke Seite der Gleichung (2) nur positive Coefficienten aufweisen.

2. Die ganze rationale Function $f(x)$, deren Coefficienten zunächst auch complexe Werthe besitzen können, möge der einen Bedingung unterworfen sein, dass sie für keinen rein imaginären Werth von x verschwindet. Bezeichnen dann N bez. P die Anzahlen der Nullstellen von $f(x)$, die negativen bez. positiven reellen Theil besitzen, so ist

(5) $$N + P = n,$$

unter n den Grad von $f(x)$ verstanden. Es sei nun c eine beliebige (complexe) Constante und

(6) $$cf(x) = \rho \cdot e^{i\pi\varphi},$$

so dass ρ den absoluten Betrag und $\pi\varphi$ das Argument von $cf(x)$ bezeichnet. Der Winkel φ ändert sich stetig mit dem Werthe von x und nimmt insbesondere um

(7) $$N - P = \Delta$$

Einheiten ab, wenn x die rein imaginären Zahlen von $+i\infty$ bis $-i\infty$ durchläuft. Man erkennt dies unmittelbar, wenn man, unter Benutzung der üblichen geometrischen Darstellung der complexen Zahlen, die Änderung des Argumentes des einzelnen Linearfactors von $f(x)$ verfolgt. Nach (5) und (7) ist nun

(8) $$N = \frac{n+\Delta}{2}, \quad P = \frac{n-\Delta}{2}.$$

Die Bestimmung von Δ wird jetzt in bekannter Weise auf die eines Cauchy'chen Index[2] zurückgeführt. Allgemein hat man unter dem Index einer Grösse R, die in jedem Punkte einer in bestimmten Sinne zu durchlaufenden Linie L einen bestimmten reellen Werth besitzt, die folgendermassen zu bildende Zahl zu verstehen. Man ordne jedem Punkte von L, in welchem R unendlich wird, die Zahl 0, oder $+1$ oder -1 zu, je nachdem R beim Ueberschreiten des Punktes nicht wechselt oder von negativen zu positiven oder von positiven zu negativen Werthen übergeht. Der Index von R bezüglich der Linie L ist dann die Summe aller den Unendlichkeitspunkten von R zugeordneten Zahlen. Man setzt hierbei stillschweigend voraus, dass R nur in einer endlichen Zahl von Punkten unendlich werdend das Zeichen wechselt, und dass $\frac{1}{R}$ in der Umgebung dieser Punkte stetig ist.

Dies in Erinnerung gebracht, sei z eine reelle Veränderliche und

(9) $$cf(-iz) = U + iV,$$

wo U und V ganze Functionen von z mit reellen Coefficienten bezeichnen. Wird nun

(10) $$\frac{V}{U} = R(z)$$

---

[2] Journal de l'école polytechnique, XV. (1837). Der Cauchy'sche Index ist als specieller Fall in dem von Kronecker eingeführten Begriff der Charakteristik von Functionensystemen (Monatsberichte der kgl. preussischen Akademie der Wissenschaften 1869) enthalten.

gesetzt, so hat man

$$\varphi = \frac{1}{\pi} \operatorname{arctg} R(z),$$

und aus dieser Gleichung folgt, dass Δ übereinstimmt mit dem Index von $R(z)$ bezüglich der im Sinne der wachsenden z zu durchlaufenden reellen z-Achse (die als eine im Unendlichen geschlossene Linie anzusehen ist). Im Folgenden nehme ich an, dass $R(z)$ für $z = \infty$ nicht unendlich wird, was offenbar gestattet ist, da man über die Constante c willkürlich verfügen kann.

3. Es sei jetzt $R(z)$ irgend eine rationale Function von z mit reellen Coefficienten, die für $z = \infty$ endlich bleibt.

Der Index von $R(z)$ (bezüglich der im Sinne der wachsenden z zu durchlaufenden Axe der reellen Zahlen) lässt sich bekanntlich durch das Sturm'sche Divisionsverfahren oder nach Hermite durch Aufstellung einer quadratischen Form bestimmen, deren Signatur mit dem gesuchten Index übereinstimmt. Unter "Signatur" einer quadratischen Form mit reellen Coefficienten verstehe ich dabei mit Hrn. Frobenius[3] die Differenz zwischen der Zahl der positiven und der negativen Quadrate, die bei der Darstellung der Form durch ein Aggregat von möglichst wenigen Quadraten reeller Linearfunctionen auftreten.

Man wird zu dieser Hermite'schen Bestimmungsweise des Index von $R(z)$ auf folgendem Weg geführt. Bezeichnet

(11) $$\Theta(z) = y_0 + y_1 z + y_2 z^2 + \cdots + y_{m-1} z^{m-1}$$

eine ganze rationale Function von z, deren Coefficienten als willkürliche Parameter angesehen werden, so stellt das Integral

(12) $$F_m = \frac{1}{2\pi i} \int R(z)[\Theta(z)]^2 dz,$$

erstreckt durch eine alle Pole von $R(z)$ einschliessende Curve, eine quadratische Form der Parameter $y_0, y_1, \ldots, y_{m-1}$ dar, die als Coefficient von $\frac{1}{z}$ in der Entwicklung von $R(z)[\Theta(z)]^2$ nach aufsteigenden Potenzen von $\frac{1}{z}$ leicht gebildet werden kann[4]. Andererseits ist das Integral gleich der Summe der Residuen von $R(z)[\Theta(z)]^2$, die den Polen von $R(z)$ entsprechen. Es sei $Z = a$ ein einfacher Pol von $R(z)$ und

(13) $$R(a+t) = \frac{c}{t} + c_1 + c_2 t + \cdots,$$

---

[3] Ueber das Trägheitsgesetz der quadratischen Formen. (Sitzungsberichte der kgl. preussischen Akademie der Wissenschaften. 1894).

[4] An Stelle des Integrals (2) kann man mit gleichem Erfolge auch das Integral $\frac{1}{2\pi i} \int R(z) \cdot \frac{[\Theta(z)]^2}{(z-\alpha)^{2m}} dz$, erstreckt um die Stelle $z = \alpha$, betrachten, unter α einen reellen Werth verstanden, für welchen $R(z)$ endlich bleibt. Für dieses Integral spielt $z = \alpha$ dieselbe Rolle, wie $z = \infty$ für das Integral (12). Im Zusammenhang hiermit steht die unmittelbar einleuchtende Thatsache, dass der Index von $R(\frac{az+b}{cz+d})$ gleich ist dem Index von $R(z)$, falls a, b, c, d reelle Constanten bedeuten, deren Determinante $ad - bc$ positiv ist.

dann ist das auf $z = a$ bezügliche Residuum

$$c \cdot [\Theta(a)]^2.$$

Wenn a reell ist, so liefert der Pol $z = a$ den Beitrag $+1$ oder -1 zu dem Index von $R(z)$ je nachdem c positiv oder negativ ist. Wenn a imaginär ist und $\bar{a}$ den zu a conjugirten Pol bezeichnet, so ist die Summe der auf a und $\bar{a}$ bezüglichen Residuen

$$c[\Theta(a)]^2 + \bar{c}[\Theta(\bar{a})]^2 = (P + iQ)^2 + (P - iQ)^2 = 2P^2 - 2Q^2,$$

wo P und Q reelle Linearfunctionen sind. Hieraus folgt -- zunächst unter der Voraussetzung, dass $R(z)$ nur einfache Pole besitzt — der Satz:

Bezeichnet n die Zahl der Pole von $R(z)$, so lässt sich die quadratische Form F_m als ein Aggregat von n Quadraten darstellen, wobei zugleich die Differenz zwischen der Zahl der positiven und der Zahl der negativen Quadrate gleich dem Index von $R(z)$ ist.

Dieser Satz gilt aber auch für den Fall, dass $R(z)$ Pole von beliebiger Multiplicität besitzt, wo dann unter n die Zahl der Pole, jeder mit seiner Multiplicität gezählt, zu verstehen ist[5]. Es sei, um dies zu beweisen, $z = a$ ein λ-facher Pol von $R(z)$ und

$$R(a+t) = \frac{c}{t^\lambda} + \frac{c_1}{t^{\lambda-1}} + \cdots + \frac{c_{\lambda-1}}{t} + \cdots,$$

$$\Theta(a+t) = \Theta_0(a) + \Theta_1(a)t + \Theta_2(a)t^2 + \cdots,$$

wo $\Theta_0(a)$, $\Theta_1(a)$,... lineare Formen der Parameter $y_0, y_1, \ldots, y_{m-1}$ bezeichnen. Das $z = a$ entsprechende Residuum lautet dann:

$$c_{\lambda-1}\Theta_0^2 + 2c_{\lambda-2}\Theta_0\Theta_1 + \cdots + c(2\Theta_0\Theta_{\lambda-1} + 2\Theta_1\Theta_{\lambda-2} + \cdots).$$

Je nachdem nun λ gerade oder ungerade ist, lässt sich dieses Residuum in die Gestalt

$$\Theta_0\psi_0 + \Theta_1\psi_1 + \cdots + \Theta_{\mu-1}\psi_{\mu-1} \qquad (\lambda = 2\mu)$$

oder

$$\Theta_0\psi_0 + \Theta_1\psi_1 + \cdots + \Theta_{\mu-1}\psi_{\mu-1} + c\Theta_\mu^2 \qquad (\lambda = 2\mu + 1)$$

setzen, wo $\psi_0, \psi_1, \ldots$, lineare Functionen der Parameter bedeuten.

---

[5] Dass die auf die Sturm'schen Reihen bezüglichen Deductionen mit den geeigneten Modificationen auch noch gültig bleiben, wenn die in Betracht kommenden ganzen Functionen mehrfache Linearfactoren besitzen, bemerkt Kronecker in seiner Abhandlung: Zur Theorie der Elimination einer Variablen aus zwei algebraischen Gleichungen (Monatsberichte der kgl. Akademie der Wissenschaften, 1881).

Ist a reell, so sind die Coefficienten von $\Theta_0, \Theta_1, \ldots \psi_0, \psi_1, \ldots$ ebenfalls reell und das Residuum kann in die Form

$$\left[\frac{1}{2}(\Theta_0 + \psi_0)\right]^2 - \left[\frac{1}{2}(\Theta_0 - \psi_0)\right]^2 + \cdots + \left[\frac{1}{2}(\Theta_{\mu-1} + \psi_{\mu-1})\right]^2$$
$$- \left[\frac{1}{2}(\Theta_{\mu-1} - \psi_{\mu-1})\right]^2 \quad (\lambda = 2\mu)$$

oder

$$\left[\frac{1}{2}(\Theta_0 + \psi_0)\right]^2 - \left[\frac{1}{2}(\Theta_0 - \psi_0)\right]^2 + \cdots + \left[\frac{1}{2}(\Theta_{\mu-1} + \psi_{\mu-1})\right]^2$$
$$- \left[\frac{1}{2}(\Theta_{\mu-1} - \psi_{\mu-1})\right]^2 + c\Theta_\mu^2 \quad (\lambda = 2\mu + 1)$$

gebracht werden, in welcher es als Aggregat von λ Quadraten reeller Linearformen erscheint.

Dabei treten, wenn λ gerade ist, genau so viele positive wie negative Quadrate auf; dagegen tritt, wenn λ ungerade ist, ein positives oder ein negatives Quadrat mehr auf, je nachdem c positiv oder negativ ist. Die Discussion des Falles, wo $z = a$ complex ist, ist in ähnlicher Weise zu erledigen, und man erkennt so die allgemeine Gültigkeit des obigen Satzes.

4. Ist $m > n$, so besitzt die quadratische Form F_m eine verschwindende Determinante, da sich die Form als Aggregat von n Quadraten, also als eine Form von weniger als m linearen Verbindungen der Parameter y_0, y_1, y_{m-1} darstellen lässt. Hingegen ist die Determinante der Form F_n von Null verschieden. Man kann dies entweder dadurch beweisen, dass man die Übereinstimmung dieser Determinante mit der Resultante von Zähler und Nenner der in reducierter Form geschriebenen rationalen Function $R(z)$ zeigt (vgl. unten Nr. 6), oder auch auf folgendem Wege:

Würde die Determinante von F_n verschwinden, so könnte man solche nicht sämmtlich verschwindende Werthe von $y_0, y_1, \ldots, y_{n-1}$ finden, für welche $\frac{\partial F_n}{\partial y_0}$, $\frac{\partial F_n}{\partial y_1}, \ldots \frac{\partial F_n}{\partial y_{n-1}}$, d.h. also die Integrale

(14) $$\frac{1}{2\pi i} \int R(z) \cdot \Theta(z) \cdot z^\lambda dz \quad (\lambda = 0, 1, \ldots, n-1)$$

sämmtlich Null sind. Wenn nun

(15) $$R(z) \cdot \Theta(z) = G(z) + R_1(z)$$

ist, wo $G(z)$ eine ganze rationale Function von z und

(16) $$R_1(z) = R(z) \cdot \Theta(z) - G(z) = \frac{k'}{z} + \frac{k''}{z^2} + \cdots$$

eine für $z = \infty$ verschwindende rationale Function bezeichnet, so ist für das Verschwinden jener Integrale erforderlich, dass

$$k' = k'' = \cdots = k^{(n)} = 0$$

ist, dass also $R_1(z)$ mindestens von der $(n+1)$ten Ordnung für $z = \infty$ verschwindet. Da aber $R_1(z)$ nur an den Polen von $R(z)$, also höchstens n Mal unendlich werden kann, so muss $R_1(z)$ identisch verschwinden. Die hieraus folgende Gleichung $R(z) \cdot \Theta(z) = G(z)$ ist aber unmöglich, da $\Theta(z)$ höchstens vom $(n-1)$ten Grade ist und $R(z)$ n Pole besitzt.

5. Aus dem Vorhergehenden ergiebt sich nun folgendes Verfahren zur Bestimmung des Index von $R(z)$:

Es sei

(17) $$R(z) = c + \frac{c_0}{z} + \frac{c_1}{z^2} + \frac{c_2}{z^3} + \cdots$$

die Entwicklung von $R(z)$ in der Umgebung von $z = \infty$. Der Factor von $\frac{1}{z}$ in der Entwicklung des Productes aus $R(z)$ und

(18) $$[\Theta(z)]^2 = \sum_{i,k} y_i y_k z^{i+k} \quad (i, k = 0, 1, \ldots m-1)$$

ist dann

(19) $$F_m = \sum_{i,k} c_{i+k} y_i y_k \quad (i, k = 0, 1, \ldots m-1),$$

und die Determinante der Form F_m stellt sich dar in der Gestalt:

(20) $$D_m = \begin{vmatrix} c_0, & c_1, & \cdots & c_{m-1} \\ c_1, & c_2, & \cdots, & c_m \\ \cdot & \cdot & & \cdot \\ \cdot & \cdot & & \cdot \\ \cdot & \cdot & & \cdot \\ c_{m-1}, & c_m, & \cdots & c_{2m-2} \end{vmatrix}.$$

In der Reihe der Determinanten

(21) $$D_1, D_2, D_3, \ldots$$

sind nun alle Glieder von einem bestimmten, etwa D_{n+1}, ab gleich Null, während D_n von Null verschieden ist. Es giebt dann n die Zahl der Pole von $R(z)$ an, und der Index von $R(z)$ ist gleich der Signatur der Form F_n.

Die Signatur der Form F_n kann man in jedem Falle aus den Vorzeichen der nicht verschwindenden unter den Determinanten $D_1, D_2, \ldots D_{n-1}$ ablesen.[6] In dem

---

[6] Frobenius, l.c. pag. 410.

Falle, wo keine dieser Determinanten verschwindet, lässt sich F_n, wie bekannt und übrigens leicht zu zeigen ist, in der Form

$$F_n = D_1 u_0^2 + \frac{D_2}{D_1} u_1^2 + \cdots + \frac{D_n}{D_{n-1}} u_{n-1}^2$$

darstellen, wo u_i eine reelle Linearform von $y_i, y_{i+1}, \ldots y_{n-1}$ ist. Der Index von $R(z)$ ist dann also gleich der Differenz zwischen der Zahl der positiven und der Zahl der negativen Glieder der Reihe

$$D_1, \frac{D_2}{D_1}, \frac{D_3}{D_2}, \cdots \frac{D_n}{D_{n-1}}.$$

Dieser Fall tritt insbesondere ein, wenn der Index von $R(z)$ seinen Maximalwerth n besitzt. Denn es ist dann F_n eine definite positive Form und ebenso F_{n-1}, F_{n-2}, ... F_1, da die letzteren Formen durch Nullstellen einiger der Parameter $y_0, y_1, \ldots y_{n-1}$ aus F_n entstehen. Die Determinante einer definiten positiven Form ist aber stets positiv. Es gilt hiernach der Satz:

Der Index von $R(z)$ hat stets und nur dann seinen Maximalwerth n, wenn die Determinanten $D_1, D_2, \ldots D_n$ positiv sind.

6. Es sei jetzt $R(z)$ in der Gestalt gegeben

(22) $$R(z) = \frac{b_0 z^\nu + b_1 z^{\mu-1} + \cdots + b_\nu}{a_0 z^\mu + a_1 z^{\nu-1} + \cdots + a_\nu},$$

wo der Coefficient a_0 von Null verschieden vorausgesetzt wird. Der Grad ν des Nenners von $R(z)$ ist grösser oder gleich n, je nachdem Zähler und Nenner von $R(z)$ einen gemeinsamen Theiler haben oder nicht. Man kann nun die Determinante D_m (20) umformen in eine Determinante, in welcher die Coefficienten $a_0, \ldots a_\nu, b_0, \ldots b_\nu$ die Elemente bilden. Diese Umformung lässt sich mit Hülfe des folgenden Satzes ausführen, den man leicht aus dem Multiplicationstheorem der Determinanten ableitet.

Es seien

(23) $$P_1, P_2, \ldots, P_m, \ldots$$

gewöhnliche Potenzreihen von z, die durch Multiplication mit

(24) $$P = k + k_1 z + k_2 z^2 + \cdots$$

in die neuen Potenzreihen

(25) $$P'_1, P'_2, \ldots, P'_m, \ldots$$

übergehen mögen, so dass also allgemein $P'_m = P \cdot P_m$ ist. Trennt man dann von jeder der Reihen $P_1, P_2, \ldots P_m$ (bez. $P'_1, P'_2, \ldots P'_m$) die ersten m Glieder ab und bezeichnet mit Δ_m (bez. Δ'_m) die Determinante der so entstehenden m ganzen Functionen $(m-1)$ten Grades von z, so ist

(26) $$\Delta'_m = k^m \cdot \Delta_m.$$

Diesen Satz wende ich nun auf folgenden Fall an. Es sei

$$\frac{b_0 + b_1 z + b_2 z^2 + \cdots}{a_0 + a_1 z + a_2 z^2 + \cdots} = c + c_0 z + c_1 z^2 + \cdots,$$

und die Reihen (23) mögen, wie folgt, angenommen werden:

$$P_1 = 1,\ P_2 = c + c_0 z + c_1 z^2 + \cdots,\ P_{2\lambda+1} = z^\lambda P_1,\ P_{2\lambda+2} = z^\lambda P_2\ (\lambda = 1, 2, \ldots),$$

während die Reihe (24) mit

$$P = a_0 + a_1 z + a_2 z^2 + \cdots$$

identificirt werden soll. Die Reihen (25) lauten dann:

$$P'_1 = P,\ P'_2 = b_0 + b_1 z + b_2 z^2 + \cdots,\ P'_{2\lambda+1} = z^\lambda P'_1,\ P'_{2\lambda+2} = z^\lambda P'_2\ (\lambda = 1, 2, \ldots).$$

Ersetzt man noch in der Gleichung (26) den Index m durch $2m$, so giebt nun diese Gleichung die gewünschte Umformung der Determinante D_m. Es kommt nämlich:

(27) $$a_0^{2m} \cdot D_m = R_m,$$

wo R_m die Determinante

(28) $$R_m = \begin{vmatrix} a_0, & a_1, & \ldots & a_{2m-1} \\ b_0, & b_1, & \ldots & b_{2m-1} \\ 0, & a_0, & \ldots & a_{2m-2} \\ 0, & b_0, & \ldots & b_{2m-2} \\ \cdot & \cdot & \cdot & \cdot \\ \cdot & \cdot & \cdot & \cdot \\ 0 & 0 & \ldots & a_m \\ 0 & 0 & \ldots & b_m \end{vmatrix}$$

bedeutet. Diese verschwindet sicher, sobald $m \geq \nu$ ist, da dann die Elemente der letzten Verticalreihe sämmtlich Null sind. Man hat hiernach zur Bestimmung des Index (und zugleich der Zahl n der Pole) der rationalen Function (22) so zu verfahren: Man bildet die Reihe der Determinanten

$$R_1, R_2, \ldots R_\nu.$$

Wenn R_n das letzte nicht verschwindende Glied dieser Reihe ist, so giebt n die Zahl der Pole, oder, was dasselbe ist, den Grad des Nenners von $R(z)$ an, wenn $R(z)$ in reducirter Gestalt geschrieben wird. Der Index von $R(z)$ wird sodann aus den Vorzeichen der nicht verschwindenden Glieder der Reihe $R_1, R_2, \ldots R_n$ abgeleitet.

7. Insbesondere ergiebt sich jetzt leicht der unter Nr. 1 angegebene Satz. Es sei

(29) $$f(x) \equiv a_0 x^n + a_1 x^{n-1} + \cdots + a_n = 0$$

eine Gleichung mit reellen Coefficienten. Dann ist

(30) $$i^n f(-iz) = (a_0 z^n - a_2 z^{n-2} + \cdots) + i(a_1 z^{n-1} - a_3 z^{n-3} + \cdots)$$

und die in Nr. 2 mit Δ bezeichnete Zahl ist der Index von

(31) $$R(z) = \frac{a_1 z^{n-1} - a_3 z^{n-3} + \cdots}{a_0 z^n - a_2 z^{n-1} + \cdots}.$$

Die Gleichung (29) hat nun, wie aus (8) in Nr. 2 hervorgeht, stets und nur dann ausschliesslich Wurzeln mit negativen reellen Theilen, wenn $\Delta = n$ ist. Hieraus folgt, dass Zähler und nenner von $R(z)$ theilerfremd sein müssen. Denn andernfalls würde $R(z)$ dargestellt werden können als ein Quotient, dessen Nenner vom Grade $n' < n$ ist und der Index von $R(z)$ wäre höchstens gleich n'.

Die nothwendige und hinreichende Bedingung dafür, dass die Gleichung (29) nur Wurzeln mit negativen reellen Theilen besitzt, ist also die, dass die Form

(32) $$F_n = \frac{1}{2\pi i} \int R(z) [\Theta(z)]^2 dz$$

eine definite positive Form von $y_0, y_1, \ldots y_{n-1}$ ist. In Folge des Umstandes, dass $R(z)$ eine ungerade Function von z ist, lässt sich F_n in zwei Formen zerlegen, von denen die eine nur die Parameter $y_0, y_2, y_4, \ldots$, die andere nur die Parameter $y_1, y_3, y_5 \ldots$ enthält. In der That sei

(33) $$H(z) = \frac{a_1 z^{\lambda-1} - a_3 z^{\lambda-2} + \cdots}{a_0 z^\lambda - a_2 z^{\lambda-1} + \cdots}$$

$(\lambda = \dfrac{n}{2}$ oder $\dfrac{n+1}{2}$, je nachdem n gerade oder ungerade ist),

so ist offenbar

$$R(z) = z \cdot H(z^2).$$

Ferner fasse man in $\Theta(z)$ die Glieder mit geraden und die mit ungeraden Potenzen von z zusammen, setze also

$$\Theta(z) = \Theta_0(z^2) + z\Theta_1(z^2).$$

Führt man nun in dem Integral (32) $z^2 = \zeta$ als neue Integrationsvariable ein und schreibt dann wieder z an Stelle von ζ, sofindet man die in Rede stehende Zerlegung

(34) $$F_n = \frac{1}{2\pi i} \int H(z)[\Theta_0(z)]^2 dz + \frac{1}{2\pi i} \int zH(z)[\Theta_1(z)]^2 dz.$$

Die hierin enthaltene Thatsache, dass der Index von $R(z)$ gleich der Summe der Indices von $H(z)$ und $zH(z)$ ist, lässt sich übrigens, beiläufig bemerkt, auch unmittelbar aus dem Begriff des Index ableiten. Stellt man jetzt nach Nr. 5 und 6 die Bedingung auf, dass F_n oder was auf dasselbe hinauskommt, jedes der beiden Integrale (34) eine positive definite Form darstellt, so wird man nach einer leichten Umformung der zu bildenden Determinanten auf den Satz von Nr. 1 geführt.

8. Durch die Gleichung (8) von Nr. 2 und die in Nr. 6 entwickelte Methode zur Bestimmung des Index einer rationalen Function ist allgemein die Aufgabe gelöst, die Anzahl derjenigen Wurzeln einer Gleichung $f(x) = 0$ zu bestimmen, die einen negativen reellen Theil besitzen, unter der Voraussetzung, dass die Gleichung durch keinen rein imaginären Werth von x befriedigt wird. (Die letztere Beschränkung kann man übrigens fallen lassen, wenn man festsetzt, dass jede rein imaginäre Wurzel mit der Multiplicität $\frac{1}{2}$ sowohl als Wurzel mit negativem wie mit positivem Theil gezählt werden soll.) Diese Aufgabe ist, wie die Substitution von $-ix$ an Stelle von x zeigt, nicht wesentlich verschieden von der anderen, die Zahl der Wurzeln einer Gleichung $n$ten Grades

$$(35) \qquad f_1(x) + if_2(x) = 0,$$

wo $f_1(x)$ und $f_2(x)$ ganze Functionen mit reellen Coefficienten bezeichnen, zu bestimmen, welche einen positiv-imaginären Bestandtheil besitzen. Diese Zahl wird ebenfalls durch die erste Formel (8) also durch $\frac{n+\Delta}{2}$ angegeben, unter Δ den Index von $\frac{f_2(x)}{f_1(x)}$ verstanden.

Mit der letzteren Aufgabe beschäftigt sich Herr Hermite in zwei Abhandlungen[7], auf die ich hier verweise. Zum Schluss bemerke ich noch Folgendes: Aus dem Begriff des Index geht unmittelbar hervor, dass eine rationale Function $\frac{f_2(x)}{f_1(x)}$ stets und nur dann den Index $\pm n$ besitzt, wenn der Nenner $f_1(x)$ in n Punkten der reellen Axe verschwindet (wobei $x = \infty$ als Nullstelle von $f(x)$ anzusehen ist, falls $f_1(x)$ nur den $(n-1)$ten Grad erreicht) und wenn zugleich $f_2(x)$ in je zwei aufeinanderfolgenden dieser Punkte Werthe von entgegengesetzten Vorzeichen annimmt. Hieraus folgert man weiter, dass der Maximalwerth $\pm n$ des Index von $\frac{f_2(x)}{f_1(x)}$ stets und nur dann eintritt, wenn jede der Gleichungen $f_1(x) = 0$, $f_2(x) = 0$ n reelle von einander verschiedene Wurzeln besitzt und zugleich die Wurzeln der einen Gleichung durch die der anderen getrennt werden. Insbesondere haben also die n Wurzeln der Gleichung (35) stets und nur dann sämmtlich positiv-imaginären oder sämmtlich negativ-imaginären Bestandtheil, wenn die Wurzeln der Gleichungen $f_1(x) = 0$, $f_2(x) = 0$ die eben erwähnte Beschaffenheit besitzen[8].

Zürich, den 12. December 1894.

---

[7] Crelle's Journal Bd. 52, pag. 39, Bulletin de la société mathématique de France, Bd. 7, pag. 128.

[8] Vgl. Biehler, Crelle's Journal Bd. 87, pag. 350, Laguerre, ib. Bd. 89, pag. 339.

International Series of Numerical Mathematics

Edited by
K.-H. Hoffmann, Technische Universität München, Germany
H.D. Mittelmann, Arizona State University, Tempe, CA, USA
J. Todd, California Institute of Technology, Pasadena, CA, USA

International Series of Numerical Mathematics is open to all aspects of numerical mathematics. Some of the topics of particular interest include free boundary value problems for differential equations, phase transitions, problems of optimal control and optimization, other nonlinear phenomena in analysis, nonlinear partial differential equations, efficient solution methods, bifurcation problems and approximation theory. When possible, the topic of each volume is discussed from three different angles, namely those of mathematical modeling, mathematical analysis, and numerical case studies.

Titles previously published in the series

ISNM 85   **H. Brass, G. Hämmerlin (Eds.):** Numerical Integration III. Proceedings of the Conference held at the Mathematisches Forschungsinstitut, Oberwolfach, November 8-14, 1987, 1988 (ISBN 3-7643-2205-5)

ISNM 86   **R.P. Agarwal, Y.M. Chow, S.J. Wilson (Eds.):** Numerical Mathematics Singapore 1988. Proceedings of the International Conference on Numerical Mathematics held at the National University of Singapore, May 31-June 4, 1987, 1988 (ISBN 3-7643-2255-1)

ISNM 87   **J.P. Penot (Ed.):** New Methods in Optimization and their Industrial Uses. State of the art, recent advances, perspectives. Proceedings of the Symposia held in Pau, October 19-29, 1987, and Paris, November 1987, 1989 (ISBN 3-7643-2286-1)

ISNM 88   **J.F. Rodriguez (Ed.):** Mathematical Models for Phase Change Problems. Proceedings of the European Workshop held at the Obidos, Portugal, October 1-3, 1988, 1989 (ISBN 3-7643-2309-4)

ISNM 89   **K. Gürlebeck, W. Sprössig:** Quaternionic Analysis and Elliptic Boundary Value Problems, 1990 (ISBN 3-7643-2382-5)

ISNM 90   **C.K. Chui, W. Schempp, K. Zeller (Eds.):** Multivariate Approximation Theory IV. Proceedings of the Conference at the Mathematical Research Institute at Oberwolfach, Black Forest, February 12-18, 1989, 1989 (ISBN 3-7643-2384-1)

ISNM 91   F. Kappel, K. Kunisch, W. Schappacher (Eds.): Control and Estimation of Distributed Parameter Systems. 4th International Conference on Control of Distributed Parameter Systems, Vorau, July 10-16, 1988, 1989 (ISBN 3-7643-2345-0)

ISNM 92   H.D. Mittelmann, D. Roose, (Eds.): Continuation Techniques and Bifurcation Problems, 1990 (ISBN 3-7643-2397-3)

ISNM 93   R.E. Bank, R. Bulirsch, K. Merten (Eds.): Mathematical Modelling and Simulation of Electric Circuits and Semiconductor Devices, 1990 (3-7643-2439-2)

ISNM 94   W. Haussmann, K. Jetter (Eds.): Multivariate Approximation and Interpolation, 1990 (3-7643-2450-3)

ISNM 95   K.-H. Hoffmann, J. Sprekels (Eds.): Free Boundary Value Problems, 1990 (3-7643-2474-0)

ISNM 96   J. Albrecht, L. Collatz, P. Hagedorn, W. Velte (Eds.): Numerical Treatment of Eigenvalue Problems, Vol. 5, 1991 (3-7643-2575-5)

ISNM 97   R.U. Seydel, F.W. Schneider, T.G. Küpper, H. Troger (Eds.): Bifurcation and Chaos: Analysis, Algorithms, Applications, 1991 (3-7643-2593-3)

ISNM 98   W. Hackbusch, U. Trottenberg (Eds.): Multigrid Methods III, 1991 (3-7643-2632-8)

ISNM 99   P. Neittaanmäki (Ed.): Numerical Methods for Free Boundary Problems, 1991 (3-7643-2641-7)

ISNM 100 W. Desch, F. Kappel, K. Kunisch (Eds.): Estimation and Control of Distributed Parameter Systems, 1991 (3-7643-2676-X)

ISNM 101 G. Del Piero, F. Maceri (Eds.): Unilateral Problems in Structural Analysis IV, 1991 (3-7643-2487-2)

ISNM 102 U. Hornung, P. Kotelenez, G. Papanicolaou (Eds.): Random Partial Differential Equations, 1991 (3-7643-2688-3)

ISNM 103 W. Walter (Ed.): General Inequalities 6, 1992 (3-7643-2737-5)

ISNM 104 E. Allgower, K. Böhmer, M. Golubitsky (Eds.): Bifurcation and Symmetry, 1992 (3-7643-2739-1)

ISNM 105 D. Braess, L.L. Schumaker (Eds.): Numerical Methods in Approximation Theory, Vol. 9, 1992 (3-7643-2746-4)

ISNM 106 S.N. Antontsev, K.-H. Hoffmann, A.M. Khludnev (Eds.): Free Boundary Problems in Continuum Mechanics, 1992 (3-7643-2784-7)

ISNM 107 V. Barbu, F.J. Bonnans, D. Tiba (Eds.): Optimization, Optimal Control and Partial Differential Equations, 1992 (3-7643-2788-X)

ISNM 108 H. Antes, P.D. Panagiotopoulos: The Boundary Integral Approach to Static and Dynamic Contact Problems. Equality and Inequality Methods, 1992 (3-7643-2592-5)

ISNM 109 A.G. Kuz'min: Non-Classical Equations of Mixed Type and their Applications in Gas Dynamics, 1992 (3-7643-2573-9)

ISNM 110	**H.R.E.M. Hörnlein, K. Schittkowski (Eds.):** Software Systems for Structural Optimization, 1992 (3-7643-2836-3)
ISNM 111	**R. Burlisch, A. Miele, J. Stoer, K.H. Well:** Optimal Control, 1993 (3-7643-2887-8)
ISNM 112	**H. Braess, G. Hämmerlin (Eds.):** Numerical Integration IV. Proceedings of the Conference at the Mathematical Research Institute at Oberwolfach, November 8-14, 1992, 1993 (3-7643-2922-X)
ISNM 113	**L. Quartapelle:** Numerical Solution of the Incompressible Navier-Stokes Equations, 1993 (3-7643-2935-1)
ISNM 114	**J. Douglas, U. Hornung (Eds.):** Flow in Porous Media. Proceedings of the Oberwolfach Conference, June 21-27, 1992 (ISBN 3-7643-2949-1)
ISNM 115	**R. Bulirsch, D. Kraft (Eds.):** Computational Optimal Control, 1994 (ISBN 3-7643-5015-6)
ISNM 116	**P.W. Hemker, P. Wesseling (Eds.):** Multigrid Methods IV. Proceedings of the Fourth European Multigrid Conference, Amsterdam, July 6-9, 1993 (ISBN 3-7643-5030-X)
ISNM 117	**R.E. Bank, R. Bulirsch, H. Gajewski, K. Merten (Eds.):** Mathematical Modelling and Simulation of Electrical Circuits and Semiconductor Devices, 1994 (ISBN 3-7643-5053-9)
ISNM 118	**W. Desch, F. Kappel, K. Kunisch (Eds.):** Control and Estimation of Distributed Parameter Systems: Nonlinear Phenomena. International Conference on Control and Estimation of Distributed Parameter Systems, Vorau, July 18-24, 1993, 1994 (ISBN 3-7643-5098-9)
ISNM 119	**R. Zahar (Ed.):** Approximation and Computation: A Festschrift in Honor of Walter Gautschi, 1995 (ISBN 0-8176-3753-2)
ISNM 120	**W. Hackbusch:** Integral Equations: Theory and Numerical Treatment, 1995 (ISBN 3-7643-2871-1)

MATHEMATICS WITH BIRKHÄUSER • MATHEMATICS WITH BIRKHÄUSER

ZAMP

Zeitschrift für angewandte Mathematik und Physik
Journal of Applied Mathematics and Physics
Journal des Mathématiques et de Physique appliquées

Editors:
U. Kirchgraber, Zürich
(Differential Equations / Applied Mathematics)
M. Renardy, Blacksburg, VA
(Differential Equations/ Fluid Mechanics)
I.L. Ryhming, Lausanne,
(Fluid Mechanics)
M. Sayir, Zürich,
(Mechanics of Solids)

Assistant Editor:
K. Nipp, Zürich

Editorial Office:
ZAMP, P.O.Box 274
CH-4024 Basel
Switzerland

Co-Editors:;
F. Bark, Stockholm
F.G. Blottner, Albuquerque, NM
S.R. Bodner, Haifa
F.H. Busse, Bayreuth
G.-Q. Chen, Chicago, IL
T.K. Fannelöp, Zürich
J. Glimm, Stony Brook, NY
M. Hiller, Duisburg
T. Maxworthy, Los Angeles, CA
P.A. Monkewitz, Lausanne
A. Moser, Zürich
P.M. Naghdi, Berkeley CA
N. Rott, Palo Alto, CA
N.M. Temme, Amsterdam
H. Thomann, Zürich
J. Waldvogel, Zürich

Abstracted/Indexed in:
Applied Mechanics Reviews, Boundary Elements Abstracts & Newsletter, Chemical Abstracts, Computing Reviews, Current Contents, Engineering Index, Mathematical Reviews, Mechanics, Metal Abstracts, Previews of Heat & Mass Transfer, Physical Abstracts, Science Abstracts, Zentralblatt für Mathematik, Mathematics Abstracts, DB MATH

Subscription Information:
1996 subscription, volume 47 (6 issues), ISSN 0044-2275
sFr. 878.–/DM 1.038.– (plus postage & handling)
Single copy
sFr. 178.–/DM 208.– (plus postage & handling)

First published in 1950
1 volume per year, 6 issues per volume,
approx. 1000 pages per volume, 17x24 cm
Back volumes are available.

The *Journal of Applied Mathematics and Physics* (ZAMP) publishes papers of high quality in fluid mechanics, mechanics of solids and differential equations / applied mathematics. The readers of ZAMP will find not only articles in their own special field but also original work in neighbouring domains. This will lead to an exchange of ideas; concepts and methods which have proven to be successful in one field may well be useful to other areas. ZAMP attempts to publish articles reasonably quickly. Longer papers are published in the section "Original Papers", shorter ones may appear under "Brief Reports "where publication is particularly rapid. The journal includes a "Book Review" section and provides information on activities (such as upcoming symposia, meetings, special courses) which are of interest to its readers.

Selected articles:
Z. Bai, W. and R.T. Shield, Oriental: Load-deformation relations in second-order elasticity / J.F. Collet, Nice: Convergence of relaxation approximation to a scalar nonlinear hyberbolic equation arising in chromatography / S. Furta, Moscow: On non-integrability of general systems of differential equations / J. Hale and W. Huang, Atlanta: Periodic solutions of singularly perturbed delay equations / K. Liu, Hangzhou and Z. Liu, Duluth: On the type of C_0-semigroup associated with the abstract linear viscoelastic system / K. Miller, Wichita: Stationary corner vortex configurations / O. O'Reilly, Berkeley: A properly invariant infinitesimal theory of an elastic Cosserat point / A. Saasen, Stavanger: The dispersion relation at large frequencies for linear surface gravity waves on a Maxwell fluid exhibiting a spatial decay / R. Srinivasan, Bangalore: Accurate solutions for steady plane flow in the driven cavity I. Stokes flow

Birkhäuser Verlag AG
P.O. Box 133, CH-4010 Basel / Switzerland
FAX: +41 / 61 / 205 07 92
e-mail: farnik@birkhauser.ch
http//www.birkhauser.ch

Prices are subject to change without notice. 8/96

Birkhäuser
Birkhäuser Verlag AG
Basel · Boston · Berlin